Quantitative Methods to Assess Capacity of Water Treatment to Eliminate Micro-Organisms

KWR Watercycle Research Institute Series

Quantitative Methods to Assess Capacity of Water Treatment to Eliminate Micro-Organisms

Wim A.M. Hijnen

KWR Watercycle Research Institute
PO Box 1072, 3430 BB Nieuwegein, The Netherlands
Tel. +31(0)306069511

Publishing
London • New York

Published by **IWA Publishing**
Alliance House
12 Caxton Street
London SW1H 0QS, UK
Telephone: +44 (0)20 7654 5500
Fax: +44 (0)20 7654 5555
Email: publications@iwap.co.uk
Web: www.iwapublishing.com

First published 2011

Cover design by www.designforpublishing.co.uk
Typeset by MPS Limited, a Macmillan Company, Chennai, India, www.macmillansolutions.com

British Library Cataloguing in Publication Data
A CIP catalogue record for this book is available from the British Library

Library of Congress Cataloging-in-Publication Data
A catalog record for this book is available from the Library of Congress

ISBN13: 9781843393764
ISBN: 184339376X

Reality is the simple truth

To Jannie, Niek, Koen and Floor

Contents

 Quantitative Methods to Assess Capacity of Water Treatment to Eliminate Micro-Organisms. Edited by Wim A.M. Hijnen. ISBN: 9781843393764. Published by IWA Publishing, London, UK.

Preface

Important motive to write this book was to put in an overview a number of major milestones of the last decade achieved in a research project on production of microbiological safe drinking water which is part of the joint research programme of the Dutch Water Works (BTO research). Safe drinking water is an important hygienic condition for a healthy living. In the developed countries this is experienced as natural and there is a large consumer trust. The challenge of the work in this book was to develop methods and collect scientific information which supports this confidence. The knowledge presented is a contribution to the development of the concept of safe drinking water with Quantitative Microbial Risk assessment and Water Safety Plans in both the developed and developing countries.

The studies presented in the book were financed by the joint research programme of the Dutch Drinking Water Companies (BTO) and carried in close collaboration with those Companies. Studies on slow sand filtration were supported financially and operationally by Waternet. The work was done in the frame work of the thematic progamme 'Microbial Safe Drinking Water' excuted under the overall responsibility of Prof. Dr. Gertjan Medema (KWR Watercycle Research Institute, TU Delft University). Part of the research has been used for a Ph.D project under the supervision of Prof. Dr. Arie Havelaar of the Institue for Risk Assessment Sciences (IRAS) of the University Itrecht (The Netherlands). The author is gratefull to Gertjan Medema and Arie Havelaar as the chief advisors in the overall work and to Dick van der Kooij (KWR Watercycle Research Institute) as the initial advisor of the project for their support and inspiring discussions.

 Quantitative Methods to Assess Capacity of Water Treatment to Eliminate Micro-Organisms. Edited by Wim A.M. Hijnen. ISBN: 9781843393764. Published by IWA Publishing, London, UK.

Summary

RATIONALE

Ever since the recognition of the important role of water in the transmission of pathogenic micro-organisms in the 19th century, microbiological safety of drinking water has been a major research issue for microbiologists in drinking water industry. In central water supply the safety of drinking water is controlled by means of treatment and water quality assessment. Chapter 1 describes the major historical developments in microbiological water quality assessment and regulations related to the transmission of waterborne pathogens of faecal origin. Standards for bacteria of faecal origin such as *Escherichia coli* and coliforms in drinking water as indicators for the presence of these pathogens have been implemented in legislations world wide. Additional drinking water standards for other indicators of faecal pollution such as enterococci and spores of sulphite-reducing clostridia (SSRC) have been implemented in regulations of selected parts of a number of countries in the world among which also the Netherlands and other European countries.

Recent epidemiological studies and outbreak reports in the United States (US) and the United Kingdom (UK) have demonstrated the importance of persistent pathogens such as *Cryptosporidium* and *Giardia* and some enteroviruses and of the pathogenic bacteria *Campylobacter* and *E. coli* O157 for drinking water safety. Inadequacies in water treatment in combination with the high virulence of these pathogens assessed in dose-response studies and the failure of faecal indicators to detect them in drinking water were important causes of the growing concern for these waterborne pathogens.

A major scientific breakthrough was the introduction of the estimation of the probability of infection given a certain dose of a selected pathogen. This enabled the introduction of quantitative health-related targets in the microbial risk management of drinking water which can be verified by means of Quantitative Microbial Risk Assessment (QMRA).

Intrinsically the safety of Dutch drinking water was assumed to be higher than the drinking water in the US and the UK because of comprehensive water treatment with multiple barriers and standards with persistent

 Quantitative Methods to Assess Capacity of Water Treatment to Eliminate Micro-Organisms. Edited by Wim A.M. Hijnen. ISBN: 9781843393764. Published by IWA Publishing, London, UK.

(*Clostridium* spores). The few reported waterborne outbreaks in the Netherlands were related to cross connections with sewage and not to inadequate drinking water quality. Though this assumption is regarded as plausible, in their aspirations to distribute a high quality drinking water Dutch Water Companies started an investigation in the joint research program to verify this. Simultaneously Dutch Authorities anticipated on these developments by implementing an accepted infection risk level of 1 person per 10.000 consumers per year (10^{-4}) for drinking water in the revised Drinking Water Decree. This health-related target should be verified by means of QMRA. Important step in this QMRA is the exposure assessment by monitoring pathogens in the source water and determination of the elimination (inactivation and removal) of these pathogens in water treatment.

OBJECTIVES

Main objective of the current study was to develop a general strategy to assess elimination capacity of water treatment processes for pathogens. For this purpose the following aspects were investigated:

- the potential use of faecal indicators Coli44 (incl. *E. coli*) and SSRC (incl. *C. perfringens*) as process indicators to assess pathogen elimination in full-scale water treatment plants.
- the value of comparative challenge tests with pre-cultured organisms for the assessment of elimination capacity of full-scale processes, to study the effect of process conditions and to validate the use of process indicators.
- the use of literature data to assess elimination capacity of water treatment processes for pathogens and the effect of process conditions on this.

INDICATOR BACTERIA TO ASSESS ELIMINATION CAPACITY OF FULL-SCALE TREATMENT

Research after the use of faecal indicators as process indicators to determine removal of pathogens in full-scale water treatment is described in the Chapters 2, 3 and 4. Drinking water Companies have generated a large amount of data on microbiological quality of the source water and the drinking water and the water after selected steps in treatment. These data proofed to be useful in assessment of the decimal elimination capacity (DEC; logremoval) of the first processes of water treatment inclusive the variation of this capacity, but the data were unsuited to do the same for the overall treatment (Chapter 2). To achieve the latter a method had to be developed with which these indicator bacteria could be detected in the drinking water. For this purpose larger water volumes were examined either with the standard membrane filtration method used to determine the concentration of indicator bacteria in water or with a special developed device, the MF-sampler. With the MF sampler the limit of detection of the standard method was lowered by a factor of 1000 or more (Chapter 3). The large volume sampling method (LVS) method yielded reproducible results with a high recovery of 75–100% assessed with dosage experiments and could easily be implemented in the daily routine of the drinking water practice. With this LVS method the concentration of Coli44 and SSRC in the water during water treatment of ten facilities was measured in a collaborative investigation with the drinking water Companies (Chapter 4).

Both indicators were detected in the finished waters with SSRC in higher concentrations than Coli44. The study yielded information on DEC of individual processes and overall treatment facilities for both indicators and its variability. A variation in elimination was observed for individual processes over time and for similar processes at different locations. Moreover, the results clearly indicated that elimination of environmental micro-organisms under full-scale conditions is lower than expected from studies presented in literature applying challenge tests with pre-cultured micro-organisms. These results emphasize the need to assess DEC of full-scale processes as local specific as possible preferably with environmental micro-organisms.

LIMITATIONS OF THE INDICATOR BACTERIA METHOD

The indicator bacteria method with LVS appeared to have a number of limitations. First of all the attention is mainly focused on the elimination of bacteria and protozoa and no information is collected on the elimination capacity of treatment for viruses. At a number of processes (slow sand filtration, surface water infiltration and post disinfection) the concentrations of indicator bacteria were too low to enable DEC assessment or some processes (UV disinfection) were not applied during the course of the study. Furthermore, the method is not suited to control DEC of processes in the daily drinking water routine. For this purpose knowledge is required on the process conditions directly affecting the DEC. Finally, the predictive value of Coli44 and SSRC for the elimination of *Campylobacter* bacteria and both protozoa, *Cryptosporidium* and *Giardia* had to be elucidated.

Because of these limitations and additional questions alternative strategies were applied to determine the DEC of processes. Challenge tests were performed with set ups at different scales (pilot plant and laboratory) and different pre-cultured micro-organisms, among which bacteriophages as a process indicator for virus removal.

ELIMINATION BY DISINFECTION

In the framework of the study two comparative studies have been conducted on the efficacy of disinfection processes. In Chapter 5 the susceptibility of spores of *Clostridium perfringens* and oocysts of *Cryptosporidium parvum* has been compared for ozonation in a continuous flow set up as model for a full-scale ozonation installation. The study showed that SSRC can be applied as process indicator for inactivation of *Cryptosporidium* by full-scale ozonation. The susceptibility of both organisms was in the same order of magnitude with an index pathogen/process indicator (IP/PI) ratio of 0.8. The inactivation data assessed with the applied continuous flow laboratory set up were predictive for inactivation of SSRC under full-scale ozone conditions. The inactivation in the gas feed chamber was significant and proportional to the calculated Ct-values (mg/l.min). When Ct of the process is controlled with the ozone concentration the dose-effect relation is in agreement with the Chick-Watson inactivation kinetics. Setting Ct in the range of 0.5–1.5 (normal for the Dutch situation) with flow and contact time at a constant low ozone dose showed to be ineffective to control the inactivation capacity of the process.

The susceptibility to Ultraviolet radiation (UV) of micro-organisms relevant for drinking water safety has been assessed and compared in Chapter 6 on the basis of literature data. Challenge test studies with a laboratory set up (collimated beam) and continuous flow systems as well as studies with environmental micro-organisms organisms in full-scale UV systems were used. This evaluation showed a higher inactivation of pre-cultured organisms in the collimated beam set up compared to the inactivation of environmental organisms of the same species in continuous flow systems. This discrepancy is caused by factors related to the micro-organisms and to the process conditions. These factors were evaluated on the basis of the available literature. A required UV fluence table for relevant micro-organisms was described in which the required fluence for bacteria and bacterial spores was corrected for this discrepancy between laboratory data and field observations. For viruses such a correction seemed to be not necessary and due to a lack of data such correction was not possible for the protozoa.

ELIMINATION BY FILTRATION

Filtration processes applied in drinking water production focused on the elimination of micro-organisms are slow sand filtration and soil passage. In Chapter 7 an investigation is presented on the elimination capacity of slow sand filtration assessed under different conditions. The results demonstrated that DEC for Coli44 determined under full-scale conditions was 2–3 log and in the same order of magnitude as DEC determined for *E. coli* in pilot plant filters operated in a pilot plant as dummy of the same treatment facility. DEC assessed in a column test was clearly lower, possibly as a result of the lack of a schmutzdecke in the column and the difference in hydraulic conditions of the filter beds. In a filter with a schmutzdecke the DEC for *E. coli* was 1–2 log higher than in the same filter without this layer. The effect of the schmutzdecke on the removal of the bacteriophage MS2 as a conservative model for viruses was marginal. These dosed organisms were removed with 1.5–2.0 log by these filters. Coli44 appeared to be a safe process indicator for the elimination of *Campylobacter* bacteria.

Removal of environmental SSRC by full-scale slow sand filtration ranged from −0.2–1.8. In a column test with short columns (0.4 m) removal of dosed spores of *C. perfringens* was higher (2.3–3.2 log). Oocysts of *Cryptosporidium* were removed in these columns even more (>5.3–>6.5 log). These data implicate that SSRC is not a proper process indicator for *Cryptosporidium* elimination by slow sand filtration. Hypothesis for the variable and sometimes low elimination of *Clostridium* spores by these filters was accumulation, survival and delayed transport. Further research after this phenomenon was done in a pilot plant filter as dummy of the full-scale filters (Chapter 8). Elimination of spores of *C. perfringens*, UV inactivated oocysts of *Cryptosporidium parvum* and centric diatoms (*Stephanodiscus hantzschii*) was determined with a prolonged challenge test with a filter operated three years in advance. Centric diatoms are persistent organisms with a size similar to the size of oocysts which were present in the influent in considerable higher concentrations than environmental SSRC. The DEC calculated from the experiment for oocysts, spores and diatoms was 4.7, 3.6 and 1.8 log, respectively. The duration of the test of three months was too short to demonstrate the occurrence of accumulation and delayed transport for spores and oocysts. On the basis of the low DEC for centric diatoms naturally present in the influent in sometimes high concentrations and the spatial distribution of spores of *C. perfringens* in the filter bed determined at the end of the dosage experiment it was concluded that this process is valid for both persistent organisms. However, most likely not for persistent organisms which are sensitive to predation such as oocysts. Mass balance calculations revealed a large loss of oocysts numbers in the filter bed. In the zooplankton population observed in the sand species were identified capable of ingesting oocysts. Because the notified predators were observed in the filter bed and in the filtrate, delayed transport and breakthrough of viable oocysts in predators as a vector was suggested as an issue for further research.

The elimination capacity of soil passage under field conditions is usually not simple to assess. In Chapter 9 a challenge test is described with columns filled with soil material from two surface water infiltration sites in the Netherlands loaded with the local surface water spiked with MS2 bacteriophages, *E. coli* and spores of *C. perfringens* and oocysts of *Cryptosporidium* and *Giardia*. From both sites quantitative information was available on the elimination of dosed MS2 phages and indigenous bacteriophages and indicator bacteria during soil passage. Comparison of the results under both conditions (filed and columns) showed clearly that column tests are not applicable for the assessment of the elimination capacity of soil passage under field conditions.

The experiment described in Chapter 9 with different organisms and two soil types demonstrated that besides size of micro-organisms and soil grains, surface properties of organism and soil and soil characteristics such as uniformity and conductivity affect the transport of micro-organisms. The delayed and second breakthrough peaks of oocysts and spores in the same soil columns and the low DEC of environmental Coli44 and SSRC, emphasize the importance of desorption in transport of micro-organisms in soil. From the results it was concluded that both straining and adsorption/desorption are mechanisms which, dependent on the conditions in different ratio contribute to the elimination of micro-organisms in soil passage. The IP/PI ratio for *Clostridium* spores and protozoan (oo)cysts assessed from this experiment was variable and higher or lower than 1.0 meaning that SSRC is not a proper process indicator for the removal of these index pathogens by soil passage. This was also concluded from the challenge test with granular activated carbon (GAC) columns on pilot plant scale described in Chapter 10. MS2 phages were not removed and both *E. coli* and SSRC removal in these columns was limited (≤ 0.5 log). Protozoan (oo)cysts, however, were removed with a capacity of 1.3–2.7 log (average of 2.0 log).

Beside differences in transport behaviour also the difference in survival rate of spores and protozoan (oo)cysts hampers the use as process indicator for filtration processes with risk of accumulation. Clostridal spores survive in sand and water for more than ten years (Chapter 11).

GENERIC METHODOLOGY

On the basis of the discussion in Chapter 10 a generic methodology was described to assess the elimination capacity of local water treatment processes for index pathogenic micro-organisms. This information is the input for the mandatory quantitative microbial risk assessment which has to be performed for drinking water production locations using surface water or vulnerable groundwater sites as the source water.

The most optimal methods are the assessment of elimination of indigenous index pathogens and indicator bacteria (Coli44 and SSRC) in treatment by large volume sampling. Because for index pathogens this is not feasible for most locations and certainly not for the total treatment, Coli44 is considered as an environmental process indicator for the elimination of *Campylobacter* by most processes and SSRC as a process indicator for *Cryptosporidium* and *Giardia* for coagulation/floc-removal, rapid granular filtration and ozonization. Further research is required to determine the necessity and possibility to add monitoring of bacteriophages as a specific process indicator for virus elimination. Because for some processes monitoring for elimination of indicator bacteria is not possible due to low influent concentrations challenge tests with pre-cultured micro-organisms are necessary to collect additional quantitative data. Quantitative data on elimination of these pre-cultured in challenge tests can be applied for DEC of full-scale processes when the tests were performed in a pilot plant operated as a dummy of the full-scale plant. Data from challenge tests on laboratory scale with multiple organisms (batch inactivation tests or filtration columns tests) must be regarded as indicative for DEC under full-scale. The value of these tests lies in the assessment of the index pathogen/process indicator ratio (IP/PI) and the elucidation of specific conditions on the elimination process. When quantitative data on elimination of indicator bacteria in the full-scale treatment and from additional challenge tests are not available, literature is an important potential source of quantitative information of elimination of micro-organisms in water treatment. Additionally these data can be used by the translation of challenge tests results to full-scale conditions.

GENERAL CONCLUSION AND RECOMMENDATIONS FOR FURTHER RESEARCH

A methodology has been developed to assess the Decimal Elimination Capacity (DEC) of full-scale water treatment processes for waterborne pathogenic micro-organisms. The methodology is tailored to the natural conditions of the water treatment and to the daily monitoring practice of microbiological water quality as much as possible.

An issue of further research is the determination of the DEC of full-scale treatment for viruses using environmental bacteriophages as potential process indicators. In relation to this the application of ultrafiltration in the recently developed hemoflow method has to be evaluated. With this method multiple micro-organisms (i.e. bacteriophages and faecal indicator bacteria) are determined in large volumes of water. For a number of water treatment processes additional knowledge on the elimination capacity and the IP/PI ratios has to be collected. These processes are granular activated carbon filtration, postdisinfection with chlorine dioxide and ultrafiltration. The effect of the intended change in legislative water quality monitoring from spores of sulphite-reducing clostridia to specific *C. perfringens* inclusive vegetative cells, an analysis without pasteurization, on the use of this parameter as a process indicator is an issue of future attention. Predation in biological filters is potentially a beneficial elimination mechanism in water treatment for water safety provided that the predators are no vectors of transport for the ingested pathogens. This is an issue of importance for further research.

Chapter 1

Microbiologically safe drinking water

Wim A.M. Hijnen

KWR Watercycle Research Institute

ABSTRACT

The management of the microbiological drinking water quality in the Netherlands has been changed in the past ten to fifteen years. One of the major changes was the shift from curative quality management to a more preventive strategy. In the first part of this first Chapter the historical perspectives of the microbiological quality of drinking water from the late 19th Century untill the beginning of the beginning of the 21th is described. Motives for the change from curative to preventive managements were epidemiological studies on waterborne diseases, scientific knowledge on the dose-response data of some waterborne pathogenic micro-organisms and shortcomings in water quality control. In the Netherlands this resulted in the introduction of a health based microbial target and the requirement to assess this target with Quantitative Microbial Risk Assessment (QMRA) in the revised Drinking Water Decree in 2001. Important requirement for QMRA is quantitative information on the capacity of drinking water to eliminate (remove or inactivate) micro-organisms.

Overall objective of the present study was to develop a generic methodology to collect this information. The line of approach in this study was to develop a method as closely related to the natural conditions in water treatment and to the daily practice of microbiological water quality monitoring. This is described and motivated in the second part of this Chapter where the hypothesis was introduced that beside their role as indicator of faecal pollution, faecal indicator bacteria can be used as process indicators for the elimination of different waterborne pathogens. In the last part of this Chapter the overall objective is specified in a number of sub-goals on the basis of research needs. In the outline paragraph the separate Chapters are mentioned focused on the sub-goals of the study.

 Quantitative Methods to Assess Capacity of Water Treatment to Eliminate Micro-Organisms. Edited by Wim A.M. Hijnen. ISBN: 9781843393764. Published by IWA Publishing, London, UK.

WATERBORNE DISEASES AND DEVELOPMENTS IN CENTRAL WATER SUPPLY

Throughout the ages, the use of fresh water by mankind for drinking was dictated by the way that they lived. As nomads, people used water available from the rivers and streams in their immediate vicinity. When they settled, and civilizations developed, more advanced centralised water supplies with mains were installed. Well known are the aqueducts from the Roman civilization. Already in older civilizations like that of the Egyptians, Persians and Pakistanis evidence of centralized water supply have been found (Wijmer, 1992). After the Roman civilization vanished from Northern Europe, their achievements in water supply disappeared for the common people. Only for the rich people, churches and monasteries a kind of central water supply from wells remain in practice. It was only since the 19th century, the century of the industrial revolution, that central drinking water supply for the more settled communities at that time was developed. Crucial for this development was the growing knowledge about waterborne diseases. Even after more than hundred years this development is valued highly as shown by a recent held internet poll by the British Medical Journal. Sanitation and clean water was seen as the major milestone in medical advances since 1840 (http://www.bmj.com/cgi/content/full/334/suppl_1/DC3).

Centralised public drinking water supplies were introduced in the 19th century and one of the major drivers for this development was public health. John Snow was the first (1849) to relate the use of contaminated water with the incidence of cholera in the population, one of the major infectious diseases at that time (Sheppard, 1995). After several outbreaks of cholera in 1832/33, 1848/49, 1853, 1854, 1855, 1859 and 1866, the role of drinking water in this severe public health problem was also recognised in the Netherlands. A national Dutch committee 'Tot onderzoek van drinkwater' (Weelden and Mingelen, 1868) came to the conclusion that a 'cholera-germ' originating from the faeces of infected people was the cause of the distribution of this disease. Pasteur (1822–1895) and Robert Koch (1843–1910) were the first scientists who showed that micro-organisms can cause diseases. Koch developed a method to cultivate these micro-organisms on a solid medium (Koch's 'Plattengussverfahren'). With this microbiological method Koch isolated the bacteria which caused cholera later known as *Vibrio cholerae* (Koch, 1884). He introduced the following postulates to identify a micro-organism as a pathogen:

- The micro-organism must be found in all organisms suffering from the disease, but not in healthy organisms.
- The micro-organism must be isolated from a diseased organism and grown in pure culture.
- The cultured micro-organism should cause disease when introduced into a healthy organism.
- The micro-organism must be re-isolated from the inoculated, diseased experimental host and identified as being identical to the original specific causative agent.

To demonstrate the significant role of centralised drinking water supply in improvement of public health, the decrease in frequency of typhus in the Dutch population, another waterborne disease, was correlated with the increase in number of inhabitants which were connected to a central drinking water supply (Figure 1.1).

Nowadays the quality of drinking water in the Netherlands is usually valued highly by consumers with regard to both perception and safety. The consumer trust in Dutch drinking water was not always obvious in recent history, as observed for instance around in the end of 19th century in Rotterdam (Wijmer, 1992). Confidence of the consumers in the city of Rotterdam in the distributed drinking water was undermined by the distribution of turbid water caused by growth of organisms in the distribution system. The public suspected a relationship with the drinking water and the occurrence of diarrhoea although this was not supported by evidence. This situation got worse due to a local waterborne outbreak of typhus. A point source sewer leakage into the deteriorated drinking water distribution network was found to be the cause and due to this incident the two Water Company managers responsible voluntarily resigned (Wijmer, 1992).

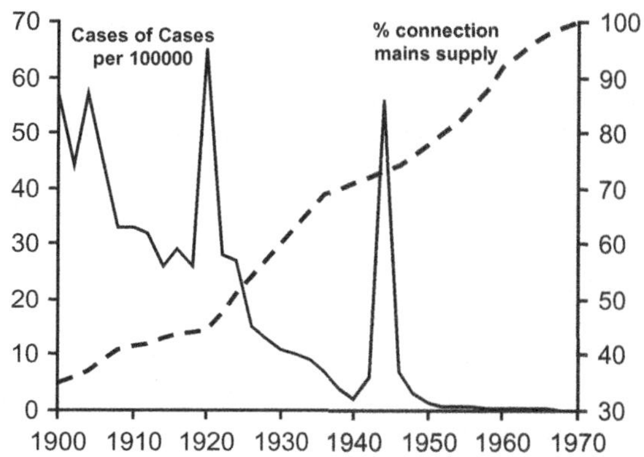

Figure 1.1 Frequency of typhus and the percentage of inhabitants connected to central drinking water supplies in the period 1900–1970 (from Van der Kooij, 2002; sources: CBS and VEWIN)

In the recent period of approximately 1960 until the present day only three waterborne outbreaks have been reported in the Dutch drinking water practice related to the consumption of faecal contaminated drinking water. The first one related to contamination with sewage occurred in 1962 when 5 cases of typhoid fever were reported in Amsterdam (Anonymous, 1962). The second one was a larger outbreak with 609 reported cases of mainly gastroenteritis caused by wastewater from a Marine vessel pumped into the drinking water mains by a mal-connection (Huisman and Nobel, 1981). The last reported outbreak dates from 2001, with an excess of gastroenteritis cases in a new residential area with the application of a dual water system with grey water. A cross connection in this system caused the consumption of partially treated river Rhine water (Fernandes *et al.*, 2007).

During the late 20th century a number of events, observations, scientific achievements and developments in society influenced the common practice associated with the production of microbiologically safe drinking water in the developed countries. Waterborne outbreaks and systematic epidemiological studies in US and other developed countries showed that consumers in these countries may be exposed to pathogens in drinking water. Due to increased urbanization surface water pollution increased and not only the amount of known waterborne pathogens increased but also the variety including more persistent species such as some viruses and protozoa (*Cryptosporidium* and *Giardia*). Moreover, it was demonstrated that current water treatment is not an absolute barrier against pathogens, and compliance with the coliform standard is not always sufficient to safeguard public health against these waterborne pathogens. In the following parts of this introduction these developments will be highlighted.

EPIDEMIOLOGICAL STUDIES ON WATERBORNE DISEASES

Major pathogenic micro-organisms causing waterborne diseases in the 19th century were bacteria (*V. cholerae*, *Salmonallae* and *Shigella* spp.). Epidemiological studies during the last period of the 20th century in the United States of America (Craun, 1988; Craun, 1990; Craun *et al.*, 1998; Herwaldt *et al.*, 1992 and Moore *et al.*, 1994) and United Kingdom (Galbraith *et al.*, 1992) showed that major causative pathogens of outbreaks of waterborne diarrhoea in the current time, however, are no longer bacteria but viruses and pathogenic protozoa (*Cryptosporidium* and *Giardia*). The outbreak of cryptosporidiosis in Milwaukee (US) was one of the largest reported outbreaks (MacKenzie *et al.*, 1994). Moreover, when bacteria are involved other species than *V. cholerae*, *Salmonallae* or *Shigella* spp. have become of importance such as *Campylobacter* spp. and enteropathogenic

E. coli O157. A recent local waterborne outbreak in Canada was related to the consumption of contaminated drinking water and both *C. jejuni* and *E. coli* O157 were identified as the main pathogens (Hrudey *et al.*, 2003). Epidemiological studies in the Europe also demonstrated a prominent role for *Cryptosporidium* in waterborne outbreaks (Medema *et al.*, 2006a) and a shift towards *Campylobacter* as the causative agent in the reported bacterial outbreaks. The prevalence of these new waterborne pathogenic bacteria in Dutch surface water was demonstrated recently by de Roda Husman *et al.* (2001) and Heijnen and Medema (2006).

Epidemiological surveys on waterborne outbreaks are valuable in assessing the emerging pathogens for drinking water safety. Another type of epidemiological study is the randomised controlled trial (RCT). With such a study Payment *et al.* (1991, 1997) demonstrated that a relatively high percentage of annual gastrointestinal illnesses among the consumers (14–40%) in a water supply were water-related. In these studies, however, consumers were not blinded to the type of water they received. Follow up RCT studies which were double- and triple-blinded (Hellard *et al.*, 2001 and Colford *et al.*, 2005) showed no significantly increased numbers of gastrointestinal illnesses as a result of the consumption of less treated drinking water. Nevertheless, the level of endemic disease due to public drinking water systems remains difficult to quantify and well documented waterborne outbreaks are still the main source of epidemiological information used by microbiologists to develop new microbial regulations on drinking water.

CONTROL OF WATERBORNE DISEASES: WATER TREATMENT

The rationale behind treating water prior to distribution for potable use was the idea that it would protect the public from waterborne diseases like cholera. After further exploration with his 'colony count' method during the Hamburg Outbreak in 1892, Koch (1893) demonstrated that slow sand filters were effective barriers against cholera and were able to reduce the number of bacteria expressed in colony counts to less than 100 per ml. Frankland and Frankland (1894) summarized the scientific knowledge on water microbiology that was available at that time. They concluded that the number of bacteria in water is related to the pollution rate of the water and that slow sand filtration is capable to reduce the number of bacteria in water with more than 90%.

Slow sand filtration was a first step in water treatment introduced by John Gibb in 1804. But in the twentieth century it became clear that the process has its limitations (rapid clogging, breakthrough at low temperatures and incomplete removal of undesirable organic compounds). Physical/chemical processes like coagulation and sedimentation as well as chemical disinfection with chlorine were added to treatment practices of drinking water. John Snow was one of the first to use chlorine in order to disinfect drinking water (White, 1999). It was not until the beginning of the 20th century that the use of chlorine as a disinfectant became common practice in drinking water supply. In 1906, the first ozonation plant for disinfection was built in France. The driving force for this development was not only public health, but also the growing public demands for clear and aesthetically sound water. Besides these disinfection practices, from 1920 facilities started to use sedimentation, filtration and chlorination. Under the influence of ongoing environmental pollution granular activated carbon (GAC) filtration was introduced to remove organic micro-pollutants. The discovery of the production of tri-halomethanes (Rook, 1974) and other disinfection by-products (DBP) by chlorine caused a decline in the use of disinfection with chlorine in the Netherlands. Ozonation as an alternative for chlorination was introduced, because this disinfectant was effective against the chlorine resistant *Cryptosporidium* (Finch *et al.*, 1993). Due to the discovery of the production of the mutagenic disinfection by-product bromate by ozone in bromide containing waters (Haag and Hoigné, 1983), this disinfectant is still not used worldwide in water industry. Though Havelaar *et al.* (2000) demonstrated in a case study that the microbial benefits of ozone disinfection can outweigh the health risks of bromate by a factor of 10 on the basis of a single measure of disease burden, the Disability adjusted life-year (DALY). During the last decade advances in technology and scientific knowledge resulted in application of membrane filtration and UV disinfection in drinking water production. The reason for the renewed interest for UV is that despite earlier communications, recent studies showed high efficacy against *Cryptosporidium* (Clancy *et al.*, 1998).

CONTROL OF WATERBORNE DISEASES: WATER QUALITY

Monitoring for pathogenic micro-organisms. As described previously, microbiologically safe drinking water was one of the major driving forces in the development of the drinking water treatment. During the last period of the 19th century, after the methodology breakthrough primarily caused by the work of Koch, considerable effort was put into the assessment of the microbiological quality of drinking water. The work was primarily focussed on the direct isolation and identification of pathogenic bacteria because it was assumed that bacteria in water might cause diseases. Scientific knowledge evolved to the general understanding at the end of the 19th century that not all bacteria present in water are pathogenic (Frankland and Frankland, 1894). Moreover, direct monitoring of pathogenic micro-organisms in drinking water proved to be difficult and not practical for monitoring the safety of drinking water. Though knowledge increased over the last hundred years this is still the current situation. There is a great diversity of pathogenic micro-organisms of faecal origin in surface water and usually the concentrations are low and variable. Some of these micro-organisms can not be cultured, such as noroviruses. For others, methods are complicated with low recovery efficiencies and not selective for the species significant to human health (*Cryptosporidium*). The introduction of molecular methods to detect micro-organisms in water offers additional possibilities to detect non-culturable pathogenic species with relatively quick methods. Currently this new analytical technology is still in the stage of development.

Faecal indicator bacteria: coliforms and E. coli. Although the intensive microbiological studies during the last part of the 19th century did not result in a practical strategy to assess safety of water with direct pathogen monitoring, an alternative strategy was developed. *Bacterium coli-commune* (corresponding approximately to *Escherichia coli*) was isolated by Escherich (1885) from the faeces of a patient suffering from cholera. It was subsequently found that this bacterium, capable of fermenting lactose with the production of acid and gas, was a normal inhabitant of the intestinal tract of man and many other animals and is excreted with the faeces in the environment. Schardinger (1892) and Smith (1895) independently introduced the examination of potable water for *Escherichia coli* to demonstrate pollution with faecal matter, a serious health threat as illustrated by the presence of cholera bacteria in faeces of infected persons. *E. coli* turned out to be a member of a larger group of bacteria defined as coliforms, gram-negative non-spore forming and rod-shaped bacteria capable to ferment lactose with gas formation within 48 hours at 35°C in selective media with briljant green and bile-salts. This group of coliform bacteria is used in common microbiologically drinking water quality monitoring for faecal contamination all over the world for most of the last century. The widespread presence of coliforms in the environment made it clear that probably not all coliforms were of faecal origin. Eijkman (1904) was the first to apply higher incubation temperature (46°C) to select for coliforms strictly related to faecal contamination and hence restricting it largely to the detection of *B. coli-commune* (*E. coli*). Later the determination of coliform species producing gas from lactose at 44.5°C within 24 hours was introduced as faecal coliforms or thermotolerant coliforms in water quality monitoring. Only recently in European regulation the use of the coliform group in water sanitation was further restricted *E. coli*, because regrowth of thermotolerant coliforms has been demonstrated in practice (Caplenas and Kanarek, 1984).

Additional faecal indicators: faecal streptococci and C. perfringens. Around the turn of the century (19th–20th) there were other bacteria identified as bacteria related to sewage pollution, such as faecal streptococci (Houston, 1899) and *Bacterium enteritidis sporogenes* later known as *Clostridium perfringens* (Klein, 1897–1898). The genus *Enterococcus* was first described by Thiercelin (1899). Sherman (1937) was the first one to classify the genus Streptococcus in the Pyogenes, Viridans, Lactic and Enterococcus groups (Godfree *et al.*, 1997). The first descriptions of *Clostridium perfringens* were given by Achalme (1891–1897) and Welch and Nuttall (1892) based on observations in tissue autopsies from patients. They described the observed organism as a lactose-fermenting spore-forming anaerobic bacterium. In 1899 Klein and Houston (1898–1899) proposed the use of *Clostridium welchii* or *Clostridium perfringens* as indicator for faecal pollution. According to Bonde (1962), they were the first to introduce the term 'bacteria of indication' for faecal pollution, for the coli-aerogenes group, faecal streptococci and *Bacterium enteritidis sporogenes* (*C. perfringens*).

Wilson and Blair (1925) showed a relationship between the presence of anaerobic sulphite-reducing spore-forming bacteria (observed in glucose-sulphite-iron agars) and the presence of *E. coli* in water. In addition, Wilson and Blair (1931) suggested that since *C. welchii* or *C. perfringens* is essentially a faecal organism and since its spores may persist long after other indicators of contamination such as coliform bacteria have disappeared, this anaerobe may well serve as an indicator of intermittent pollution.

Prescott *et al.* (1945) reviewed the developments of coliforms and 'other intestinal bacteria which have been used as indices of pollution', the sewage streptococci and spore-forming, lactose-fermenting anaerobes. This review of the potential use of both alternative organisms as faecal indicators was mainly focussed on data comparison with the coliform test. Based on the reviewed literature they came to the conclusion that streptococci 'adds little to the information furnished by the test of coliform organisms' and that the presence of *C. perfringens* indicate a faecal pollution 'so remote as to be of little or no sanitary significance'.

Later, Bonde (1962) reviewed the historical developments on both alternative indicators in the early days of developments in water microbiology. He stated as follows: 'The Royal Commission on sewage disposal in 1898 (UK; Klein, Houston, Adeney, Frankland, MacConkey) initiated research which revealed that concentrations of both indicators, faecal streptococci and *B. enteritidis sporogenes* (*C. perfringens*), were lower than the concentrations of *E. coli* and the results indicated insufficient reliability of both microbiological parameters.' Moreover, further developments with respect to coliform monitoring (Eijkman, 1904 and MacConkey, 1905) and determination of *E. coli*, forced the enteritidis test of Klein (1895) as well as the faecal streptococci as indicators into the background of microbiological water quality monitoring. In his studies, Bonde (1962, 1977) found similar results: no relationship between the presence of *C. perfringens* spores and coliforms/*E. coli*. Counts of *C. perfringens* in environmental water samples were sometimes up to a factor of 100 lower than *E. coli* numbers. He observed a better quantitative relationship between *Streptococcus faecalis* and *E. coli*, but still the *E. coli* numbers were usually higher. Despite these findings he reasoned in his extended thesis that the relationship of *C. perfringens* with *E. coli* was not to be expected. Furthermore, in his opinion *C. perfringens* meets the specifications for a proper indicator to a higher degree than does *E. coli*. The organism must be considered to be a faecal organism, excreted always together with a possible faecal pathogenic organism and present in environmental water samples in numbers often as numerical as the *E. coli* numbers. The indicator is more persistent towards environmental stress and disinfectants than the pathogens and can be identified with rapid and unambiguous methods of determination. Cabelli (1977) supported this, by mentioning the ubiquity of *C. perfringens* in nature, primarily because they are spore-forming bacteria, as an advantage. In the more recent literature the use of spores of sulphite-reducing clostridia (SSRC) including *C. perfringens* has been used by several authors to assess the environmental pollution with sewage of remote or intermittent nature and in situations where resistance to disinfectants and environmental stress is at a premium (i.e. Fujioka and Shizumura, 1985; Sartory, 1988; Sorensen *et al.*, 1989; Hill *et al.*, 1993; Coffey *et al.*, 1999; Robles *et al.*, 2000 and Buchholtz ten Brink *et al.*, 2000).

Quantitative approach of microbiological parameters. Klein and Houston (1898, 1899) showed the superiority of microbiological methods to chemical methods in assessing water safety as described by Bonde (1962). Pollution of water with sewage could be detected by *B. coli* at a ratio of 1 : 20.000 and with the enteritis test for *Bacterium enteritidis sporogenes* (*C. perfringens*) at a ratio of 1 : 500.000, despite the lower prevalence of this indicator in natural waters. With chemical methods it was hardly possible to detect one part of sewage diluted in 1000 parts of pure water. The plate method using solid media limited the amount of water for examination to 1 ml. Phelps (1907 ref. in Prescott *et al.*, 1945) suggested that the reciprocal of the highest dilution in the liquefied coliform test in which a positive result is obtained should be used as the expression of the number of coliforms present in the examined water sample. This approach was further explored and resulted in a quantitative assessment of indicator bacteria in water samples by means of the Most Probable Number (MPN).

The use of the membrane filtration to filter larger volumes than 1 ml followed by incubation on solid growth media, was introduced in 1953 (Bush, 1953). Windle Taylor *et al.* (1955) attempted the use of membrane filtration for coliform counting and demonstrated high number of false positives at 37°C. By increasing the incubation temperature to 44°C and using resuscitation, he found close agreement between the results of the dilution in liquid media (MPN) and this membrane filter method. False positives and overgrowth was mentioned by Bonde (1977) as major drawbacks associated with the use of membrane filtration. With the development of more selective culture media and improvement of the general water quality, these drawbacks were reduced to a large extent. The use of membrane filtration as a quantitative microbial method in order to examine larger volumes than 1 ml has become more and more common practice and is currently the most commonly used method in microbial water quality assessment.

Dutch legislation on microbiological quality of water. Nowadays, standards for coliforms and also for the other faecal indicators in drinking water are implemented in drinking water legislation worldwide. Due to methodological optimization and increased knowledge on coliforms in the environment, standards in the Dutch Drinking Water Decree changed from coliforms to thermotolerant coliforms in the Drinking Water Decree of 1984 (Anonymous, 1984; Table 1.1).

Table 1.1 The Maximum Tolerable Concentrations (MTC; colony forming units or cfu) of indicator bacteria in the Dutch drinking water as prescribed in 'normbladen; N' before 1960 (Anonymous, 1942, 1956), in the first Dutch Drinking Water Decree of 1960 and the adapted Decree in 1984

	N1028/N3034 1942/1956	Decree 1960	Decree 1984
Coliforms[a]	≤1 per 50 ml	0 per 200 ml	0 per 300 ml
Therm. lactose-ferm. bacteria[a]	0 per 50 ml	0 per 50 ml	–
Thermotolerant coliforms[a]	–	–	0 per 300 ml
Faecal enterococci[b]	0 per 20 ml	0 per 10 ml	0 per 100 ml
Spores of sulphite-reducing clostridia[c]	–	–	0 per 100 ml

Methods of assessment: [a]Anonymous, 1981; [b]Anonymous, 1982; [c]Anonymous, 1985a

Introduction of the other indicator bacteria, faecal streptococci and *Bacillus enteritidis sporogenes* (*C. perfringens*), to examine drinking water was suggested previously by de Graaf (1922). It was more than 50 years later until both indicators were implemented in Dutch legislation. These standards for indicator bacteria in drinking water implemented in the Dutch Decree of 1984 are described in detail by Havelaar (1983) and Van der Kooij (2002) and were based on the harmonisation of drinking water regulations in the European Community (80/778/EEG, Pb. L 229/11). The background of implementation of streptococci and SSRC was explained by Havelaar (1983) as follows:

- the higher persistence of faecal streptococci and spores of sulphite-reducing clostridia (SSRC) against environmental stress compared to coliforms;
- the higher specificity of faecal streptococci as faecal indicator;
- the high resistance of SSRC to common used chemical disinfectants comparable to persistent pathogenic protozoa.

Additional to the these new drinking water standards in the Dutch Drinking Water Decree of 1984 was the requirement for source water monitoring in order to indicate the treatment required to obtain safe drinking water from different sources of surface water with different state of pollution (Table 1.2). Again this regulation was based on a European directive related to the required quality of surface water used for production of drinking water (Anonymous, 1975).

Table 1.2 Microbial standards for source water quality (n/100 ml) and water treatment

Class	Description and requirements	Coliforms	Therm. coliforms Coli44	Faecal streptococci
I	*Low level of contamination: simple treatment (Rapid granular filtration and disinfection)*	100	20	10
II	*Moderate domestic/industrial contamination: Coagulation/floc removal, rapid granular filtration, oxidation/disinfection*	10.000	2.000	1.000
III	*Heavily domestic/industrial contamination: as II plus Cl_2, Granular Activated Carbon*	100.000	20.000	10.000

Nomenclature of indicator organisms. As described before Klein and Houston (1899) were the first the use the title 'bacteria of indication of faecal pollution' for three different bacteria implemented in Dutch Drinking Water Decree. Mossel (1982) introduced the term 'marker' organisms for indicating potentially unsafe food samples and described an ecological based nomenclature for two groups, the first group named index organisms to provide information on the risk of occurrence of pathogens. The second group named indicator organisms used for the purpose of assessing the risk of inadequate bacteriological quality of a more general nature. This nomenclature was adopted by Ashbolt *et al.*, 2001 and extended with the term 'process indicator' (Table 1.3). This classification adds to the specification of the different groups of micro-organisms that are significant for microbiological sanitation of water.

Table 1.3 The three identified groups of faecal micro-organisms used in water quality control (from Ashbolt *et al.*, 2001)

Group	Definition
Index and model organism	A group/or species indicative of pathogenic presence and behaviour respectively, such as *E. coli* as an index for *Salmonella* and F-RNA coliphages models of human enteric viruses
Faecal indicator	A group of organisms that indicates the presence of faecal contamination; pathogens may be present
Process indicator	A group of organisms that demonstrates the efficacy of a water treatment process

One unambiguous and universal group/or species of organism for one specified group of waterborne pathogen (viruses, bacteria, parasites), however, is still not available. It is doubtful if this ever will be found. Differences in persistence, nutritional requirements and environmental metabolic conditions as well as in their morphological and physical characteristics between the potential groups and their target pathogens, but also within both groups of organisms can be large.

Furthermore, the rate of environmental distribution between both groups is different. Pathogens are only excreted by infected individuals and the extent of excretion depends on seasonal variation and presence of epidemics, but the non-pathogenic candidates for the three classes are excreted at a higher and stable level as part of the normal faecal microbiota. Therefore, the choice for the use of a variety of potential candidates which meets the following requirements

1. being present in the presence of pathogens in higher numbers;
2. being equally or more persistent in the environment than the pathogens;
3. showing no multiplication in the environment;
4. and being determined with rapid, simple, reliable and unambiguous methods, is, to a large extent still the best choice for microbial water quality monitoring with respect to safety.

FURTHER DEVELOPMENTS IN MICROBIAL RISK MANAGEMENT

Shortcomings of faecal indicators and water treatment. Beside the identification of new emerging pathogens in waterborne diseases as a major finding of the epidemiological studies at the end of the 20th century, there were two other major findings which initiated increased attention to the microbiologically safety of drinking water in the developed countries. First of all it was noticed that waterborne outbreaks of enteric diseases associated with pathogenic bacteria, viruses and protozoa in drinking water were reported while the coliform regulation was not violated (Dutka, 1973 and Craun *et al.*, 1997). Craun (1997) concluded that coliform bacteria are adequate indicators for the potential risk from pathogenic bacteria and viruses, but not for persistent pathogens like *Giardia* and *Cryptosporidium*. However, the failure of the coliform standard to protect against outbreaks of pathogenic bacteria like *Campylobacter* spp., has also been demonstrated twice in literature by Rosef and Mork (1985) and Rosef *et al.* (2001). Secondly, waterborne outbreaks of diarrhoea caused by protozoan parasites have been related to peak concentrations in the source water and inadequacies in water treatment (Dykes *et al.*, 1980; Badenoch, 1990; Craun, 1990 and Richardson *et al.*, 1991).

Legislation: microbial risk management. Health related developments in the drinking water industry in the eighties of the 20th century were mainly focussed on chemical issues like disinfection by-products (DBP's) and other trace chemical contaminants of toxicological significance. The increased knowledge on waterborne diseases and the awareness of the severe consequences for vulnerable parts of the population (immune-deficient persons) initiated a greater interest in microbiologically aspects of drinking water during the last decade of the 20th century. A large part of these studies and observations were carried out in the US and resulted in legislations in that country to enhance public health protection. The Surface Water Treatment Rule (SWTR) was introduced prescribing treatment requirements to guarantee a certain level of virus and protozoan (oo)cyst removal (von Huben, 1991). To balance the risks of DBP's and pathogens an Information Collection Rule (ICR) was published in the US to collect actual data on both health significant contaminants (pathogens and DBP's) in drinking water. This resulted in the Long Term 2 Enhanced Surface Water Treatment Rule which regulates monitoring *Cryptosporidium* levels in the source water and the required removal or inactivation capacity for this specific pathogen in addition to *Giardia* and viruses of new and existing treatment facilities (USEPA, 2006).

Simultaneously, a major scientific breakthrough was the introduction of the estimation of the probability of infection given a certain dose of micro-organisms (Haas, 1983; Haas *et al.*, 1999; Medema *et al.*, 1996 and Teunis *et al.*, 1996). Comparable to the regulation for mutagenic contaminants prescribing Maximum Tolerable Concentrations (MTC) for the risk of cancer during a life time exposure, the US Environmental Protection Agency (EPA) used these dose-response relationships to calculate the MTC values for viruses, *Giardia* and *Entamoeba histolytica* in drinking water for an annually accepted infection risk of 1 infection per 10.000 consumers, drinking 2 litres per day (Regli *et al.*, 1991). For the viruses this MTC-value was based on the most infectious virus type known at that time, the rotavirus. This development can be regarded as a major breakthrough in microbial risk management in drinking water. In the US where this was first described, however, these MTC-values are not implemented in the current legislation. US authorities do not communicate an accepted risk level with Public.

Developments in the Netherlands: Quantitative Microbial Risk assessment. In the Drinking Water Decree (Anonymous, 1984, 2001) as well as in the revised EU-regulations (Anonymous, 1980a) it is prescribed that drinking water may not lead to an unacceptable microbial health risk. This formulation was chosen with the awareness that absolute absence of pathogenic micro-organisms in drinking water can not be guaranteed. Based on the available knowledge is was assumed that the standards for indicator bacteria, and coliforms specifically, were enough to assure that the drinking water did not lead to unacceptable health risks.

In the Netherlands the number of registered waterborne outbreaks is very low and only related to cross contamination of the water mains due to human failures as presented before. Dutch drinking water was regarded as intrinsically safer than drinking water in the US and other Western countries with reported waterborne outbreaks because of the following reasons:

- The ground water used for Dutch drinking water is abstracted from fine sandy soil aquifers protected by clay layers;
- The surface water sources used for Dutch drinking waters are relative highly polluted and therefore stabilized (reservoir storage or dune passage) prior to a treatment plant with multiple barriers;
- Microbial standards for drinking water in the Netherlands with multiple indicators implemented in legislations since 1984 (thermotolerant coliforms, enterococci and SSRC; Table 1.1) are more stringent than in other European countries (lower MTC-values) and even more stringent compared to non European countries using solely coliforms or *E. coli*.

Though these assumptions are regarded as plausible, the Dutch Drinking Water Companies in their aspiration to distribute a high quality drinking water started a research funded by the joint research program to verify these assumptions.

Assessment of the microbiological water quality with the current microbial standards (Table 1.1) is a curative control strategy with a delay time of at least one day; consumers can be exposed to pathogens during this delay time while the consumed water was potentially contaminated with pathogens. Moreover, the increasing knowledge on waterborne outbreaks and the inadequacy of the coliform or *E. coli* standard demonstrate the need for more preventive strategies. Hazard Analysis and Critical Control Points (HACCP) is such a tool, introduced in the food industry and also proposed (Havelaar, 1994 and Anonymous, 2003) and applied in the drinking water industry (Hijnen *et al.*, 2001). Later this approach is implemented in the more generic approach of Water Safety Plans (WSP; WHO, 2004). A related approach is the use of Quantitative Microbial Risk Assessment (QMRA). The introduction of the MTC values for pathogenic micro-organisms in drinking water based on an annual risk of infection first presented by Regli *et al.* (1991), is a basic requirement for quantitative risk management. Implementation of QMRA in the revised Drinking Water Decree (Table 1.4) was initiated by the Dutch Authorities (Medema and Havelaar, 1994a) in close cooperation with the Dutch Drinking Water Companies (Van der Kooij *et al.*, 1995 and Medema, 1999). Besides the preventive character of this new approach, it offers a framework for new and emerging pathogens such as *Cryptosporidium* and noroviruses. Moreover, it enables balancing the microbial risks with toxicological risks of disinfection processes (by-products; Havelaar *et al.*, 2000) and with risks associated with other routes of exposure.

Quantitative Risk Assessment for pathogenic micro-organisms in drinking water was described by Haas *et al.* (1999), Teunis *et al.* (1997), Haas and Eisenberg (2001) and Medema *et al.* (2003) and is part of a total system approach as presented in Figure 1.2. The major steps in the process of QMRA are (i) quantitative information of the relevant pathogenic micro-organisms in the source water and (ii) quantitative knowledge about the elimination (removal and inactivation) capacity of water treatment processes for pathogenic micro-organisms and the factors influencing elimination and (iii) calculation of the exposure on the basis of these data, drinking water consumption and the dose-response curves.

The scope of the present study is on the second step of QMRA, assessment of the elimination capacity of a process or a treatment as a chain of processes. The three basic processes in water treatment are coagulation/floc removal, filtration (underground or in filters) and disinfection.

Table 1.4 The Maximum Tolerable Concentrations (MTC) of indicator bacteria and pathogenic micro-organisms in the drinking water after treatment as prescribed in the Dutch Drinking Water Decree using surface water as the source

Parameter	Maximum value	Unit	Remarks
Cryptosporidium			Note 1
Escherichia coli	0	CFU/100 ml	CFU 5 colony Forming Units
Enterococci	0	CFU/100 ml	
(entero)viruses[a]			Note 1
Giardia			Note 1

[a]The purpose of the brackets in (entero)viruses was to indicate that possible other virus groups that may be of concern for drinking water provision are included in the Dutch Drinking Water Decree

Note 1:

According to article 4, paragraph 1, micro-organisms should not be present in tap water in such concentrations that public health is jeopardized. Specific micro-organisms such as viruses and protozoa (e.g. Cryptosporidium *and* Giardia*) can not be detected at the very low concentrations at which exposure is relevant to the consumer's health. Instead, the owner who employs surface water as source for drinking water production should in consultation with the Inspectorate carry out a quantitative risk assessment based on data regarding the source water quality and treatment efficiency. The theoretical infectious risk resulting from the risk assessment should comply with a provisional standard of one infection per 10,000 individuals per year. Verification of the (provisional) infectious risk standard should be carried out for enteroviruses,* Cryptosporidium *and* Giardia *but also concerns other pathogenic micro-organisms. If the assessed infection risk is greater than aforementioned standard, the owner should consult with the VROM-Inspector about necessary measures. The Inspectorate can decide if a risk assessment should be carried out for vulnerable groundwater supplies. The expression << provisional standard >> is used to indicate the value that is verified in practice. Adjustment of this value can therefore not be precluded.*

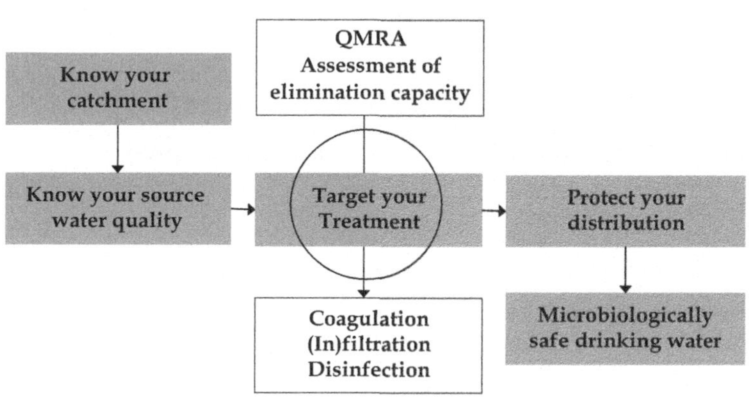

Figure 1.2 The total system approach for the (microbial) safety of drinking water

Dutch drinking water production systems with a QMRA obligation. QMRA is mandatory for Drinking Water Companies in the Netherlands with treatment facilities using surface water as the source and for Drinking Water Companies using ground water subtracted from areas vulnerable to contamination as appointed by the inspectorate. The major surface water sources used for drinking water production at eight major Dutch production facilities are the River Rhine and Meuse (Table 1.5).

Table 1.5 The source water and treatment before 2000 of eight major Dutch drinking water production facilities with a QMRA requirement

Number and Source water	Pre-treatment Open systems[a]	Treatment (all covered systems)
1. River Meuse	IR	CFR – O_3 – RGF – GAC – PD
2. River Meuse	IR	CL_2 – CFR – RGF- PD
3. River Meuse	IR	CFR – O_3 – RGF – GAC – PD
4. Lake IJssel	SR	CL_2 – CFR – RGF – GAC – PD
5. Drentsche Aa	SR	CFR – RGF – GAC – RGF – SSF
6. River Rhine	IR	RGF – SR – O_3 – SF – GAC – SSF
7. Polder[b]	DP – SR	RGF – SF – O_3 – GAC – SSF
8. River Meuse	DP – SR	RGF – SF – SSF

[a]*Re-contaminated by wild life;* [b]*small fraction river Rhine water*

The treatments of these facilities in operation before 2000 are presented in Table 1.5. The surface waters are pre-treated with either open pre-treatment systems like impoundment reservoirs (IR) or small reservoirs (SR) or by dune passage (DP) with open reclamation. These pre-treatment steps are followed by covered unit processes in different process schemes: coagulation/floc removal (CFR), rapid granular filtration (RGF), ozonation (O_3), chlorination (Cl_2), granular activated carbon filtration (GAC), slow sand filtration (SSF), powdered activated carbon (PAC), lime stone softening (SF) and post-disinfection (PD) with chlorine or chlorine dioxide.

Source water monitoring: required elimination capacity. Monitoring of the microbiological quality of the source water was made mandatory in the former Dutch Drinking Water Decree in 1984 to prescribe the level of water treatment (Table 1.3). In the further developments implemented in the revised Water Decree of 2001 source water monitoring was intensified resulting in monitoring for index pathogens and for *E. coli* and *C. perfringens* as process indicators (Table 1.6).

Table 1.6 Monitoring requirements (annual number of samples) of the source water and the treated water in the revised Drinking Water Decree (2001) at a production of 180.000 $m^3.d^{-1}$

Micro-organisms	Groundwater (GW)	Surface water (SW)	Treated water
E.coli	13	13	GW:52 SW:365
Coliforms	13	–	–
Enterococcci	–[a]	13	–
C. perfringens	–	13	GW:18 SW: 544
Cryptosporidium	–	18	–
Giardia	–	18	–
Enteroviruses	–	18	–
Bacteriophages	–	18	–

[a]*– = no requirement*

In the period 2001–2006, Water Companies have been occupied to comply with these new regulations. Their activities were based on an extended interpretation of this new regulation developed by a collaborative committee with representatives of Water Companies, Kiwa Water Research, Water Authorities and Inspectorate and the National Institute of Public Health and the Environment RIVM (Wetsteyn, 2005).

During the period of these new developments surveys have been performed to assess the prevalence of pathogenic micro-organisms in Dutch surface waters used as the source for drinking water production. In an extensive study the distribution of the protozoan parasites, *Cryptosporidium* and *Giardia* was monitored by Hoogenboezem *et al.* (2001a) and additional samples were analyzed for enteric viruses. This revealed that (oo)cysts were ubiquitous in the river Rhine and Meuse, two of the most important rivers for the Dutch drinking water production (Table 1.7).

Table 1.7 The range of required elimination capacities (log) to meet the infection risk for the index pathogens of 10^{-4}

Index pathogens	Dutch river waters	After reservoir storage (24 weeks)
Cryptosporidium	5.8–7.0[a]	4.8–5.9[b]
Giardia	6.0–7.8[a]	4.3–5.9[b]
Enteroviruses	6–7[c]	2.8–3.6[d]
Campylobacter	4.5–8.0[d,e]	6.5–7.3[d,e]

[a]Hoogenboezem et al. (2001a); [b]Ketelaars et al. (1995); van Breemen and Waals (1998); [c]Theunissen et al. (1998); [d]de Roda Husman et al. (2001a); [e]Wubbels (1996)

From the observed concentrations the required elimination of a treatment for *Cryptosporidium* ranged from 5.8 up to 7.0 log and for *Giardia* from 6.0 up to 7.8 log to comply with the annual infection risk of 10^{-4} per person. These estimations were calculated from the average surface water concentrations corrected for recovery using dose-response data (Regli *et al.*, 1991; Gerba and Rose, 1993 and Rose *et al.*, 1993), the set point of the maximum tolerated annual infection risk of 10^{-4} per person and the probability of infection assessed with the exponential model (Haas, 1983, 1993).

Theunissen *et al.* (1998) summarized the enterovirus concentrations at the intake points of surface waters of the Dutch drinking water companies and estimated a required elimination of 6–7 log (MTC-value for rotavirus of 1.83×10^{-7}/l for infection risk level of 10^{-4}). The additional enterovirus concentrations collected by Hoogenboezem *et al.* (2001a) and de Roda Husman *et al.* (2001a) indicated similar required elimination capacities.

Because the Dutch surface waters are not pristine but heavily to moderate polluted there is a pre-treatment prior to the final and covered production stage of the drinking water. These pre-treatment stages consisted of some kind of open storage reservoirs with or without preceded artificially recharge in the sand dunes. In these pre-treatment systems the level of pathogens in the surface water is reduced to some extent but for some micro-organisms introduction by waterfowl (i.e. *Campylobacter*) or other wild life may occur. Medema (1999) showed that waterfowl contributes to the introduction of pathogenic protozoan (oo)cysts, and Fallacara *et al.* (2001) demonstrated that in approximately 50% of the faecal shedding of free-living waterfowl sampled at six different parks with water reservoirs in Ohio *C. jejuni* was isolated. *Campylobacter* concentrations in surface waters showed a large variation and consequently the required elimination varied considerably. Introduction of this pathogen was observed in open reservoirs and collection ponds for recollection of infiltrated river water in concentrations of 0.2–40 per 100 ml (Medema and Schets, 1994b). In the same study, 25% of gull and duck faeces sampled in the neighbourhood of these reservoirs and ponds were positive for *Campylobacter*. Additionally, 84% of the water samples after the large storage reservoirs of the Biesbosch contained

Campylobacter (de Roda Husman *et al.*, 2001a) and based on the measured concentrations the required DEC value ranged from 6.5 to 7.3 log (Table 1.7).

For *Giardia* cysts and *Cryptosporidium* (oo)cysts, a reduction in numbers of 2.3 and 1.4–1.9 log, respectively was observed in the storage reservoirs Biesbosch with an average residence time of 24 weeks (Ketelaars *et al.*, 1995 and van Breemen and Waals, 1998). The reduction of enteroviruses in the Biesbosch reservoirs was in the same order of magnitude (2.1 log; de Roda Husman *et al.*, 2001a). Thus, the required DEC values for protozoan (oo)cysts and viruses for the treatment facilities after reservoir storage are approximately 2 log lower than based on the concentrations observed in the untreated river water (Table 1.7). The required elimination for both protozoan parasites is in the same order of magnitude as calculated by LeChevallier and Norton (1992) for US surface waters.

Need for information on removal of pathogens. In development and implementation of new water treatment technologies throughout the 20th century, improvement of water quality was a major driving force with the protection of public health as the main rationale. The performance of these processes was commonly assessed by monitoring the removal of physical/chemical parameters such as suspended solids, turbidity and dissolved organic carbon and by assessing the microbiological quality of the water with colony counts and faecal indicator bacteria. For the microbial parameters the performance of the processes was usually judged from the quality of the produced water rather than the capacity of the process to reduce micro-organisms.

As prescribed in the revised Dutch Drinking Water Decree (2001) quantifying the elimination capacity of water treatment processes for pathogenic micro-organisms present in the source water has become a major requirement to assess microbiological safety of the drinking water. Under the influence of the growing interest in microbiological safety of drinking water in the last decade of the 20th century studies were initiated to quantify the removal of the pathogenic micro-organisms by full-scale water treatment plants (Payment *et al.*, 1985, 1993; LeChevallier *et al.*, 1991; Stetler *et al.*, 1992 and States *et al.*, 1997).

Direct monitoring of enteroviruses, *Cryptosporidium* and *Giardia* elimination in Dutch water treatment facilities is restricted to water pre-treatment water systems using river water directly to produce semi finished water (Havelaar *et al.*, 1995; Ketelaars *et al.*, 1995; Hoogenboezem *et al.*, 2001a and Oesterholt *et al.*, 2007). In the nineties of the last century it was recognized that some Water Companies with open reservoirs directly before treatment were confronted with steady and sometimes high concentrations of *Campylobacter* bacteria in the source water (Medema and Schets, 1994b). This urged these Water Companies to assess the elimination of these bacteria by their treatment (Visser *et al.*, 2004 and Dullemont, 2004, 2006). These investigations yielded quantitative information on the efficacy of rapid granular filtration, ozonation and slow sand filtration to eliminate these pathogenic bacteria in comparison with the elimination of Coli44 (Hijnen *et al.*, 1995, 1998 and Smeets *et al.*, 2005). In recent years, open reservoirs with water recontamination by waterfowl immediately before the drinking water production facility have been eliminated as much as possible. Nonetheless, these observations clearly demonstrate the possibility of collecting elimination data by direct pathogen monitoring in treatment.

This strategy of pathogen monitoring in the full-scale treatment is not applicable for most Dutch locations certainly not for the total treatment. The concentrations during water treatment (multiple barriers) and required for an infection risk of 10^{-4} are too low and furthermore the pathogen enumeration methods are too complex (Wetsteyn *et al.*, 2005). Hence, it was necessary to develop and validate alternative methods or strategies.

PROCESS INDICATORS FOR PATHOGENS REMOVAL AND A GENERAL HYPOTHESIS

Generally speaking, the pathogenic micro-organisms can be divided in resistant and less resistant organisms. The parasitic protozoa *Cryptosporidium, Giardia* and *Entamoeba histolytica* are resistant against the commonly used concentrations of chemical disinfectants (chlorine, ozone). Viruses and bacteria are more susceptible to disinfectants. Thus, the major barriers in treatment for the resistant pathogens are physical barriers like coagulation and filtration. For viruses and bacteria chemical disinfection is the barrier of primary importance.

Physical process indicators. The large *Cryptosporidium* outbreak in Milwaukee prompted research into the validity of commonly used parameters in water treatment such as turbidity and more advanced techniques like particle counting to determine removal of this pathogen. Several authors demonstrated that setting a treatment goal for turbidity of <0.1 NTU will help to reduce breakthrough of parasitic protozoa in water treatment. To a certain degree, a quantitative correlation between removal of turbidity and particles with the removal of protozoan (oo)cysts both monitored in situ, has been observed (LeChevallier *et al.*, 1992 and Nieminsky *et al.*, 1995). However, a large national survey on the use of particle counting showed no correlation between the elimination capacity of treatment for this parameter and the elimination capacity for protozoan (oo)cysts (McTigue *et al.*, 1998 and LeChevallier and Au, 2004). Furthermore, the same authors and also others who performed dosing experiments (Patania *et al.*, 1995; Coffey *et al.*, 1999; Swertfeger *et al.*, 1999; Dugan *et al.*, 2001 and Emelko, 2001) demonstrated that generally the elimination capacity for both physical parameters underestimates the elimination capacity of the treatment for both parasites. The poor quantitative relationship between the removal of turbidity and particles and the removal of micro-organisms is most likely due to the fact that there is no quantitative relationship between the concentrations of both parameters (physical and microbiological) in the source water as shown by McTigue *et al.* (1996, 1998). With both physical methods unspecific colloids in the water are monitored which are present in much higher concentrations than the pathogens. Most of these studies, however, demonstrated that both parameters are valuable to control removal of colloidal particles by the processes which certainly help in minimizing the risk of breakthrough of these persistent pathogens in water treatment.

Microbial process indicators. Information on multiple indicator bacteria in treated water, as applied in the Netherlands since 1984, offers the opportunity to quantify the removal capacity of treatment processes for susceptible and more resistant micro-organisms.

General hypothesis

The general hypothesis of the work described in this thesis is that Escherichia coli and spores of sulphite-reducing clostridia can be used as process indicators to assess the elimination capacity of water treatment processes for, respectively, pathogenic micro-organisms that are susceptible and pathogenic micro-organisms that are resistant to disinfection processes under the commonly applied conditions in water treatment.

Monitoring of these indicators both in the source waters and the finished waters are prescribed by legislation (the Dutch Drinking Water Decree 1984, Tables 1.1 and 1.2; revised Decree 2001, Tables 1.3 and 1.5). Beside this required data collection, Water Companies monitor indicator bacteria throughout their treatment facilities regularly to detect contaminations and to check process performances. Thus, the use of faecal indicators as microbiological process indicators is closely linked to the daily practice of water quality monitoring in the drinking industry and a multi year data base of full-scale systems is available. Furthermore, these indicators have been selected on the basis of requirements related to the presence of pathogenic micro-organisms. They are commonly present in the source waters and related to nominal faecal contamination and indicative for some peak events (Atherholt *et al.*, 1998 and Signor *et al.*, 2005).

For a process indicator the requirement of being presence in the presence of pathogens is not stricktly required. In addition to the three other previously mentioned requirements for proper indicator micro-organisms of faecal contamination, further requirements for an ideal process indicator are as follows:

4. being continuously present in the source water at measurable concentrations to enable site specific assessment of elimination;

5. being determined with a method that is sensitive enough to determine concentrations at all stages of water treatment;
6. being eliminated at an equal or slightly lesser extent than the pathogens by individual water treatment processes or a chain thereof.

Due to major interest in the use of coliforms and *E. coli* as faecal indicators during the last century, scientific knowledge on the validity of the first four requirements for this group of indicators is extensive. This revealed a serious limitation for the total group of coliforms which multiply at moderate temperatures in treatment steps. The failure to detect the potential presence of parasitic protozoa and pathogenic viruses in drinking water by coliform monitoring (*E. coli* included) as described before, however, emphasize that also *E. coli* does not meet the requirements for a proper microbial process indicator for all pathogens. The difference in nature of indicator bacteria, parasitic protozoa, and viruses is evident and the major reason for their different behaviour during treatment. It is hypothesized that for the pathogenic bacteria such as *E. coli* O157 and *Campylobacter*, *E. coli* is a proper process indicator. Scientific knowledge on behaviour of faecal streptococci in water treatment is limited. Moreover, the environmental persistence of these bacteria (die off and susceptibility to chemical disinfection) is in the same order of magnitude as the persistence of coliforms (i.e. Grabow *et al.*, 1983). Therefore, faecal streptococci is not of additional value to *E. coli* as a process indicator.

The low incidence of waterborne outbreaks of parasitic protozoan in the Netherlands has been attributed to the benefit of multiple barriers in the treatment. Additionally, also the required compliance with a standard for a persistent indicator bacterium like spores of sulphite-reducing clostridia SSRC in drinking water (Table 1.1 and 1.2) has been hypothesized as a possible explanation. Both micro-organisms, *Cryptosporidium* and *Clostridium*, exhibit a life-cycle with a dormant life-stage meant for environmental transmission and survival.

In the NPR 6568 (Anonymous, 1985b; note of the Dutch normalized method NEN6567) SSRC is described as a 'biological process indicator at the drinking water production, especially for the performance of treatment steps like sedimentation and filtration'. The use of the anaerobic spores as technical parameter to monitor deficiencies in filtration processes was proposed by Taylor (1949) and Willis (1957). The use of *Clostridium perfringens* as process indicator for the removal of viruses and protozoan oo(cysts) of *Cryptosporidium* and *Giardia* in drinking water treatment was studied later by Payment and Franco (1993). For the removal of enteric viruses, however, the use of bacteriophages as process indicators is more appropriate due the similar nature of those organisms (Havelaar *et al.*, 1993, 1995 and Jofre *et al.*, 1995). The first study to propagate the use spores of sulphite-reducing clostridia (SSRC) as a surrogate for oocyst removal as suggested by the NPR6568 was done on pilot plant scale by Hijnen *et al.* (1997). The use of SSRC as potential process indicator was chosen, because of the daily routine of monitoring, the availability of historical data in source water and the subsequent stages of drinking water production of this faecal indicator bacterium and the improbability of growth in the aerobic water treatment processes.

In countries like US and UK with no monitoring routine of these faecal indicators in water treatment, the aerobic spore forming bacteria (*Bacillus*) was introduced as a potential process indicator for the assessment of protozoan (oo)cyst removal in water treatment (Rice, 1996) and applied in research (Patania *et al.*, 1995; Nieminski *et al.*, 2000; Emelko, 2001 and Dugan *et al.*, 2001). Indications for regrowth of aerobic spores in filters, however, have been documented in literature (Galofre *et al.*, 2004; Mazoua and Chauveheid, 2005).

SPORES OF SULPHITE-REDUCING CLOSTRIDIA AS PROCESS INDICATOR FOR *CRYPTOSPORDIUM* AND *GIARDIA*

The genus Clostridum. Bacteria of the genus *Clostridium* can be described as gram-positive, anaerobic spore-forming, non-motile, and rod-shaped. *C. perfringens* is the species most commonly associated with faecal contamination (Cabelli, 1977) which ferments lactose, sucrose and inositol with the production of gas.

The bacterium causes 'stormy fermentation of milk', reduces sulphite to sulphide, reduces nitrate, hydrolyzes gelatine, and produces lecithinase and phosphatase.

The genus Cryptosporidium and Giardia. The genus *Cryptosporidium* is a small coccidian protozoan parasite that infects a host through the oocyst stage of its life-cycle. This is the environmentally resistant transmission stage of the parasite which may remain in the environment for long periods without loosing infectivity. After ingestion by a suitable host the wall of the oocysts is opened (excystation) and the four motile sporozoites are released which transform in either an asexual or a sexual reproduction cycle. The oocysts are the product of the sexual reproduction cycle in which spherical thin- and thick-walled oocysts are produced. The thick-walled oocysts are the oocysts which are excreted with the faeces and thus environmentally transmitted.

Giardia is a genus of protozoan parasites potentially found in water and other media. The recent taxonomy of the genus *Giardia* includes the following species and their potential hosts: *G. lamblia* (also called *G. intestinalis* or *G. duodenalis*; humans and other mammals); *G. muris* (rodents); *G. agilis* (amphibians); *G. psittaci* and *G. ardeae* (birds). As with *Cryptosporidium*, the parasite is shed with the faeces as an environmentally robust cyst that is transmitted to a new host. In the duodenum of a new host, the trophozoïte emerges from the cysts and completes a mitotic division to produce two trophozoïtes that attach to the epithelial cells by their adhesive disc and feed on the epithelial cells. During the passage of the trophozoïtes through the intestine, part of these trophozoïtes begin to encyst and leave the host as cysts.

Characterization of spores and (oo)cysts. The bacterial spore (Figure 1.3) consists of an inner part, the core, which contains the cellular substances necessary for the outgrowth of vegetative cells.

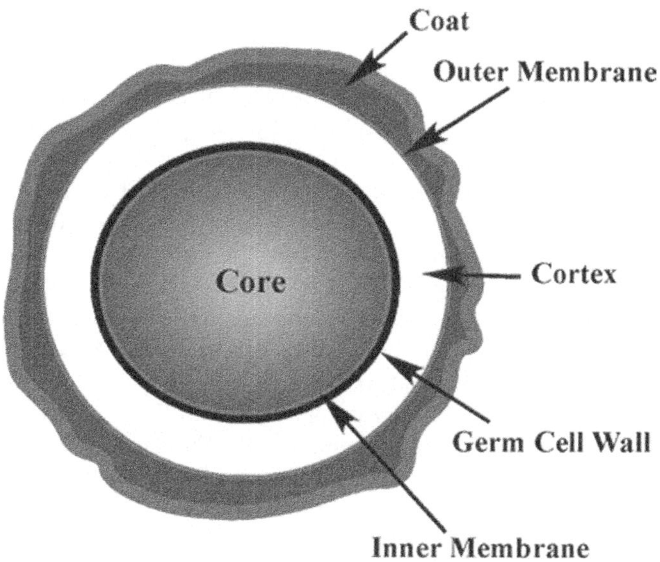

Figure 1.3 A schematic image of a bacterial spore left (http://www.micro.cornell.edu/cals/micro/research/labs/angert-lab/images/endospore.jpg) and right from Driks (2007; with permission of Copyright holder) cartoons illustrating the diversity of spore appendages. The coat is drawn as a light gray layer. The exosporium (in the top two spores) is indicated by a thin black line surrounding the coat. Appendages extend from the coat or exosporium. A thick grey layer indicates the spore cortex (peptidoglycan). Top: pili-like structures found on spores of some B. cereus strains (ref. in Driks, 2007). Middle: pin-like structures on Clostridium bifermentans spores (Yolton *et al.*, 1968). Bottom: ribbon-like structures on C. taeniosporum spores

This core is surrounded by the inner membrane, a lipid bilayer with no detectable fluidity, encaged by the cortex and primordial cell wall, a thick layer of peptidoglycan. This cortex is enclosed by the spore coats, a proteinaceous layer (Foster and Johnstone, 1990). Finally, several *Bacillus* and *Clostridium* species are surrounded by the exosporium, a highly structured, paracrystalline basal layer with an external hair – like nap, consisting of protein, glycoprotein and most likely, lipid (ref. in Walker *et al.*, 2007). The composition of the exosporium of *B. cereus* was determined (Matz *et al.*, 1970) and proved to be chemically complex, consisting mainly of protein (52%), amino and neutral polysaccharides (20%), lipids (18%), and ash (4%). Little information was found on the composition of the exosporium of *Clostridium* spores but it may be assumed that they consist of similar constituents in different ratios. The large morphological divergences of *Clostridium* spore was presented by Yolton *et al.* (1968). They showed the presence of appendages on spore surface and showed five distinct morphological spore types in 12 strains of *C. bifermentans*. External appendages in the form of ribbons, pili, feathers, brushes, tubules or swords (ref. in Walker *et al.*, 2007; see Figure 1.3) on the spore surfaces are observed frequently. Yolton *et al.* 1972 demonstrated that these appendages of *C. taeniosporum* consist for 80% of protein, 15% of glucose, rhamnose, glucosamine and 5% phosporus.

The spore coat is approximately 50% of the spore volume and 40–60% of the dry weight (Murell, 1969) and consists of hexosaminepeptides with large number of amino acids and high cystine concentrations. The density of *Bacillus* spp. (*B. cereus*, *B. subtilis*, *B stearothermophilus*) and *Clostridium perfringens* spores was assessed with several techniques by Tisa *et al.* (1982) and ranged from 1.120 to 1.380 kg/m^3 (average of 1.253 $\pm$ 0.074). The circular size of the spores of *C. perfringens* in a special prepared suspension (removal of vegetative cells by washing with lysozyme, trypsine and sodium dodecyl sulphate) was determined at approximately 1 μm (0.79–0.98 μm) with transmission electron microscopy (TEM) (Novak *et al.*, 2002). Individual ovoid spores of three strains of *B. subtilis* were sized by TEM; length ranged from 1.39 $\pm$ 0.14 – 1.42 $\pm$ 0.18 μm and the width from 0.68 $\pm$ 0.05 – 0.8 $\pm$ 0.11 μm (Leuschner *et al.*, 1999).

Comparable to the bacterial spores, the characteristic wall of the oocysts is responsible for their environmental persistence. This wall consists of an outer wall of acidic glycoprotein, a central rigid layer composed of complex glycolipid/lipoprotein layer followed by a thick layer of filamentous glycoprotein. Both the central and thick wall provides rigidity and elasticity of the oocyst (Harris and Petry, 1999). Generally, similar components have been described for the spore wall. No quantitative data of these components were found to enable a relative comparison with the spores. Additionally, no information of the presence of appendages on the oocyst wall comparable to the appendages described for bacterial spores have been found in literature. Observations of 'electrosteric' repulsion in a liquid/solid interface study by Kuznar and Emilech (2005) may point to the existence of protein containing appendages. These oocysts are spherical (Figure 1.4) and have a mean size of 4.9 (3.9–5.9) μm and a geometric mean density of 1,045.4 kg/m^3 (Medema *et al.*, 1998) and thus larger and less dense than bacterial spores.

Adam (2001) describes the wall of the *Giardia* cysts as a 0.3 to 0.5 μm thick and composed of an outer filamentous layer and an inner membranous layer with two membranes. The outer portion of the cyst wall is covered by a web of 7- to 20-nm filaments and the outer cyst wall consists of four major proteins. The predominantly sugar component is galactosamine in the form of Nacetylgalactosamine (refs in Adam, 2001). Earlier claims that the cyst wall is composed of chitin (N-acetylglucosamine) have been refuted. The cysts of *Giardia* are elliptical in shape with the mean largest (length) and mean smallest diameter (width) of 12.2 and 9.3 μm, respectively. The mean ratio between length and width is 1.3 and the geometric mean density is 1,036.2 kg/m^3 (Medema *et al.*, 1998).

Surface properties of spores and (oo)cysts. Electrophoretic mobility studies (references in Murell, 1969) indicated that the effective surface layer of spores is non-ionogenic. Any charge of the surface is due to adsorption of ions. There are other studies who demonstrated that the spores of *Bacillus* are negatively charged (Douglas, 1955 and Watanabe and Takesue, 1976). The latter authors defined the two negatively charged groups on the spore surface of *B. megaterium* to be strong acidic groups such as phosphates and the weaker acidic groups of the carboxylates.

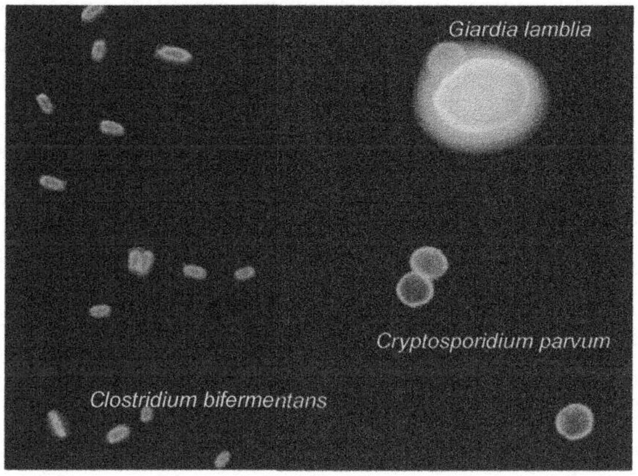

Figure 1.4 A microscopic image of spores of Clostridium bifermentans labelled with Acridine-orange and a cyst of Giardia lamblia and oocysts of Cryptosporidium parvum labelled with fluorescein isothiocyanate (FITC)

Another important surface property which is of influence of attachment of micro-organisms to surfaces is hydrophobicity. Spores of *Bacillus* and *Clostridium* were found to be more hydrophobic than the vegetative cells of these species (Wiencek *et al.*, 1990). This increased hydrophobicity was attributed by the same authors to the relative abundance of protein in the outer coat and exosporium compared to the peptidoglycan on gram-positive vegetative cell-surfaces. The hydrophobic nature of SSRC was also deduced from the sticky behaviour of these spores observed in plastic Petri dishes used for enumeration (Hijnen, unpublished results).

Electrophoretic mobility of oocysts studied by Brush *et al.* (1998) showed differences in surface charge of oocysts, depending on the pre-treatment of the used oocysts. Oocysts pretreated (purified) with deionized water exhibit no surface charge whereas oocysts pre-treated with surfactants (defatting agents) were negatively charged. Drozd and Schwartzbrod (1996) found that oocysts were low in hydrophobicity and additionally, Hsu and Huang (2002) found an increasing hydrophobicity at decreasing pH. Brush and colleagues demonstrated that the hydrophobicity of oocysts can change as they age.

In conclusion, there are similarities and differences between the oocysts and clostridia spores which influence their similarity in resistance and transport behaviour during water treatment processes. Furthermore, comparative studies are required to elucidate the validity of the use of SSRC as a process indicator.

Environmental fate: resistance, activation, germination and outgrowth. The outer structure in combination with the dehydrated core components in a lattice of divalent metal ions and dipicolinic acid is the basis of the metabolic dormancy and heat resistance of the spores (Gould and Hurst, 1969). UV resistance is established by complex formation of spore DNA with spore-specific, acid-soluble, low molecular-weight proteins (Setlow, 1988). The general view on bacterial spore survival in nature is that they are extremely persistent (Gould, 2006). Longevities of thousands of years have been demonstrated in literature. A recent comparative survival study showed that spores of *C. perfringens* were more persistent than oocyts of *C. parvum* assessed by *in vitro* excystation and exclusion of propidium iodide (Medema *et al.*, 1997).

Besides this protective structure to maintain environmental dormancy, spores have also a sensory mechanism for favourable environmental conditions which can initiate germination, the pre-stage of vegetative outgrowth, within seconds. Both the sporulation and germination stage of the genera *Bacillus* and *Clostridium* is induced on exposure to stressful environmental conditions, and these differ between genus and species. *Bacillus* spp. have their active life-cycle under aerobic conditions, whereas *Clostridium* spp. germinate and grow under anaerobic conditions. The redox conditions of the environment seem to be not essential for the germination (references in Lewis, 1969). Clostridia spores do geminate under aerobic conditions and Bacillus spores can germinate under anaerobic conditions. Germination of the spores is preceded by activation, defined by Keynan and Evenchik (1969) as breaking dormancy and overcome the inability to germinate. Heat treatment of the spores is the most described activator, but also low pH, reducing agents, calcium, dipicolinic acid, ionizing radiation, chemicals and ageing. The latter activator is presented as a similar activator as heat treatment by Keynan and Evenchik.

The germination of spores is defined as a series of events triggered by specific germinants leading to irreversible loss of spore resistance properties (Johnstone *et al.*, 1994). Possible germinants are amino acids, sugars, alcohols, dodecylamine, enzymes, hydrostatic pressure, mechanical actions and heat. *Bacillus* and *Clostridium* spores differ in their need for germinants as described by Gould (1969) and the latter spores need generally more complex germinant mixtures than *Bacillus* spores.

After germination a spore can be changed into a mature vegetative cell with all the characteristics and needs for multiplication. When the environmental conditions are not appropriate i.e. the required nutrients are not available, re-sporulation may occur without the evolution to a mature vegetative cell (Strange and Hunter, 1969).

Being a parasitic micro-organism *Cryptosporidium* needs a host for multiplication. This is not necessary for bacterial spores and the potential propagation of these spores in water treatment is a critical aspect and observed for the aerobic spores as previously described. Before proposing anaerobic spores as technical parameter, Willis (1956) was negative about the use of these anaerobic spores because of their prevalence in sand from drinking water filters. Later Willis (1957) proposed that 'filters may be a reservoir of anaerobes which cause intermittent appearance of sulphite-reducers in the treated water throughout the year'. He found anaerobic bacilli, defined as bacteria forming black colonies in anaerobic incubated Wilson-Blair medium, frequently in the sand of filters with no relationship with the depth. Vegetative cells of this bacterium were predominantly present, indicating the occurrence of growth in the filter bed. He emphasized the value of the sulphite-reduction test for the indication of deficiencies in the filtration process as previously reported (Taylor, 1949), but again doubted the use of the parameter for the classification of the safety of the water. This observation of vegetative anaerobic bacilli in the filter beds is controversial, however, with respect to the assumption of growth. The way to determine vegetative cells of sulphite-reducing clostridia in the samples was to omit the pasteurization prior to the cultivation. Because of the following arguments the assumption that this procedure gives valid evidence for the occurrence of outgrowth of these anaerobic bacterial spores is disputable:

- the assumption that all black colonies in the non pasteurized samples can be classified as vegetative bacilli (or more precise clostridia because of the sulphite-reduction) is not correct without excluding of the presence of other non-spore forming bacteria that can reduce sulphite in the used medium and produce black colonies such as *Salmonellae*, *Proteus*, *Escherichia freundii*, *Citrobacter* spp. (formerly *Paracolobactrum*) and certain species of *Erwinia*, *Flavobacterium* and *Achromobacter* (Angelotti *et al.*, 1962);
- secondly, the susceptibility of spores to heat is variable and depends on the state of the sporulation and species present in the samples (Bonde, 1962). In general the optimum growth range of *Clostridium* spp. influences the thermoresistance of the produced spores. Spores formed by psychrotrophic spp. at low temperature are less resistant to heat than spores formed by thermophylic spp. at higher temperatures (Roberts and Hitchins, 1969).

Thus, subtracting the number of SSCR assessed with pasteurization from the number of SSRC assessed without pasteurization not necessarily reveals the concentration of vegetative *Clostridium* cells but can also be regarded as the number of heat susceptible *Clostridium* spores.

The outgrowth of *Clostridium* spores is preceded by the activation and germination. Germination of anaerobic spores occurs only under certain conditions. As previously described germination of the anaerobic spores may occur under aerobic conditions but outgrowth is restricted to the anaerobic environment. Gibbs (1964) described that activation of dormant clostridial spores occurs only in the presence of mixtures of amino acids. Evidence for outgrowth of *Clostridium* spores in the sediment-water interface of a lake was presented by Molongoski and Klug (1976). The fraction of anaerobic bacteria in the total bacterial population of the sediment was 0.001% and with *Clostridium* as the dominant genus. This lake was a hyper-eutrophic lake. Thus, the large numbers of vegetative 'anaerobic bacilli' in filters reported by Willis (1957) can be regarded as debatable results with respect to the conclusion on *Clostridium* growth in the aerobic and mostly oligotrophic environment of water treatment. No information of that kind was found in his work, but if outgrowth of *Clostridium* in the filter bed had occurred, these filters were possibly supplied with water rich in organic matter and operated discontinuously with anaerobic periods as a result. Both at the time of introduction of the SSRC parameter in the Dutch Drinking Water Decree in 1984 and also during the subsequent period until now, indications of growth of SSRC in water treatment have not been documented.

Analytical methods for Clostridium spore. To assess the concentration of spores of sulphite-reducing clostridia in water the environmental spores must complete activation and germination before they can be determined by outgrowth in an appropriate medium incubated under the optimal conditions. The method to assess concentrations of SSRC in water and sediments used in the current study, is described in the Dutch NEN6567 (Anonymous, 1985a) and based on the work of Havelaar *et al.* (1983). The anaerobic or micro-aerophilic incubation method of the sample in solidified Perfringens Agar Base medium in between the bottom and the cover of a petri-dish was a method developed during laboratory practices in Italy (personal communications, Havelaar, 2007). A comparison with the methods of strict anaerobic incubation revealed small differences (Havelaar *et al.*, 1983). It was concluded that this simple and cheap incubation method could be included in the prescribed enumeration method (Anonymous, 1985a).

RESEARCH NEEDS

Assessment of the elimination capacity of water treatment facilities for micro-organisms has not been common practice in drinking water production. To enable QMRA in practice of a water company, methods or strategies have to be developed to assess this elimination capacity of the local specific operated treatment plant.

Information on variation is of interest because waterborne outbreaks of diarrhoea caused by protozoan parasites have been related to peak concentrations in the source water or inadequacies in the treatment (Dykes *et al.*, 1980; Badenoch, 1990; Craun, 1990 and Richardson *et al.*, 1991). Beside obvious inadequacies in treatment, normal operated processes in a treatment exhibit a variation in the elimination of micro-organisms. LeChevallier and Norton (1992) for instance showed that the removal of particles of $>5\,\mu m$ (size of *Cryptosporidium* oocysts) by individual unit filters in the filtration stage of a treatment may vary by as much as 1.000-fold. These variations will influence the overall efficiency of a treatment stage. Variation in the removal of (oo)cysts of *Cryptosporidium* and *Giardia* by a conventional treatment (coagulation/floc removal plus filtration) was observed by Hashimoto *et al.* (2001). Elimination capacity for *Cryptosporidium* and *Giardia* ranged from 2.0–3.2 and 1.7–3.1 log, respectively.

The variability of elimination of processes for *Cryptosporidium* has also been demonstrated by spiking tests on pilot plant scale or in laboratory experiments. Emelko (2001) intensively studied the removal of *C. parvum* oocysts by filtration processes under different filtration conditions. Elimination assessed under stable operation, ripening, sub-optimal coagulant dosing, early and late breakthrough ranged between 5.5 and <0.5 log. Also with spiking experiments on laboratory scale (dose/response data with a continuous flow system) a high variability of the efficiency of ozone disinfection for *C. parvum* in natural waters was demonstrated

(Oppenheimer *et al.*, 2000). The average Chick/Watson inactivation constant at 10°C was 0.21 with a range of 0.08–0.46 l/mg.min.

This variation in the elimination of micro-organisms is caused by the multiple variables involved in the mechanisms responsible for the elimination in the different treatment processes (Table 1.8).

Table 1.8 Elimination mechanisms significant for the different treatment processes related to the applied processes and the abiotic conditions influencing (+) the elimination

Processes:	**Inactivation**		**Removal/interception**		
Mechanisms:[a]	**Die off**	**Disinfect.**	**Coag./Sed.**	**Strain.**	**Att./det.**
Soil passage	+			+	+
Storage	+		+		
Coag./flocc.			+		+
Disinfection		+			
Filtration				+	+
Abiotic conditions:					
– Temperature	+	+	+		+
– DOC		+			
– UV		+			+
– Fte		+	+		+
– pH		+	+		+
– EGV					+
– Hydraulics		+	+		+
– Contact time	+	+	+		+
– Granular mat.[b]					
– Grains charact.[c]				+	
– Chemical type		+	+		
– Dose[d]		+	+		

[a]Disinfection, coagulation and sedimentation, straining, Attachment and Detachment; [b]Type of granular material; [c]size and uniformity; [d]assessment and control

Inactivation which is passive die off or active destruction of the organisms, is an elimination mechanism significant for water treatment processes such as infiltration and soil passage, storage in reservoirs and disinfection. Especially the efficiency of the disinfection process is influenced by a lot of abiotic conditions. Removal and interception of micro-organisms as mechanism, caused by coagulation and sedimentation, straining and attachment/detachment is also under the influence of partly the same abiotic conditions.

Due to these observations and considerations it is of importance for risk management to integrate the level of variability in both source water concentrations and in elimination capacity in the variability of the infection risk assessed with QMRA. Additional statistical tools to integrate uncertainty calculations in QMRA have been described (Teunis *et al.*, 1997; Evers and Groennou, 1999; Haas *et al.*, 1999 and Medema *et al.*, 2003). Moreover, in the framework of WSP it is of importance to determine the causes of these fluctuations since that can lead to measures in process design or operation to minimize infection risks (Smeets, 2008).

As a consequence of these considerations on variability of elimination capacity, data on treatment efficiency for QMRA must be as site specific and actual as possible.

Development of a method to determine the elimination of faecal indicators. Assessing the elimination capacity of treatment processes and a train of processes at a treatment facility is one of the current challenges in drinking water industry (Medema, 1999). The general hypothesis is that *E. coli* and SSRC can be used as process indicators for susceptible and the more resistant pathogens, respectively. The standard microbial methods faecal indicators can be used to assess elimination capacity of the first steps in the treatment train. The detection limit of the current methods, however, is too low to detect *E. coli* and SSRC after subsequent processes. Thus, a first need is a method with a lower detection limit to determine the concentrations of *E. coli* and SSRC in the water in the last stages of treatment.

Comparative studies on process indicators and pathogens. The major missing information to decide whether both *E. coli* and SSRC are usefull process indicators for viruses, bacteria and protozoan (oo)cysts are comparative data on the elimination in water treatment assessed under the same conditions. There are a number of morphological, physical and physiological differences between both indicator bacteria and the target pathogens which will affect the similarity of organism removal during treatment. To elucidate the influence of these differences on the elimination of indicator and pathogen, comparative studies are necessary, ideally under full-scale conditions or otherwise with spiking studies using multiple micro-organisms under well defined experimental conditions. These comparative studies are needed to answer the following questions:

- Is the elimination of the proposed process indicators *E. coli* and SSRC in water treatment predictive for the elimination (removal and inactivation) of pathogens and, if so, for which pathogens?
- To what extent are comparative (in)filtration and inactivation experiments predictive for the observed elimination under field conditions and what is the influence of the experimental scale on the predicted elimination?
- Can these comparative experiments be used to elucidate the process conditions influencing the variation of elimination of micro-organisms?

Validation and extension of the data base with literature data. Collecting site specific information on elimination of process indicator (*E. coli* and SSRC) supplemented with comparative spiking studies, is a relative large effort and is most likely not feasible for each specific situation. The scientific knowledge on elimination of micro-organisms by water treatment processes has been extended during the last decade and potentially includes valuable and applicable information. There is a need for exploration of the literature and systematically assemble relevant and reliable quantitative information in a central data base which enables to obtain a default value of the elimination capacity of a treatment process and the conditions of influence on elimination. These data can be used in the tiered approach of QMRA (Medema *et al.*, 2003) for assessing a point estimate of the annual risk of infection. When the calculated infection risk is close to the required safety level of 10^{-4} and the available full-scale data on process indicator removal is limited, a distribution of the risk of infection can be calculated by more comprehensive statistical methods using the range of elimination capacities collected in this data base (Smeets, 2008).

CONCLUSIONS AND OBJECTIVES FOR RESEARCH PRESENTED IN THIS BOOK

The developments in the supply of microbiologically safe drinking water over the last century (1900–2000) have been described showing that most developments occurred at the turn of both centuries. The period with the highest impact on the production and control of microbiologically safe drinking water was around 1900. During the 20th century the fraction of the population in the Netherlands connected to a central drinking water steadily increased to almost 100% with a decreasing incidence of outbreaks of waterborne diseases.

At the end of the 20th century, however, epidemiological studies showed that outbreaks of waterborne diseases in the developed countries were not totally banned. Outbreak investigations showed shortcomings in both microbiological water quality monitoring and water treatment with respect to more persistent pathogens such as *Cryptosporidium,* a pathogen considered as a new challenge to the drinking water industry (Medema, 1999). Developments in the Netherlands resulted in a shift from reactively end product quality monitoring to proactively quantitative microbial risk assessment QMRA. QMRA is a new method to assess the safety of drinking water by means of calculation of the risk of infection to the consumers. This risk of infection is estimated from the dose/response curves for the individual pathogen on the one hand and the exposure assessment on the other. The exposure assessment is derived from the concentration of pathogens in the source water, the elimination capacity of the water treatment and the drinking water consumption. A major lack in the process of QMRA is quantitative information on the elimination capacity of water treatment processes for pathogenic micro-organisms and the variation in elimination. Teunis *et al.* (1997) showed that the uncertainty in the estimated removal efficiency of treatment processes dominates over uncertainties in the estimates of the other factors used in QMRA.

The objective of the studies presented in this book is to develop and evaluate methods for the assessment of the elimination capacity of full-scale water treatment processes particularly for bacteria and protozoan (oo)cysts. Elimination of both environmental and spiked micro-organisms under full-scale, pilot plant and laboratory conditions is determined and compared in order to evaluate the suitability of these methods for the assessment of the elimination capacity of processes and the variability of this elimination capacity.

The objective of the study is divided in a number of sub-goals

1. Evaluation of Coli44 (incl. *E. coli*) and SSRC (incl. *C. perfringens*) as process indicators to quantify the elimination capacity of full-scale water treatment for the index pathogens *Campylobacter* and *Cryptosporidium/Giarda*, respectively.
2. Develop and apply large volume sampling to assess the concentrations of indicator bacteria in the final stages of full-scale water treatment.
3. Compare the elimination of *E. coli* and *C. perfringens* with the elimination of pathogens in a selection of processes by comparative spiking experiments and literature data.
4. Determine the value of challenge tests under pilot plant or laboratory scale for the assessment of the elimination capacity of full-scale processes.
5. Explore and analyse literature data on micro-organism removal in water treatment as tool for assessment of the elimination capacity and the process conditions affecting removal.
6. Develop a general strategy to quantify the elimination of pathogenic micro-organisms in water treatment processes.

Chapter 2

Indicator bacteria concentrations in water treatment and assessment of elimination capacity

Wim A.M. Hijnen[a], Y.C. Drost[a], G.J. Medema[a], D. van der Kooij[a] and A.H. Havelaar[b]

[a]*KWR Watercycle Research Institute;* [b]*University Utrecht*

ABSTRACT

The use of thermotolerant coliforms (Coli44) and spores of sulphite-reducing clostridia (SSRC) as process indicators to assess the elimination capacity of water treatment and unit processes for pathogenic micro-organisms susceptible and resistant to disinfection was explored. Historical data sets of routine monitoring programs revealed that both indicators were detected in the finished water after treatment in variable frequencies. Coli44 was detected incidentally, whereas SSRC was observed after every treatment in percentages of samples ranging from 0.8 up to 23%. With data collected after unit processes efficacy of these unit processes was calculated to eliminate both process indicators. The percentage of positive samples, however, decreased rapidly during treatment. This affects the accuracy of the calculated decimal elimination capacity (DEC). Decreasing the detection limit of the standard membrane filtration method by filtering larger sample volumes enabled more accurate assessment of DEC. The required detection limit of the analytical method, however, to evaluate the required DEC of a total water treatment for pathogenic micro-organisms is ≤1 organism per 100 l which requires further research for new methods.

INTRODUCTION

Chapter one describes developments in microbiologically safe drinking water over the years. An important development in management of microbiological quality of drinking water in the last decade is the introduction

 Quantitative Methods to Assess Capacity of Water Treatment to Eliminate Micro-Organisms. Edited by Wim A.M. Hijnen. ISBN: 9781843393764. Published by IWA Publishing, London, UK.

of health-based targets. In the revised Dutch drinking water decree (Anonymous, 2001) a mandatory quantitative microbial risk assessment (QMRA) is implemented for drinking water produced from surface water to demonstrate compliance with an annual accepted infection risk of 10^{-4} for pathogenic micro-organisms. An important step in this QMRA is the determination of the removal and inactivation of pathogenic micro-organisms. The objective of the current study is to develop and apply methods for the assessment of the elimination capacity (removal and inactivation) of water treatment processes. Direct measurement of elimination of pathogens is not feasible. The general hypothesis is that the elimination of the faecal indicator bacterium *E. coli* and spores of sulphite-reducing clostridia (SSRC) can be used for assessment of elimination of susceptible and resistant pathogens, respectively. First step in the study was to explore the use of the standard microbiological methods in routine water quality monitoring for the assessment of the elimination capacity of treatment processes. Routinely collected data of concentrations of indicator bacteria in full-scale water treatment were evaluated. Additionally, the use of an analytical method with lower detection limit by examining larger volumes with the standard membrane filtration technique was investigated in a pilot plant study.

ANALYZING HISTORICAL MICROBIAL WATER QUALITY DATA

Routinely collected microbiological data. Analysis data on indicator bacteria in water monitored over periods of three years by the Water Companies with standard methods were collected and evaluated. The data were monitored in the eight full-scale systems mentioned in Chapter 1 (Table 1.5). Total coliforms (Coli37) and thermotolerant coliforms (Coli44), faecal streptococci (FS) and spores of sulphite-reducing clostridia (SSRC) were monitored mandatory in the source waters and the drinking water with prescribed methods (Anonymous, 1985a,b,c). Each Water company also measured all or selected indicators at some selected sampling points in treatment.

Process conditions. The general operational and design conditions of the evaluated processes presented in Table 1.5 of Chapter 1 were gathered during this survey: coagulant and disinfectant dosages, pH, turbidity, temperature, contact time, flow rate. For disinfection processes the Ct-values $(mg.l^{-1})$.min calculated from the residual disinfectant concentration $(mg.l^{-1})$ and the average contact time (min).

PILOT PLANT STUDY ON DISINFECTION

Pilot plant study Zevenbergen. Water treatment plant Zevenbergen of Water Supply Company Brabant Water Ltd. uses river water after storage in open reservoirs for the production of drinking water. Treatment includes coagulation/flotation, chlorination, dual media filtration and GAC filtration. The use of ozone as an alternative for chlorine was studied on pilot plant scale (Figure 2.1). The specifications of the treatment stages are described elsewhere (Hijnen *et al.*, 1997). In this study the detection limit of the standard SSRC analysis was decreased by examining larger sample volumes to assess concentrations of SSRC in the last stages of treatment.

Determination of the decimal elimination capacity (DEC). The collected historical data on indicator bacteria in the water from the eight surface water treatment locations showed that concentrations in the source waters were high. Percentage of detection in the examined samples was 100%. After the first applied processes, however, percentages of positive samples (%PS) decreased and after the subsequent processes in an increasing number of the samples indicator bacteria were absent (zero count). Therefore a general strategy was introduced in order to calculate the elimination capacity of processes and a total treatment for these micro-organisms.

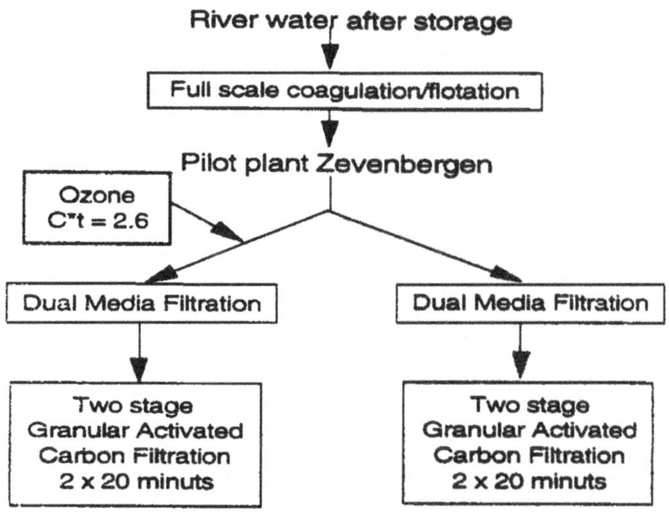

Figure 2.1 Scheme of the pilot plant at the water treatment plant Zevenbergen

For the first processes with high percentages of positive samples the actual decimal elimination or DE value is calculated from the concentration of indicator bacteria in the influent (C_{in}) and effluent (C_{out}) of the process measured at the same time (paired data) with the following equation

$$DE = Log_{10}C_{in} - Log_{10}C_{out} = Log_{10}\frac{C_{in}}{C_{out}} \tag{1}$$

The Decimal Elimination Capacity (DEC) is the 50-percentile or median values from the calculated DE values of a process.

When in the outlet or in both the inlet and the outlet of a process, the majority of the samples >80% were negative for the indicator bacteria, DEC is determined with the average concentration. In order to include the information of the 'negative' samples (zero counts) in these calculations all individual samples are considered as part of a large volume examined during a specified period of time. The average concentration is calculated with the following equation

$$\overline{C} = \frac{\sum cfu}{N * v} \tag{2}$$

where $\overline{C}$ is the average concentration in the inlet or the outlet of the process, Σcfu the total number of colony forming units in N samples with volume v (l). These average concentrations are used to calculate the Decimal Elimination Capacity with

$$DEC = Log_{10}\frac{\overline{C}_{in}}{\overline{C}_{out}} \tag{3}$$

where C_{in} and C_{out} are the arithmetic average concentrations in the water. This method, where the observations are used as unpaired data, is called the ratio-estimation method (Drost *et al.*, 1997). For statistical analysis of the data Excel (Microsoft) was used.

ACCEPTABLE QUALITY LEVEL OF DRINKING WATER

On the basis of the standards for indicator bacteria in drinking water of the Dutch Drinking Water Decree (Anonymous, 1984; MTC-values and sampling frequencies) critical mean concentrations $N_c.l^{-1}$ (95% confidence level) can be calculated that will generate annually 0, 1 and 5% positive samples of indicator bacteria in the drinking water, as suming Poisson distribution (Van der Kooij *et al.*, 1995). These N_c values for an acceptable quality level of 1% of samples non complying with the standard for thermotolerant coliforms (Coli44 incl. *E. coli*) daily monitored in 300 ml is $0.018.\ l^{-1}$. For SSRC and FS weekly monitored in 100 ml the N_c is $0.035.l^{-1}$.

FAECAL INDICATORS IN SURFACE WATER TREATMENTS

Concentrations of indicator bacteria in the source waters of the eight treatment facilities showed a high variation, both between the locations but also at the individual locations (Figure 2.2; arranged from high to low concentrations). Coliform concentrations were usually higher than the concentrations of FS and SSRC. At facilities 1 and 2, however, SSRC concentrations in the source water were higher than coliform concentrations.

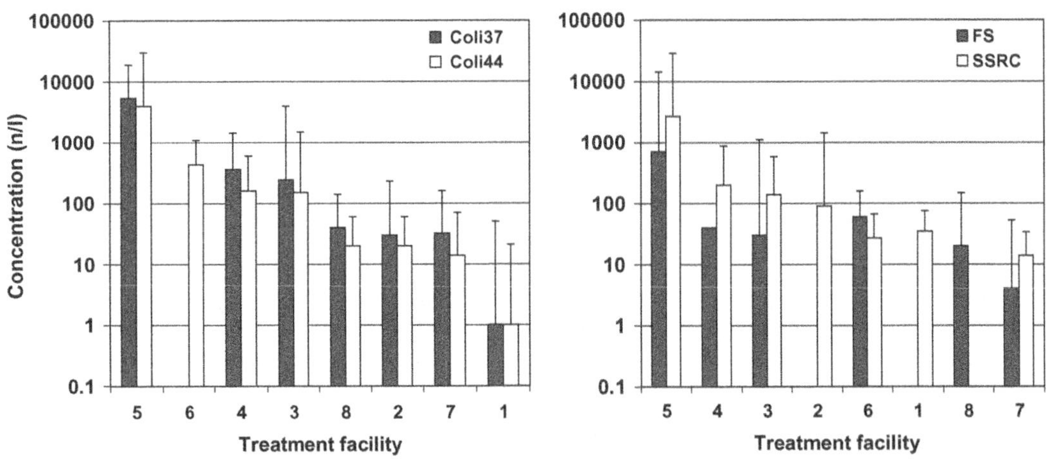

Figure 2.2 Concentrations (arithmetic mean) of Coli37, Coli44, FS and SSRC in the source waters of eight treatment facilities calculated from the routine monitoring programs (n = 150–10,000) over a period of 3 year and arranged in decreasing order of Coli44 and SSRC concentrations (error bar 90-percentile-values interpolated from the collected data)

At four of the five locations using main disinfection with either chlorine or ozonation (Table 2.1) indicator bacteria were monitored after this process. These results demonstrate that main disinfection processes are no absolute barrier against indicator bacteria. In percentages of the total samples ranging from 0–33.6% coliforms and faecal streptococci were detected after disinfection.

Table 2.1 Percentage (%) of positive samples (number of samples) for faecal indicator bacteria in the water after main disinfection at two locations with chlorine and two locations with ozonation with the corresponding Ct-values (mg/l.min.)

Organisms	Location 4	Location 2	Location 3	Location 7
	Cl_2 (15–100)[a]	Cl_2 (9–12)	O_3 (2)	O_3 (1.7–2.2)
Coli37	2.4 (82)[b]	0.4 (2013)	1.8 (1520)	33.6 (411)
Coli44	2.4 (83)	0.1 (2013)	0.6 (1519)	12.2 (582)
FS	0.0 (82)	1.2 (415)	1.3 (310)	8.3 (72)
SSRC	32.9 (82)	–	30.4 (312)	65.2 (135)

[a]range of applied Ct values; [b]number of samples

Despite higher Ct values of chlorination and a significantly lower number of samples at location 4, the percentage of positive samples for coliforms (Coli37 and Coli44) after this chlorination was higher than the percentage of positive samples with these indicator bacteria after chlorination at location 2. The difference between the performance of ozonation at location 3 and 7 with approximately the same Ct values was even larger which indicates that the efficacy of these full-scale disinfection processes may vary considerably. A second important but expected result was the higher resistance of SSRC. In 30.4 up to 65.2% of the samples these anaerobic spores were detected after disinfection.

The data of the finished waters revealed that 100% compliance with the microbiological standard (Water Decree, 1984) was not achieved during the evaluated period of three years (Table 2.2). The first four plants used post-disinfection and at the other plants post-disinfection was not used. A high level of non compliance for Coli37 (data not presented) in the finished water at plants with main disinfection (1, 2, 3, 4 and 7) was observed. This was caused by regrowth of these bacteria in the filtration processes GAC or SSF applied at the end of the treatment facilities. During summer periods C_{out} of this indicator was regularly higher than C_{in} (negative DE value). Consequently, this parameter is unsuitable as a process indicator for elimination assessment.

Table 2.2 The average concentrations of Coli44 and SSRC in the finished waters with or without post-disinfection of the eight treatment facilities, the percentage of positive samples (%PS) and DEC values

Facility (DIS)[a]	Coli44 (N_c = 0.018 cfu l^{-1})			SSRC (N_c = 0.035 cfu l^{-1})		
	C_{out} (n/l)	%PS	DEC	C_{out} (n/l)	%PS	DEC
1 (O_3)	0.0009[b]	0.09	4.1	0.051[b]	2.6	3.0
2 (Cl_2)	0.05[b]	0.02	2.7	0.32[b]	1.9	2.8
3 (O_3)	0.160[c]; <0.0008[b]	1.6; 0.0	3.5; >5.3	1.375[b]	5.6	2.1
4 (Cl_2)	<0.013[c]; <0.03[b]	0.0; 0.0	>4.3; >3.9	0.117[b]	1.0	3.2
5	0.090	1.2	5.0	0.037	1.9	4.9
6	0.029	2.0	4.3	0.320	23	3.0
7 (O_3)	0.002	0.1	4.2	0.008	0.8	3.4
8	0.013	1.0	3.5	0.014	1.4	3.4

[a]DIS = main disinfection; Cl_2/O_3 = chlorine/ozone; [b]data after post-disinfection with chlorine dioxide; [c]data before post-disinfection

The percentage of non compliance with the standards for Coli44 (Table 2.2) and faecal streptococci (data not presented) was higher after treatment plants without a main disinfection, except for location 3. Concentrations of SSRC in the finished waters (only measured after post disinfection when applied) were usually higher than Coli44 concentrations again showing the higher resistance of these indicator bacteria with a higher probability of breakthrough. The finished water concentrations of Coli44 after five facilities were below the N_c for Coli44 of 0.018 $CFU.l^{-1}$ for an acceptable quality level (AQL) of 1%. For SSRC, however, finished water concentrations were usually above the N_c of this parameter of 0.035 $CFU.l^{-1}$ and percentages of positive samples were higher than 1% up to 23% at location 6.

DECIMAL ELIMINATION CAPACITY (DEC)

DEC of the treatments. Using these data DEC of the facilities was calculated using the arithmetic mean concentrations in source and finished water (zero counts are include as indicated before). DEC of the facilities for these indicator bacteria ranged from 2 to 5 log (Table 2.2). Application of main disinfection not necessarily revealed a higher DEC value. The highest DEC value was determined for location 5 without either main or post-disinfection. In agreement with the former conclusion on higher breakthrough of SSRC, DEC of the facilities for SSRC was lower than DEC for the more susceptible Coli44.

Decimal Elimination Capacity of unit processes. From the three year routine data of indicator bacteria samples monitored during treatment of some facilities, indicator bacteria concentrations in the water after the first processes was high enough to assess the actual decimal elimination (DE) with variation presented by the 50- (median) and 10-percentiles assessed by interpolation (Table 2.3).

Table 2.3 DEC and median and the 10-percentile value of DE of some processes at a number facilities for thermotolerant coliforms (Coli44) and SSRC based on three year routinely collected data

Process	Coli44			SSRC		
		DE			DE	
	DEC	Median	P10	DEC	Median	P10
CFR	–[a]	–	–	1.2	0.9	0.1
CFR	0.7	0.8	0.5	0.6	0.6	0.1
RGF	1.1	1.2	0.8	–	–	–
RGF	–	–	–	1.4	1.3	0.7
RGF	0.9	0.7	0.2	–	–	–
O_3	–	–	–	1.0	0.7	0.0
O_3	–	–	–	–	0.8	0.3
O_3	1.6	1.0	0.5	0.4	0.5	0.0

[a]*no data collected*

These data demonstrate that in 10% of the time the actual elimination in the processes for faecal indicator bacteria is 0.3–0.8 log lower than the average elimination (DE). The impact of not including the zero counts in the DE calculations on the average elimination capacity was illustrated from these data. DEC values with zero counts included in the calculation were in 5 out of 9 processes higher than the median DE value (Table 2.3). For one process DEC was similar to the median and at 3 processes DEC was lower than the median.

LARGE VOLUME SAMPLING

Lower detection limit of the standard analytical method. A simple option to decrease the detection limit of the standard analytical method is by increasing the sample volume as demonstrated for SSRC in a pilot plant study (Hijnen *et al.*, 1997). The removal of these spores by coagulation/flotation followed by dual media filtration and granular activated carbon filtration (GAC) was monitored with 1 and 10 liter samples filtered through 47 mm (0.45 μm) membranes. By increasing the sample volume the percentage of positive samples after dual media filtration increased from 22 to 84% and after the subsequent GAC-filtration (no 1 liter samples examined) the %PS was 58% (Figure 2.3).

The cumulative distribution of the results of SSRC in 1 and 10 liter samples is depicted in Figure 2.4.

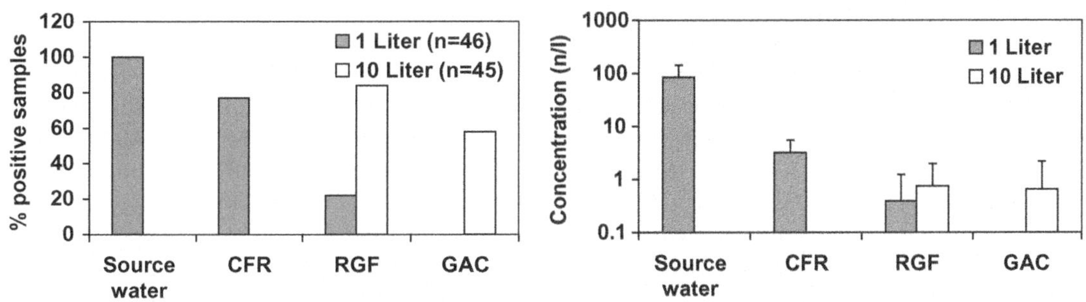

Figure 2.3 Percentage of positive samples (left) and (right) average concentration of SSRC at the different stages of the pilot plant Zevenbergen (GAC = granular activated carbon filtration) (error bar = SD) (data from Hijnen *et al.*, 1997)

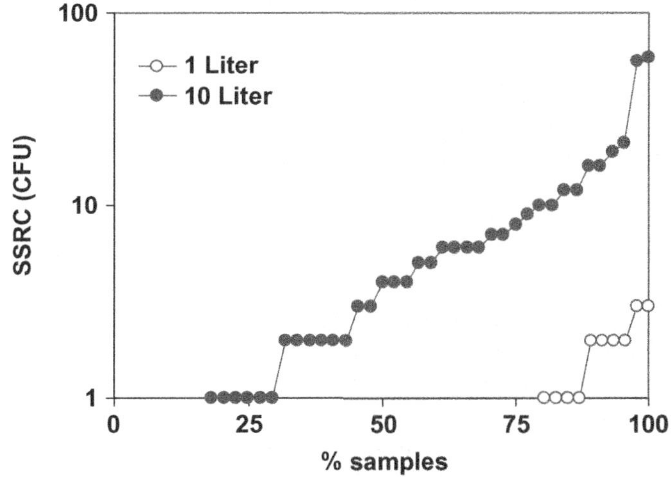

Figure 2.4 Cumulative distribution of SSRC numbers (colony forming units; CFU) in GAC filtrate monitored in 1 and 10 liter samples (from Hijnen *et al.*, 1997)

DISCUSSION

DEC assessment of a treatment or process for indicator bacteria. Quantification of the elimination capacity of water treatment facilities for faecal indicator requires presence of these organisms in both the source and finished waters. At all facilities these micro-organisms were present in the source water. On the basis of the source water concentrations and the detection limits of the standard analytical methods of 1 CFU per 100 or 300 ml (10 or 3.3 CFU per l), the maximum assessable DEC of a treatment for coliforms (total Coli37 and thermotolerant coliforms Coli44) is approximately <1–3 log depending on source water quality. With the data of faecal streptococci and SSRC (incl. *C. perfringens*) this maximum assessable DEC is <1–2.4 log. The required DEC values for viruses, *Cryptosporidium* and *Campylobacter* for an annual infection risk level of 10^{-4} estimated from source water concentrations was 6–8 log (Chapter 1). Consequently, with the routine data of indicator bacteria monitoring with the standard membrane filtration method the required DEC for index pathogens is not assessable.

The evaluation of historical routinely collected data on indicator bacteria in the finished water of drinking water facilities revealed that in a small number of samples these indicators were observed. In a first attempt to describe elimination capacity of water treatment processes with the routinely collected data on faecal indicators Drost *et al.* (1994) compared a number of possible options focussed on the question how to include the zero counts in data sets as valuable observations in the calculations of elimination. Zero counts were replaced by the value of the detection limit (A), by the average concentration (B) or by the actual concentration plus 1 (C). Using real data sets and simulated data sets, DEC derived from these data manipulations were compared with DEC assessed with the ratio-estimation method (E) described before and with DE values using only positive samples (D; excluding zero counts). Method assessment was done by calculation of the residual values (RES) of the actual concentration in the outlet and the concentration calculated from the linear regression coefficient a of $C_{out} = a.C_{in}$:

$$RES = C_{out} - \hat{C}_{out}$$

Conclusion and recommendation of these tests was to use paired data of C_{in} and C_{out} if >20% of these pairs were both positive. Otherwise the ratio-estimation method E was recommended as second best followed by the method where zero counts are replaced by the average concentration of the actual data set.

Calculation of the arithmetic average concentrations of faecal indicator bacteria in source and finished water (zero counts included), enables assessment of higher DEC values of 2–5 log as demonstrated in Table 2.2. These DEC values are lower than the required DEC-values of 6–8 log for viruses, *Campylobacter*, *Cryptosporidium* and *Giardia* for an annual 10^{-4} infection risk in the produced drinking water.

A second study to explore calculations of micro-organism removal in water treatment was conducted with the (statistical) methods B and D described above and additional mathematical methods (Evers and Groennou, 1999). Using SSRC data from full-scale plants the study concluded that a binominal mathematical method with fixed p value was best followed by statistical method B (zero = average). Nonetheless, the authors concluded that the accuracy of the determined elimination capacity depends on the quality of the collected data. From datasets with high level of zero counts no high accuracy can be expected and it was recommended to reduce the number of these zero counts by analyzing larger sample volumes.

Possibility to lower the detection limit. The data of the pilot plant study at Zevenbergen demonstrate that by simply increasing the sample volume with a factor of 10 a higher percentage of samples with SSRC can be obtained. This results in more accurate quantitative data on the DEC of processes. In the presented pilot plant study without a main disinfection step, the standard sample volume of 100 ml was increased with a factor of 100 (detection limit of 0.1 cfu l^{-1}). For an accurate assessment of concentrations of Coli44 and

SSRC in the finished water of most treatment facilities a detection limit of 0.01 cfu l^{-1} is required. Volumes of 100 liter or more, however, can not be analyzed with the standard membrane filtration technique and require an adapted method. The use of large volume sampling under full-scale conditions have been explored by others (Goyal *et al.*, 1980 and Payment *et al.*, 1989) but these techniques are based on membrane adsorption/ elution techniques with relative complex procedures compared to the standard microbial methods and have low recoveries.

Variability of DEC of unit processes. Another omission in the current data sets is the lack of information on variation in elimination. Information about the variability in removal is of interest because waterborne outbreaks of diarrhoea caused by protozoan parasites have been related to peak concentrations in the source water but also by inadequacies in water treatment (Dykes *et al.*, 1980; Badenoch, 1990; Craun, 1990 and Richardson *et al.*, 1991). Evaluation of historical data sets can help to identify such inadequacies as demonstrated for the ozonation process at facility 7 (Table 2.1). DEC of this ozonation operated at Ct values of 1.7–2.2 mg/l.min. was 1 log for the ozone susceptible Coli44. On the basis of dose/response data for *E. coli* known in literature (Finch *et al.*, 1988), a much higher DEC was expected.

Beside obvious inadequacies in treatment, normal operated processes in a treatment exhibit a variation in the elimination of micro-organisms. LeChevallier and Norton (1992) for instance showed that the removal of particles of $>5\,\mu m$ (size of *Cryptosporidium* oocysts) by individual unit filters in the filtration stage of a treatment may vary by as much as 1.000-fold. These variations will influence the overall efficiency of a treatment stage. Variation in the removal of (oo)cysts of *Cryptosporidium* and *Giardia* by a conventional treatment (coagulation/ floc removal plus filtration) was observed by Hashimoto *et al.* (2001). DEC for *Cryptosporidium* and *Giardia* ranged from 2.0–3.2 and 1.7–3.1 log, respectively.

The variability of DEC of unit processes for *Cryptosporidium* has also been demonstrated by spiking tests on pilot plant scale or in laboratory experiments. Emelko (2001) intensively studied the removal of *C. parvum* oocysts by filtration processes under different filtration conditions. DEC assessed under stable operation, ripening, sub-optimal coagulant dosing, early and late breakthrough ranged between 5.5 and <0.5 log. Also with spiking experiments on laboratory scale (dose/response data with a continuous flow system) a high variability of the efficiency of ozone disinfection for *C. parvum* in natural waters was demonstrated (Oppenheimer *et al.*, 2000). The average Chick/Watson inactivation constant at 10°C was 0.21 and a range of 0.08–0.46 l/mg.min.

Variation in elimination of micro-organisms is caused by the multiple variables involved in the mechanisms responsible for the elimination (inactivation and removal) in the different treatment processes (see Chapter 1). Due to these observations and considerations it is of importance for risk management to integrate variability in both source water concentrations and elimination in the uncertainty calculations of the infection risk assessed with QMRA. Additional statistical tools to integrate uncertainty calculations in QMRA have been described (Teunis *et al.*, 1997; Evers and Groennou, 1999; Haas *et al.*, 1999 and Medema *et al.*, 2003). Moreover, in the framework of Water Safety Plans (WSP, WHO, 2004) it is of importance to determine the causes of these fluctuations since that can lead to measures in process design or operation to minimize infection risks.

CONCLUSIONS

From the results of the studies presented in this chapter it can be concluded that monitoring for faecal indicator bacteria in full-scale water treatment facilities for the production of drinking water can be used to quantify the removal and inactivation (elimination) capacity (DEC) of unit processes and a complete train of processes. Evaluation of the historical collected monitoring data with standard analytical methods yielded a first impression of DEC of processes and treatment facilities. It also showed the limitations because of the high percentage of zero counts (no indicators detected) in the last stages of treatment. Methods have been used and evaluated to determine DEC from these datasets with a high level of zero counts, but were judged as being inaccurate. Beside this inaccuracy assessment of variability of DEC values required to calculate the uncertainty of infection risk levels with QMRA was also not possible. The first impression on elimination of indicator bacteria in water treatment showed lower DEC values than the DEC values required for index pathogens. This emphasizes the need for improvement of accuracy of DEC assessment of a treatment for both process indicators which requires an analytical method with lower detection limit.

The challenge is to develop a rapid, simple and reliable isolation method for the assessment of concentrations of indicator bacteria in volumes of ⩾100 litre treated water. First results presented in this study indicated that decreasing the detection limit of the standard membrane filtration method by increasing sample volumes is potentially an easy and successful option.

The data collected on the elimination of Coli44 and SSRC by unit processes showed not only variation in DEC of processes over time but also a significant variation in DEC of similar processes operated at different locations. This emphasizes the need for quantitative information on elimination of micro-organisms by water treatment processes for QMRA as site specific and actual as possible.

Chapter 3

Enumeration of faecal indicator bacteria in large water volumes using on site membrane filtration to assess water treatment efficiency

Wim A.M. Hijnen[a], D. Veenendaal[a], W.M.H. van der Speld[a], A. Visser[b], W. Hoogenboezem[c] and D. van der Kooij[a]

[a]KWR Watercycle Research Institute; [b]Dune Water Company of South Holland Ltd.; [c]Hetwaterlaboratorium

ABSTRACT

Sample volumes as tested in routine microbiological methods for determining the presence and absence of faecal indicator bacteria in water are too small to assess the actual concentration in the last stages of a water treatment. Consequently no accurate information can be obtained about the removal efficiency of a water treatment for micro-organisms. Therefore a method for on site isolation of faecal indicator bacteria from large volumes (100 litre or more) of treated water using membrane filtration (MF-sampling) was developed and tested. The procedures for culturing the isolated micro-organisms were similar to those applied in the routine methods for small volumes using membrane filtration (mf-method). The recovery efficiency of MF-sampling for *E. coli*, *S. faecalis* and spores of sulphite-reducing clostridia ranged from 74.6 to 100% and only for *E. coli* a slight decrease with increasing sample volume was found. Field studies revealed that MF-sampling can easily be implemented in (routine) laboratory practice for an accurate determination of the concentration of faecal indicator bacteria in treated water after various treatment stages. From these data the treatment efficiency of the involved processes and the overall treatment for those micro-organisms and the fluctuation in micro-organism removal were determined. Such data can be used to improve water treatment regarding the removal of micro-organisms and for quantitative microbial risk assessment. Validation of the use of faecal indicator bacteria as

Edited by Wim A.M. Hijnen. ISBN: 9781843393764. Published by IWA Publishing, London, UK.

a surrogate parameter for the assessment of the effects of treatment processes on pathogenic micro-organisms needs further investigation.

INTRODUCTION

Outbreaks of waterborne diarrhoea in the USA and the UK caused by persistent pathogenic protozoa *Cryptosporidium* and *Giardia* (Richardson *et al.*, 1991; MacKenzie *et al.*, 1994 and Kramer *et al.*, 1996) have increased interest in the effect of water treatment on pathogenic micro-organisms. Direct monitoring of these micro-organisms is difficult because (i) their concentrations in source and treated water are usually low and strongly fluctuating and (ii) analytical procedures are time-consuming and have a low and variable recovery efficiency. Traditionally the microbiological quality of drinking water is assessed by monitoring for non-pathogenic bacteria of faecal origin (faecal indicator bacteria). These indicator bacteria usually include total and faecal coliforms (FC) and in the European legislation also water quality criteria for faecal streptoccocci and spores of sulphite-reducing clostridia (SSRC) have been defined (Anonymous, 1998). SSRC have been included in legislation in the Netherlands since 1984 (Anonymous, 1984). Most SSRC present in raw and treated water are of faecal origin (*Clostridium perfringens*) and their presence may be indicative for the presence of persistent pathogens. More recently it has been suggested that SSRC may be used as a process indicator for the assessment of the capacity of water treatment processes to remove (oo)cysts of pathogenic protozoa and viruses (Payment and Franco, 1993 and Hijnen *et al.*, 1997).

In the Netherlands data are collected on the concentrations of faecal indicator bacteria in water in various treatment stages to obtain quantitative information on the removal capacity of water treatment for micro-organisms. The concentration of indicator bacteria can easily be determined in the source water, but after one or more treatment processes the concentration decreases and a sample volume of 100 ml as tested in the routine membrane filtration methods (mf-method) is too small to assess the actual concentration. SSRC removal efficiency in treatment processes at pilot plant scale was assessed by the enumeration of these bacteria in 10 litre samples with the mf-method (Hijnen *et al.*, 1997), but even larger sample volumes (100 litre or more) are needed to determine the efficiency of the final treatment steps and the overall treatment. Therefore the mf-method was scaled up and adapted for on site isolation of (indicator) bacteria from large volumes of water (MF-sampling) and enumeration in the laboratory with similar cultivation techniques used in the routine methods. This new isolation method was tested under laboratory and field conditions.

DEVELOPMENT OF A LARGE VOLUME SAMPLING METHOD

MF-sampling. An MF-sampling procedure was designed to filter a water volume of 100 litre within 1 h. With a tubing pump (Masterflex 07549-50 with 07019-00 pump head), using silicone tubing, stainless steel (SS) piping and pressure regulation, the supplied water was forced through a sterile Ø 142 mm 0.45 μm pore size membrane filter (Millipore HAWP 142-50) in a filter house (Schleicher & Schuell AJ6020-2). The entire system of water supply and filter house is autoclaved for 15 minutes at 121°C before sampling. The required flow (maximum of 250 l/hr) and an overpressure of 1 ATO is adjusted with a flow control system and the tubing pump. When the intended volume has passed through the membrane filter, the valve of the water supply is closed. Thereafter the membrane filter is aseptically transferred in a sterile glass petri dish (Ø 165 mm). When no further pre-treatment is required before incubation, resuscitation can be started on site by placing the membrane filter directly onto the solid medium.

Recovery tests. Stock cultures (one batch) of *Escherichia coli* (WR1), *Strepto faecalis* (WR63) and spores of *Clostridium bifermen* (CP1) in peptone/glycerol solution stored at −70°C were used for the recovery tests. The concentration of micro-organisms in these stock cultures had been determined periodically (Table 3.1) in routine quality checks of the mf-method (0.45 μm, Ø 47 mm membrane filters; Sartorius 13906-50-ACN) for enumeration of coliforms, faecal streptococci and SSRC in environmental samples.

The effect of the sampling volume on the recovery was tested using 550 litre volumes of non-sterile tap water (turbidity 0.1 NTU) without residual disinfectant stored at room temperature in an SS tank with a mixing device. These volumes were separately inoculated with the stock cultures to achieve a final concentration of approximately 1 CFU per litre. Faecal indicator bacteria were isolated from 10, 20, 25, 40 and 50 litre samples with the MF-sampler. *E.coli* and *C. bifermentans* were tested in duplicate samples; *S. faecalis* was tested in single samples.

The recovery efficiencies obtained with MF-sampling and with the routine mf-method for *E. coli* and *C. bifermentans* and for indigenous SSRC from river water were determined in separately prepared suspensions in sterile tap water. For this purpose 1 litre volumes were inoculated with stock cultures for a concentration of approximately 900 CFU of *E.coli* and 500 CFU of *C. bifermentans* per litre, respectively. From these suspensions 100 ml samples were tested directly with the mf-method using an SS filtration device (Schleicher & Schuell AS002/3). Prior to the testing with the MF-sampling procedure volumes of 100 ml of these suspensions were diluted in 5 litres of autoclaved (121°C) tap water at room temperature. For the test with indigenous SSRC a volume of 50 litre sterile tap water was inoculated with 50 ml of river water with an SSRC concentration of 10^3 to 10^4 CFU/l. Samples of 5 and 7.5 litre were tested in duplicate with the mf-method and single samples of 10 and 15 litre samples were tested with MF-sampling.

Microbiological analysis. Membrane filters were incubated on Lauryl Sulphate Agar (LSA; Oxoid MM615) for 5 ± 1 h. at 25 ± 1°C and 14 ± 2 h. at 44 ± 0.5°C to cultivate *E. coli* and the faecal coliforms (FC). FC concentrations in the samples of the field study were derived from the number (N) of yellow colonies on LSA at 37 ± 1°C for 14 ± 2 h. Typical colonies ($\sqrt{N}$) were tested in Brilliant Green Bile broth (Oxoid CM31) during 24 h. at 44 ± 0.5°C and the percentage of positive tests was used to calculate the FC concentration in the samples. Faecal streptococci were determined by incubation on KF streptococcus Agar (Difco 0496-17-2) during 48 ± 4 h. at 37 ± 1°C. In the laboratory experiments concentrations of SSRC and spores of *C. bifermentans* were determined on Perfringens-Agar-Base medium (PAB; Oxoid CM587) incubated for 48 ± 4 h. at 37 ± 1°C. For the field studies Shahidi Ferguson Perfringens agar base (Difco 0811-17-0) was used for SSRC determinations. Sample volumes of 100 ml were pasteurized in a water bath at 70 ± 1°C for 30 minutes prior to filtration. With sample volumes of 1 litre or more the membrane filters were pasteurized in the liquefied medium placed in an oven at 70 ± 1°C for 30 minutes. The cultivation procedure of the membrane filters for SSRC analysis was carried out as previously described (Hijnen *et al.*, 1997).

Field tests. The suitability of the MF-sampling procedure under field conditions was tested in two treatment facilities (Figure 3.1).

The concentration of faecal coliforms (FC) and SSRC was daily monitored by the involved water supply companies in the water after several treatment stages in a two-week period in January (water temperature of 10°C) and in February (water temperature 3–5°C) at facility 1 and 2, respectively. Samples of 1 up to 10 litre collected after the open collecting bassin (OCB) and after rapid granular filtration (RGF) at facility 1 and after chlorination (CL_2) at facility 2 were tested with the routine mf-method. Samples after CL_2 were neutralized with sodium thiosulphate (30 mg/l). The MF-sampling procedure was used to isolate both indicator bacteria from 57–100 litre samples of the SSF-filtrate at facility 1 and of the GAC-filtrate at facility 2, respectively. The results were compared with the results of the routine microbiological monitoring program conducted by the water supply companies in this period.

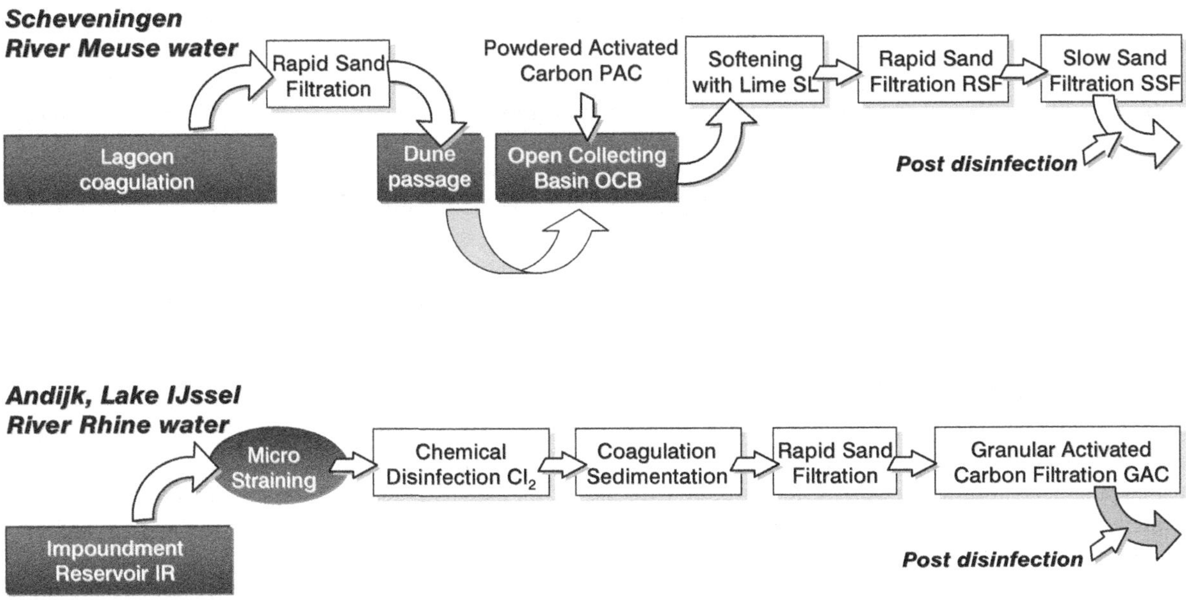

Figure 3.1 Water treatment of facility 1 (the Dune Water Company South-Holland Ltd. and of facility 2 (PWN Water Supply Company North Holland Ltd.) where MF-sampling was tested under field conditions

Statistical analysis. The similarity of the numbers of typical colonies obtained with the MF-sampling and the mf-method was tested by the student-t test and the F-test (Excel software). Normal distribution of the data was tested with the Kolmogorov-Smirnov-test (Statistical Package for Social Sciences SPSS).

RECOVERY EFFICIENCY

Recovery tests. A method with high recovery efficiency was developed for an accurate determination of the concentration of indicator bacteria in water. The average recovery efficiencies of the experiment with increasing sample volumes obtained from large volumes (550 litre) were estimated from the ratio between the measured and the calculated concentrations of the indicator bacteria in each separate sample (Table 3.1).

The calculated concentration was derived from the average concentration in the stock culture as determined with the routine mf-method. The average recovery efficiencies for the three indicator bacteria were ≥74.6% of the calculated concentrations. The colony counts increased linearly with the sample volume (Figure 3.2) with slopes of 1.06 and 1.00 for *S. faecalis* and *C. bifermentans*, respectively (Table 3.1), indicating that the recovery was independent of the sample volume. For *E.coli* a slope value of 0.787 revealed that the recovery efficiency slightly decreased with increasing sample volume. It is possible that this decrease was caused by a decline of the concentration in the suspension during the experiment. These recovery efficiencies are similar or higher than those reported for the membrane adsorption-elution techniques using cartridge filters and elution-procedures (≥30 up to 86%; Goyal *et al.*, 1980 and Payment *et al.*, 1989). The recovery efficiencies of the MF-sampling and the mf-method were similar for *E. coli*, *C. bifermentans* and for the indigenous SSRC from river water (Table 3.2; $P \geq 0.99$, 0.37 and 0.81, respectively for 95%-interval).

Table 3.1 The calculated and measured average concentrations (± standard deviation, SD) as well as the recovery efficiencies determined from the colony counts in different sample volumes from separate suspensions of *E. coli*, *S. faecalis* and *C. bifermentans* in tap water tested with MF-sampling; linear regression analysis for the relation between the sampled volumes and the colony counts (Figure 3.2)

Data of recovery tests		*E. coli*	*S. feacalis*	*C. bifermentans*
Stock cultures	CFU[a]/ml	27 (±3.3; 5[c])	46 (±11.1; 30[c])	102 (±36; 30[c])
Conc. (n/l)	Calculated	1.09 (±0.22)	0.94 (±0.23)	0.93 (±0.33)
	Measured	0.89 (±0.12)	0.703 (±0.23)	1.06 (±0.11)
Samples	N[b]	8	4	8
Recovery	%	81.9 (±11.1)	74.6 (±24.9)	114 (±11.4)
Lin. reg. anal. (Vol. vs CFU)[d]	Slope	0.787 (±0.06)	1.061 (±0.07)	1.000 (±0.09)
	r^2	0.967<	0.991<	0.950<
	p	0.0001	0.0100	0.0001

[a]Colony Forming Units; [b]N = number of observations; [c]number of tested samples from the stock cultures with mf-method; [d]linear regression analysis: volume versus CFU

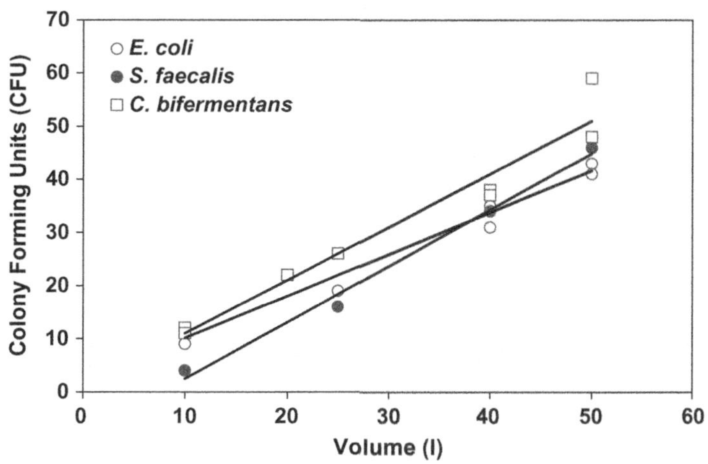

Figure 3.2 The linear regression fit of the recovery tests with faecal indicator bacteria in tap water without residual disinfectant

Table 3.2 Concentrations of *E. coli*, *C. bifermentans* and indigenous SSRC from river water (arithmetic mean±SD) in separate suspensions in sterile tap water determined with the mf-method and with the MF-sampling procedure

	Sample volume (ml)	*E. coli* (CFU/100 ml)	*C. bifermentans* (CFU/100 ml)	SSRC (CFU/l)
mf-method	100	93.3 (±12.2) (n = 3)	38.3 (±18.6) (n = 3)	8.9 (±2.1) (n = 4)
MF-sampling	5000[a]	93 (±9) (n = 3)	49.3 (±3.7) (n = 3)	9.2 (±1.2) (n = 2)

[a]prior to MF-sampling: 100 ml of a prepared suspension mixed in 5 litre sterile tap water

APPLICATION UNDER FIELD CONDITIONS

Field tests. Relatively high concentrations of FC (205 ± 94 CFU/l) were observed in the water after dune passage sampled from the open collecting basin (OCB) at facility 1 (Table 3.3). Apparently the water in this open basin was faecally recon-taminated by birds and other wild life. The SSRC concentration in this water was lower and more fluctuating (8.3 ± 16.3 CFU/l). MF-sampling of the slow sand filtrate before post-disinfection showed that treatment (PAC, SL, and RGF followed by SSF; Figure 3.1) clearly reduced these concentrations.

Table 3.3 Concentrations (n/l) of faecal coliforms (FC) and spores of sulphite-reducing clostridia (SSRC) in the water of the open collection basin (OCB) after dune passage and after powderer activated carbon, softening with lime and rapid granular filtration (PAC/SL/RGF) and slow sand filtration (SSF) of facility 1 in January 1997 (water temperature 10.4 ± 0.4°C)

No.	Faecal coliforms			Spores of sulphite-reducing clostridia		
	OCB[a]	PAC/SL/RGF[a]	SSF[b]	OCB[a]	PAC/LS/RGF[a]	SSF[b]
1	418	47.6	0.19	<1	<0.1	0.03
2	192	39	0.19	ND[c]	ND	0.01
3	208	64.4	0.19	12	0.9	ND
4	299	48.5	0.21	ND	ND	0.13
5	226	54	0.24	<1	<0.1	ND
6	111	36.3	0.21	1	0.2	0.13
7	164	34.4	0.21	1	0.1	0.16
8	103	24.6	0.14	44	<0.1	0.03
9	140	26.5	0.14	ND	ND	0.2
10	193	33.4	0.17	<1	<0.1	0.12
avg[d]	205	40.9	0.19	8.3	0.17	0.10
SD[d]	94	12.6	0.03	16.3	0.33	0.07

[a]Routine mf-method: 4 times 2.5 litre using Ø 47 mm membranes (10 litre samples); [b]MF-sampler: sample volume of 100 l; [c]ND = not determined; [d]avg = arithmetic mean and SD = standard deviation

Average FC and SSRC concentrations in the SSF-filtrate were 0.19 ± 0.03 CFU/l and 0.10 ± 0.07 CFU/l, respectively. The routine microbiological monitoring program conducted in the same period in the SSF-filtrate yielded an average concentration for FC of 0.09 ± 0.54 CFU/l (145 samples of 300 ml) and for SSRC of 0.36 ± 1.88 CFU/l (58 samples of 100 ml), respectively. These values are in the same order of magnitude as those determined with MF-sampling, but relative standard deviations clearly exceeded 100% as the result of a low percentage of positive samples (3%).

The FC concentrations in the water from the impoundment reservoir (IR) of treatment facility 2 were 4 times lower than the concentrations of these micro-organisms in the OCB of facility 1 (Table 3.4), but SSRC concentrations were 10 times higher.

Chlorination (Cl_2) strongly reduced the FC concentration to 0.01 ± 0.3 CFU/l and these indicator bacteria were not found (<0.001 CFU/l) in a total volume of 1.000 litre GAC-filtrate tested with MF-sampling. Average SSRC concentrations after Cl_2 and GAC were 29 and 0.43 CFU/l, respectively and could easily be determined by testing 1 and 100 litre samples with the mf-method and MF-sampling, respectively. No SSRC were detected in the GAC-filtrate (8 samples of 1 l) during routine microbiological monitoring in this period and also all samples after post-disinfection tested as prescribed by legislation were negative for FC and SSRC.

Table 3.4 Concentrations (n/l) of faecal coliforms (FC) and spores of sulphite-reducing clostridia (SSRC) in the water of the impoundment reservoir (IR) and after chlorination (Cl_2) and granular activated carbon filtration (GAC) at facility 2 in February 1997 (water temperature 4 ± 1°C)

No.	Faecal coliforms			Spores of sulphite-reducing clostridia		
	IR[a]	Cl_2[a]	GAC[b]	IR[a]	Cl_2[a]	GAC[b]
1	25	<0.33	<0.014	270	6.2	1.60
2	210	<0.2	<0.013	635	17.8	0.11
3	10	<0.2	<0.01	1010	43.6	0.07
4	60	<0.2	<0.01	2085	40.4	0.06
5	150	0.2	<0.01	1020	44.6	0.09
6	150	<0.2	<0.01	250	27.2	0.37
7	610	<0.1	<0.01	900	28	0.59
8	320	<0.1	<0.01	1550	28	0.66
9	70	<0.1	<0.01	1350	28	1.80
10	20	<0.1	<0.01	1050	18	0.81
avg[d]	49.5	0.01	<0.001	1079	29	0.43
SD[d]	180.5	0.3	ND[c]	562	14.3	0.53

[a]Routine mf-method: 4 times 2.5 litre using Ø 47 mm membranes (10 litre samples); [b]MF-sampler: sample volume of 100 l; [c]ND = not determined; [d]avg = arithmetic mean and SD = standard deviation

THE REMOVAL EFFICIENCY OF THE TREATMENT PROCESSES

Due to the low percentage of positive samples the routine microbiological monitoring program of both water supply companies was unfit to determine removal efficiencies of all processes or the overall treatment for FC and SSRC. Increasing the sample volume resulted in higher percentages of positive samples and revealed a more accurate determination of the actual concentrations as indicated above. For FC at facility 1 and for SSRC at facility 2, respectively this enabled the assessment of the actual Decimal Elimination (DE), which is defined as $\log C_{in} - \log C_{out}$. From these DE-values and from the average concentrations in the water, the decimal elimination capacities (DEC) of unit processes and of the overall treatment without post-disinfection for both micro-organisms in this period were calculated (Table 3.5).

Clearly, at treatment facility 1 SSRC concentrations were removed more efficiently by the combination of PAC, SL and RGF than FC concentrations. For slow sand filtration however it was the other way around. The DEC value of this process for FC was 2.3 ± 0.09 log, but the filters showed hardly any elimination of the clostridia spores. This phenomenon has also been observed in GAC filter beds in a pilot plant which were infrequently backwashed (Hijnen *et al.*, 1997). At facility 1 concentrations of SSRC were observed in the filter bed of two slow sand filters and the highest numbers were detected in the top layer (Figure 3.3). Most likely SSRC accumulated and subsequently survived in the filter bed, but under anaerobic conditions multiplication can not be excluded. These observations demonstrate the need for further research on the use of SSRC as a surrogate parameter for the assessment of the effects of filtration processes on protozoan (oo)cysts.

Chlorination in treatment facility 2 (average product of residual concentration C and the contact time t (Ct) of 40 mg.min/l) reduced the average FC concentration in the water with 3.6 log (Table 3.5).

Table 3.5 The Decimal Elimination Capacity (DEC) of unit processes and the overall treatment of drinking water facilities Scheveningen and Andijk

Facility	Treatment processes	DEC value (log)	
		FC	SSRC
1	PAC/SL/RGF	0.7±0.13[a]	1.7[b]
	SSF	2.3±0.09	0.2
	Treatment	3±0.17	1.9
2	Cl_2	3.6	1.5±0.25
	Treatment	>4.7	3.5±0.6

[a]DEC ± SD = arithmetic mean (avg) of DE-values; [b]DEC = log (avg C_{in}) − log (avg C_{uit})

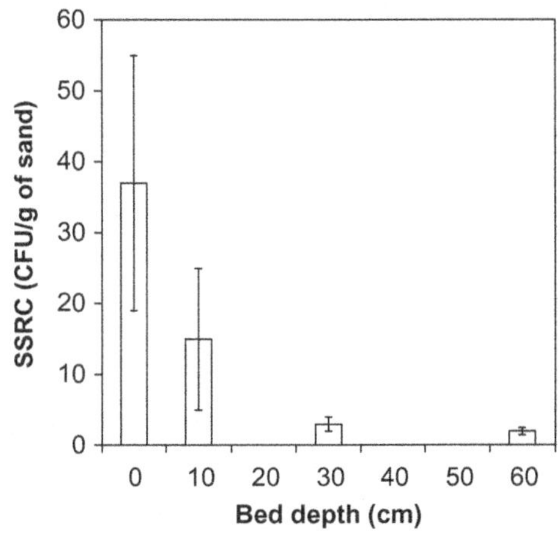

Figure 3.3 The average SSRC concentration (± SD; n = 6) in the sand of two individual slow sand filters of drinking water production plant Scheveningen

This DEC for these Ct-values is low compared to values reported in literature (Sobsey, 1989). A DEC-value of 1.5 ± 0.25 log was calculated for the inactivation of spores of clostridia by chlorination, indicating that these spores are more resistant to chemical disinfection. From the concentrations of FC and SSRC in the GAC-filtrate, assessed with MF-sampling, a DEC-value of ≥4.7 and of 3.5 ± 0.6, respectively was calculated for the overall treatment.

The DE-values of unit processes and the overall treatment also gave information about the variation in removal of the indicator bacteria during these two-week periods in winter (Figure 3.4). The DE-values of unit treatment processes and the overall treatment of both facilities for FC and SSRC showed a normal distribution ($P \geq 0.827$–0.999, 95%-interval). Table 3.5 show that the variation in DE-values of the processes at facility 1 for FC was limited ($SD < 0.2$ log). Also the inactivation of SSRC by chlorination at facility 2 showed little variation ($SD = 0.25$ log), but for the DEC-value of the overall treatment for SSRC the SD-value was 0.6 log (Table 3.5). As yet it is not clear which processes or process conditions are responsible for variations in micro-organism removal.

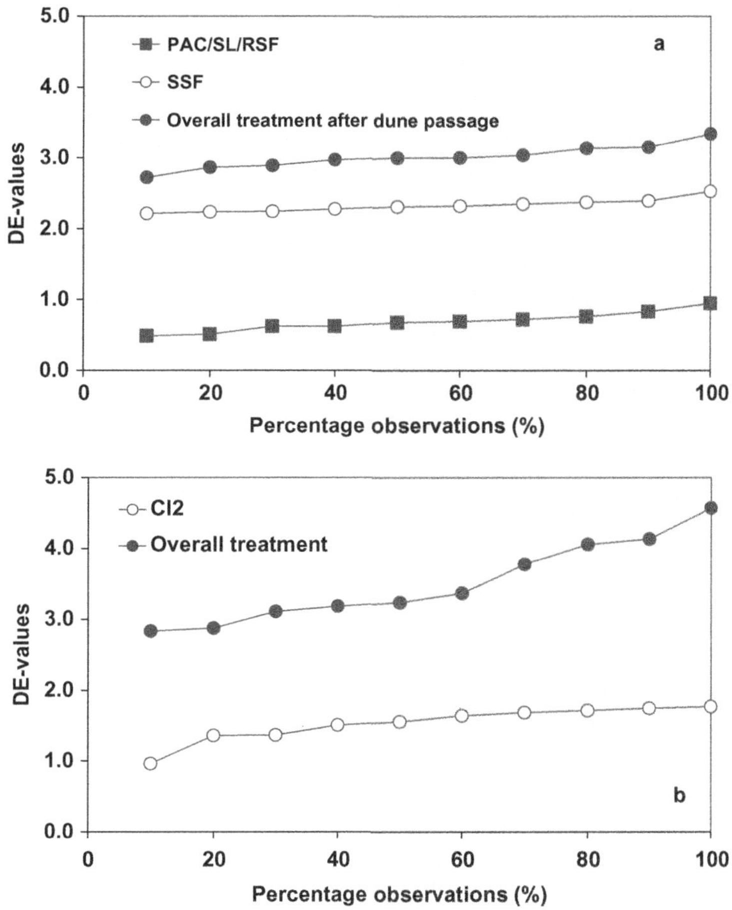

Figure 3.4 The cumulative frequency distribution of the DE-values of unit processes and of the overall treatment for FC and SSRC at facility 1 (a) and 2 (b), respectively, during a two week period in the winter period of 1997

The presented quantitative data about the removal efficiency of full-scale treatment processes for indicator bacteria obtained with MF-sampling contributes to a more accurate assessment of the risk of infection with pathogens via drinking water (Teunis *et al.*, 1997) necessary to determine the need for process optimization. Furthermore, with MF-sampling the process conditions responsible for the low DE-values may be elucidated subsequently leading to process optimization for the removal of pathogens.

CONCLUSIONS

The routine method for determining the presence and absence of faecal indicator bacteria (e.g. faecal coliforms (FC), faecal streptococci and spores of sulphite-reducing clostridia (SSRC) in relatively small volumes of water using membrane filters (mf-method) is inadequate for the assessment of micro-organism removal efficiency of all processes in water treatment. Sampling of larger volumes with specially designed equipment

(MF-sampling) in two full scale treatment plants yielded accurate determination of the concentration of faecal indicator bacteria (FC and SSRC) in the finished water (filtrate of SSF and of GAC) with a detection limit of 0.01 CFU/l or less. The same cultivation procedures are used as those applied in the routine mf-methods and the recovery efficiencies of both methods were similar.

By determining simultaneously the concentration in the source water and in the water after selected treatment stages the decimal elimination capacity (DEC) and the variation in DEC of unit processes and the overall water treatment without post-disinfection for these micro-organisms was assessed. Such data can be used to improve the microbiological safety of the drinking water by improving water treatment for the removal of micro-organisms. The significance of the observed accumulation of SSRC in slow sand filters and the relative low DEC of chlorination for FC and SSRC needs further research. This is necessary to verify the use of these indicator bacteria as surrogate parameters for the assessment of the effects of water treatment processes on pathogenic micro-organisms.

Chapter 4

Quantitative assessment of the removal of indicator bacteria in full-scale treatments plants

Wim A.M. Hijnen[a], G.J. Medema[a], D. van der Kooij[a] and A.H. Havelaar[b]

[a]KWR Watercycle Research Institute; [b]University Utrecht

ABSTRACT

The elimination of thermotolerant coliforms (Coli44) and spores of sulphite-reducing clostridia (SSRC) in full-scale water treatment was determined by large volume sampling. Objective was to determine the elimination capacity of full-scale treatment processes for micro-organisms, both vegetative bacteria and bacterial spores. In two short-periods in winter and summer, information was collected about the elimination of Coli44 and SSRC by the overall treatment, the contribution of the unit processes and the variability in elimination. Coli44 concentrations in the source waters were reduced by 3.2 to 6.3 log to an average concentration sufficiently low to achieve more than 99% compliance with the Dutch drinking water standard (0 in 300 ml; daily sampling). The elimination of SSRC was lower (1.4 to 4.2 log units) and SSRC were observed occasionally (>1%) in finished water by the routine weekly sampling of 100 ml samples. The study also yielded much information about the elimination efficacy of unit processes at the different locations, which enables process optimization and improved process control. Moreover, it is demonstrated that this quantitative information on removal of indicator bacteria by full-scale treatment systems can be used as input for quantitative microbial risk assessment.

 Quantitative Methods to Assess Capacity of Water Treatment to Eliminate Micro-Organisms. Edited by Wim A.M. Hijnen. ISBN: 9781843393764. Published by IWA Publishing, London, UK.

INTRODUCTION

In the beginning of the 20th century, bacteria that indicate faecal contamination were introduced to monitor the microbiological safety of drinking water (Eijkman, 1904). *E. coli* (or thermotolerant coliforms) and enterococci are currently used throughout the world to monitor drinking water quality. Outbreaks of cryptosporidiosis and other enteric diseases through drinking water meeting the coliform standard have been reported in developed countries (Craun *et al.*, 1997). Thus, this indicator is not adequate to predict the safety with regard to persistent pathogens like *Cryptosporidium*, *Giardia*, and enteric viruses. The microbiological standards for drinking water in the Netherlands include a standard (0/100 ml) for sulphite-reducing clostridia spores (SSRC) since 1984. The rationale was that compliance with this standard would require sufficient treatment to eliminate chlorine-resistant pathogens as well. This multiple indicator concept, in combination with other water quality issues such as disinfection by-products, has resulted in the application of ozonation and physical treatment processes rather than chlorination as principal microbial barriers in treatment.

The development of the Quantitative Microbiological Risk Assessment (QMRA) for defining the microbiological safety of drinking water (Haas, 1983; Regli *et al.*, 1991; Teunis *et al.*, 1997 and Medema *et al.*, 2003) has stimulated quantitative research into the effect of treatment processes on micro-organisms. Because the concentration of pathogens in water normally is low, much research is conducted with challenge tests on pilot plant scale with lab-cultured micro-organisms, rather than removal of indigenous micro-organisms by full-scale treatment systems. However, such studies may overestimate the removal capacities of naturally occurring micro-organisms. A relatively simple large volume sampling (LVS) technique was developed for assessing the concentration of faecal indicator bacteria after water treatment (Hijnen *et al.*, 2000a). In the present study, this method was applied to determine the elimination capacity of full-scale water treatment processes for thermotolerant coliforms (Coli44) and SSRC as process indicators for susceptible and resistant pathogenic micro-organisms. By monitoring the concentration of these indicator bacteria throughout the treatment, information was collected about the elimination capacity of overall treatment, the contribution of unit processes and the variability of elimination. The use of this information as input for process optimisation and Quantitative Microbiological Risk Assessment will be discussed.

QUANTIFYING ELIMINATION OF INDICATOR BACTERIA UNDER FULL-SCALE CONDITIONS

The treatment plants and process information. Eight full-scale water treatment plants were involved using several surface waters as source water and a number of different unit processes (Table 4.1). The following unit processes were included: coagulation/floc-removal (CFR), rapid granular filtration (RGF), ozonation (O_3), chlorination (Cl_2), granular activated carbon filtration (GAC), slow sand filtration (SSF), lime softening (LS) and post-disinfection (PD) with chlorine or chlorine dioxide. Process conditions (design and operational) of the unit processes were collected as well as some physical water quality parameters (temperature and turbidity) in order to relate these conditions with the observed removal of micro-organisms.

Large volume sampling. The data of the routine monitoring program showed that the concentration of faecal indicators in the water was generally below the limit of detection after the first one or two unit processes (Chapter 2). Therefore, an additional study with large volume sampling (LVS) was conducted. In a two-week period in winter and summer thermotolerant coliform (Coli44) and spores of sulphite-reducing clostridia (SSRC) were determined daily before and after unit processes, except for the weekends. Volumes of 1 up to 10 litre were examined with the routine membrane filtration method (Ø 47 mm; mf method). Faecal indicators in 50 up to 1,100 litres samples of finished water before post-disinfection (if applied) were determined with an in-situ filtration device, the MF-sampler (Ø 142 mm; Hijnen *et al.*, 2000a).

Table 4.1 The source water and the successive processes of the eight full-scale treatment plants

Loc.	Source water	Processes
1	River[a]	CFR – O_3 – RGF – GAC – PD
2	River[a]	CFR – RGF – O_3 – GAC – PD
3	River[a]	CFR – O_3 – RGF – GAC – PD
4	Lake	CL_2 – CFR – RGF – GAC – PD
5	Regional stream	CFR – RGF – GAC – RGF – SSF – (PD)[e]
6	River[b]	RGF – O_3 – LS – GAC – SSF – (PD)[e]
7	Local surface water[c]	O_3 – LS – GAC – SSF – (PD)[e]
8	River[d]	RGF – SSF – PD

[a]After impoundment reservoirs; [b]After CFR, RGF, soil passage, open recollection reservoir; [c]After CFR, reservoir and RGF; [d]After RGF, soil passage and LS; [e]Stand-by post-disinfection

Microbiological analysis. Thermotolerant coliforms (Coli44) were determined using Lauryl Sulphate Agar (LSA; Oxoid MM615). Concentrations of Coli44 were derived from the number (N) of yellow colonies on LSA at 37 ± 1°C for 14 ± 2 h. Typical colonies ($\sqrt{N}$) were tested in Brilliant Green Bile broth (Oxoid CM31) during 24 h at 44 ± 0.5°C and the percentage of positive results was used to calculate the Coli44 concentration. Spores of sulphite-reducing clostridia (SSRC) were determined on Perfringens-agar-base (PAB, Oxoid CM387) or on Shahidi Ferguson Perfringens agar base (Difco 0811-17-0). Sample volumes of 100 ml were pasteurised in a water bath at 70 ± 1°C for 30 minutes prior to filtration. With sample volumes of 1 litre or more the membrane filters were pasteurised in the liquefied medium placed in an oven at 70 ± 1°C for 30 minutes (cultivation procedure previously described; Hijnen *et al.*, 1997).

Calculation of the elimination capacity. The Decimal Elimination Capacity (DEC) or log-removal of the treatments or unit processes was calculated as described in Chapter 2. Actual decimal elimination (DE) was calculated from the C_{in} and C_{out} concentrations paired by date

$$DE = Log_{10}C_{in} - Log_{10}C_{out} = Log_{10}\frac{C_{in}}{C_{out}}$$

Additionally, DEC of the treatments was calculated from the arithmetic mean concentrations in the source and the finished waters with the same equation (ratio estimation method; Chapter 2). Statistical analysis of the collected data was performed with Microsoft Excel.

CONCENTRATIONS OF INDICATOR BACTERIA IN THE WATER

The concentrations of Coli44 and SSRC in the source and finished water (before post-disinfection when applied) of the eight facilities during the monitoring periods are presented in Table 4.2. The source water concentrations of both indicators varied between <1–>1000 CFU/l and were not related to the season. There was a significant positive correlation between the average concentrations of Coli44 and SSRC in the source waters ($P < 0.05$); the correlation, however, was not proportional (slope of 0.48; SSRC (y) versus (x) Coli44) and the regression coefficient was low ($r^2 = 0.43$).

Table 4.2 The average concentrations of Coli44 and SSRC (Colony Forming Units; CFU) in the source and finished water at the eight facilities in winter (N = 10) and summer (N = 10)

Loc.	Source water (n/l)		Finished water in winter			Finished water in summer		
	Winter	Summer	Vol. (l)[a]	%PS[a]	Avg[a] (CFU/l)	Vol. (l)	%PS	Avg (CFU/l)
				COLI44				
1	1.0	0.6	1285	0	<0.0008	1833	0	<0.0005
2	7	–[b]	1000	0	<0.0010	–	–	–
3	36	–	970	10	0.0010	–	–	–
4	50	439	912	0	<0.0010	497	50	0.0160
5	970	438	3907	40	0.0010	5170	10	0.0017
6	37	140	1761	0	<0.0006	–	–	–
7	2.2	59	1725	0	<0.0006	2780	0	<0.0004
8	143	62	1000	100	0.186 (0.035)[b]	1000	100	0.145 (0.063)
				SSRC				
1	10	77	3305	70	0.0030 (0.06)[b]	5113	80	0.041 (0.058)
2	33	–	900	67	0.033 (0.054)	–	–	–
3	67	–	515	100	1.045 (0.076)	–	–	–
4	1079	1331	831	100	0.425 (0.543)	920	78	0.062 (0.072)
5	1100	540	2130	80	0.062 (0.073)	3230	90	0.028 (0.019)
6	88	555	1000	100	0.181 (0.103)	984	100	0.216 (0.074)
7	26.0	15	1050	100	0.202 (0.111)	1815	100	0.398 (0.208)
8	8	7	996	90	0.202 (0.358)	1000	100	0.706 (0.500)

[a]*Vol. = total sampled volume; %PS = percentage of positive samples; Avg = weighted average calculated from the total of detected colonies in the total sampled volume;* [b]*standard deviation; – = not monitored*

The concentrations of faecal indicators after treatment determined with Large Volume Sampling (LVS) showed a distinct difference between the concentration of Coli44 and SSRC (Table 4.2). Coli44 were detected in the finished water at four locations. At two of these locations (5 and 8; Table 4.1) no main disinfection is applied and there was some indication that breakthrough of Coli44 was related to the concentration in the source water (Figure 4.1). Except for location 8, the percentage of positive sample (%PS) for Coli44 in most finished waters was ≤50% whereas %PS for SSRC in the finished waters was significantly higher (≥67%). These indicator bacteria were observed in the finished water at all locations. The weighted average concentration of Coli44 varied between <0.0004 and 0.186(±0.035) CFU/l (Table 4.2). SSRC concentrations were considerably higher and varied between 0.003(±0.003) and 1.045(±0.076) CFU/l. These results clearly demonstrate that breakthrough of persistent micro-organisms in water treatment occurs more easily compared to breakthrough of the more susceptible micro-organisms.

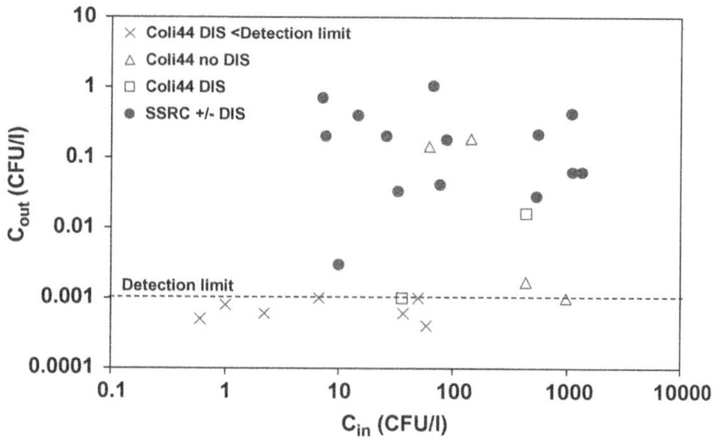

Figure 4.1 Relationship between concentrations of Coli44 and SSRC in the finished water and the source water (DIS = main disinfection in treatment)

ELIMINATION CAPACITY OF THE OVERALL TREATMENT

Coli44 were observed after the treatments of locations with no disinfection or higher source water concentrations. The data showed that the concentration in the finished water C_{out} was not correlated with the source water concentrations C_{in} for both indicators (Figure 4.1). This demonstrates that the concentration in the finished water depends on the DEC of the overall treatment. The DEC values for summer and winter periods were calculated with the ratio-estimation method using the average concentrations in source water and finished water (Drost *et al.*, 1997 and Evers and Groennou, 1999) or using the individual DE values ($n > 5$). DEC of the treatments for Coli44 ranged from 3.0–6.6 log (Table 4.3). Due to a low %PS in the finished water for most facilities DEC values for Coli44 were minimal values and based on the detection limit of the analysis. This detection limit depended on the examined water volume during the selected period (Table 4.2). For SSRC DEC values of the treatments varied between 0.9 and 4.7 log. For elimination of SSRC no influence of the season (winter or summer) was observed but DEC was significantly correlated with source water SSRC concentrations C_{in} (Figure 4.2a; $p < 0.001$). Assuming that higher source water concentrations of faecal indicators are correlated with the general quality of the source water, this correlation suggests that DEC of a treatment is higher at locations with poor source water quality.

In general treatment is in design and operational conditions (contact times and chemical dosages) tailored to meet the drinking water standards. A more general water quality parameter for the source water quality is turbidity and for most of the locations turbidity data were collected during the course of the study. These data showed a positive correlation of source water turbidity and the source water concentrations of both indicators; correlation coefficient of 0.99 ($p < 0.001$) and 0.67 ($p = 0.066$) for Coli44 and SSRC, respectively. Removal of turbidity to levels of around 0.1 NTU is one of the general quality goals used in treatment design. Hence, the higher the source water turbidity, the higher the physical treatment effort to meet this goal. The number of physical steps in the individual treatments is added up for the different locations (lime softening not included) and presented with the removal of turbidity and both indicators in Table 4.3. The use of a chemical disinfection on the locations is also indicated. DEC values of the overall treatment for Coli44 and SSRC was low at locations with two physical processes (locations 7 and 8) and high at location 5 with five physical processes (Table 4.3). The removal efficiency of turbidity ranged from 0.5–2.1 log and was positively correlated with the DEC values for SSRC ($r^2 = 0.30$; $p < 0.05$; Figure 4.2b).

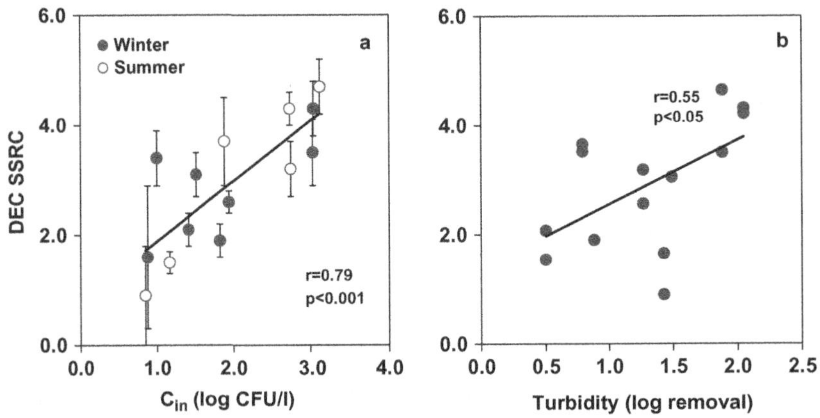

Figure 4.2 Relationship between the DEC (error bars = SD) of treatments for SSRC and the concentration of SSRC in the source waters (C_{in}; a) and (b) the average turbidity removal (DEC) in these treatments

Table 4.3 The average source water quality, turbidity removal (log) and DEC(±SD) in winter and summer for Coli44 and SSRC, arranged in increasing order of successively, number of physical processes, turbidity and Coli44 in the source waters at the different locations

Location (process)[a]	Source water (avg.)			Turb. (log)	DEC Coli44		DEC SSRC	
	Turb. (NTU)	Coli44 (CFU/l)	SSRC (CFU/l)		Winter	Summer	Winter	Summer
8 (2)	1.7	102	7	1.4	3.0 (0.2)	3.2 (0.4)	1.6 (1.3)	0.9 (0.9)
7 (2 + D)	0.2	30	20	0.5	>3.6	>5.2	2.1 (0.3)	1.5 (0.2)
3 (3 + D)	0.6	36	67	0.9	4.6	–	1.9 (0.3)	–
2 (3 + D)	1.2	7	33	1.5	>3.8	–	3.1 (0.4)	–
1 (3 + D)	1.5	0.8	44	0.8	>3.1	>3.1	3.4 (0.5)	3.7 (0.8)
6 (3 + D)	1.9	88	321	1.3	>4.8	–	2.6 (0.2)	3.2 (0.5)
4 (3 + D)	3.9	244	1205	1.9	>4.7	4.4	3.5 (0.6)	4.7 (0.5)
5 (5)	11	704	820	2.1	6.6	5.3	4.3 (0.5)	4.2 (0.3)

[a]*number of physical processes plus disinfection (D)*

For Coli44 this relationship was not significant ($p > 0.1$). These results show that the level of micro-organism removal assessed with Coli44 and SSRC as process indicators for susceptible and persistent pathogenic micro-organisms is to some degree related to the removal of suspended solids. This demonstrates the importance of physical processes (straining, attachment/detachment) in the removal of micro-organisms in treatment. Moreover, this explains the correlation of the source water concentration and the DEC of the overall treatment as observed for SSRC (Figure 4.2a).

ACCEPTABLE QUALITY LEVEL

Required treatment efficacy to meet the drinking water standards. The Dutch drinking water standards require absence of Coli44 and SSRC spores in 300 ml (daily samples) and 100 ml (weekly samples) of treated water, respectively (Anonymous, 1984). It is obvious that a high level of compliance with these standards (i.e. >99% of all samples taken in one year) requires a much lower average concentration in the drinking water than the detection limit of the method. Assuming a random (Poisson) distribution of bacteria in the water, the average concentration of Coli44 and SSRC in drinking water should be as low as 0.018 and 0.035 per litre, respectively, to obtain an acceptable quality level (AQL) of 1% positive samples per year (95% CI; Van der Kooij *et al.*, 1995). To verify these theoretical calculations, Coli44 and SSRC data of the LVS study at the locations were compared with microbiological data from three years of legislative water quality monitoring in standard volumes prior to the period of the LVS study. The average Coli44 concentration in the finished water of locations 5 and 7 assessed with LVS was 0.0013, and <0.0004 CFU/l, respectively.

This is below the required concentration of 0.018 CFU/l for an AQL1%. The percentage of positive samples non complying with the standard during the three years of monitoring at both locations was 0.6% at location 5 and 0.09% at location 7, thus showing that the elimination of Coli44 of these treatments is indeed enough to obtain an AQL1% for these indicator bacteria. In the finished water of locations 2, 3, 4 and 8 the percentage of positive samples for SSRC from the 3 year legislative monitoring program was ≥1% (1.9, 5.6, 1.0) and <1% (0.6), respectively); at the other locations the monitoring program was not according to legislation (higher volumes than 100 ml were analysed). The SSRC concentrations assessed with the MF-sampler in a short period in winter at locations 2, 3, 4 and 8 were 0.033, 1.045, 0.457 and 0.202 CFU/l, respectively, (Table 4.2) and at locations 4 and 8 in summer 0.062 and 0.706 CFU/l, respectively. These concentrations are in the same order of magnitude or higher than the concentration of 0.035/l for the AQL1%, thus supporting the theoretical calculations presented by Van der Kooij *et al.* (1995). Further support for these theoretical calculations came from a comparison of the results of the routine monitoring program at location 8 with the MF sampling in the same periods. The percentage positive samples for Coli44 and SSRC in the finished water during the sampling period of this study was 2.5 and 3.4%, respectively. The concentrations of both indicators Coli44 and SSRC assessed with the mf method in the routine program were 0.083 ± 0.53 ($n = 80$) and 0.34 ± 0.61 ($n = 58$), respectively and with the MF sampler 0.186 ± 0.035 and 0.201 ± 0.358, respectively. In conclusion both monitoring methods showed similar concentrations larger than required to comply with an AQL of 1%.

DECIMAL ELIMINATION CAPACITY OF UNIT PROCESSES

Elimination of Coli44 and SSRC by unit processes. The LVS-study yielded much information about the elimination efficiency of individual water treatment processes. To illustrate this, two locations are presented in detail, one with (5) and one without main disinfection (7). Indicator concentrations and performance of the unit processes are presented and compared with DEC of the overall treatment (Figures 4.3 and 4.4). At location 5 no disinfection is applied and the local river water is treated with coagulation and floc removal (CFR), double rapid granular filtration (RGF; sand), granular activated carbon filtration (GAC) and slow sand filtration (SSF).

After treatment, a post-disinfection is stand-by. The elimination of Coli44 and SSRC by the unit processes and the average concentrations after these processes are presented in Figure 4.3. The results show that without chemical disinfection the relatively high Coli44 and SSRC concentrations in the source water were eliminated with 5.4 up to 6.6 and 4.1 up to 4.3 log, respectively. Slow sand filtration (SSF) was responsible for the high elimination of Coli44 in the treatment. These indicator bacteria were removed with 2–4 log by this process.

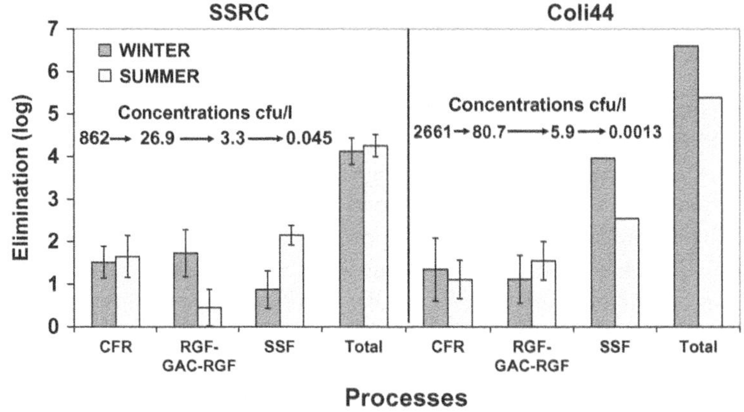

Figure 4.3 Elimination (log) of SSRC and Coli44 by the unit processes and the overall treatment at location 5 determined in winter and summer; RGF-GAC-RGF effect of combined processes; average concentrations in source water and after the unit processes above the columns; error bars is SD

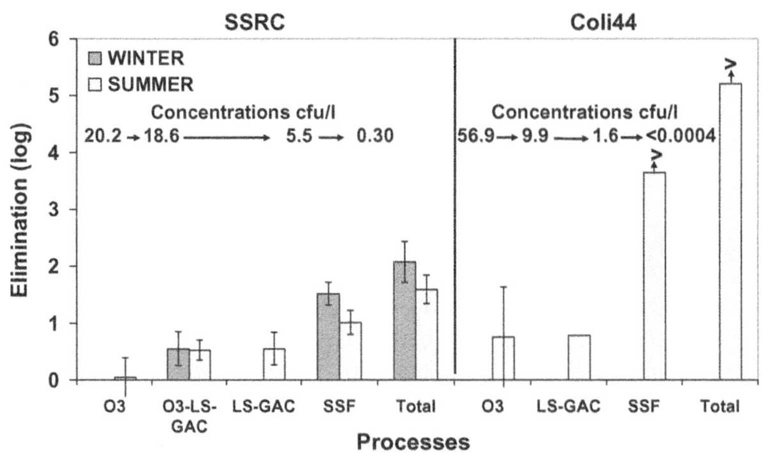

Figure 4.4 Elimination (log) of SSRC and Coli44 in the unit processes and the overall treatment at location 7 determined in winter and summer; O3-LS-GAC effect of combined processes; average concentration in source water and after the unit processes above the columns; error bars is SD

The elimination of both indicators in treatment of location 7 during winter and summer is presented in Figure 4.4. Local surface water was pre-treated with CFR, impoundment reservoir storage and RGF before the post-treatment that is investigated in this study (Table 4.1). First process of the post-treatment was disinfection with ozone (O3) followed by lime softening (LS), GAC and SSF. SSF was the most important barrier for both Coli44 and SSRC in this treatment (Figure 4.4). A relatively low inactivation of both SSRC and Coli44 was observed in the disinfection process with ozone.

Based on the collected data an overview of the assessed DEC values for SSRC and Coli44 of all applied unit processes is presented in Figure 4.5. The collected data revealed more DEC values for SSRC than for Coli44. DEC of coagulation/floc-removal (CFR) for SSRC varied 0.8 log (0.8–1.6) with no systematic difference between winter and summer. The observed relative standard deviation of DEC ranged from 20–40%. The few data collected for Coli44 (0.5–1.3 log) indicated lower removal of these indicators by CFR than SSRC.

The DEC values of rapid granular filtration (RGF) for SSRC showed a similar variation as observed for CFR (0.9 log; 0.8–1.7). These filters were more effective when they were directly applied to the source waters (locations 6 and 8; Table 4.1) than when applied after CFR as a secondary floc removal. This difference is most likely caused by differences in the operational conditions: filters at location 6 and 8 were operated at 3–5 m/h with contact times of approximately 20 minutes; the other filters were operated at 9–12 m/h (contact times of 7–15 min.). Only for the RGF at location 8 DEC for Coli44 could be calculated and this value was 1 log lower than DEC of the same filters for SSRC.

For most of the applied disinfection processes DEC for Coli44 could not be calculated due to 100% inactivation during the period of monitoring. For SSRC DEC of disinfection processes varied between 0 and 3 log with the highest value for chlorine at location 4 most likely due to the high Ct of this process (62–84 mg/l.min) compared to the relative low Ct values of the ozonation processes (1–2 mg/l.min). In summer this chlorine process was more efficient in inactivation of SSRC than in winter.

GAC filtration is applied as a polishing step for the removal of micro-pollutants but the results showed that the filtration process also removed both indicators. For Coli44 little information was collected but for SSRC DEC varied from −0.6 to 1.1 log. For slow sand filters with no backwash and long operational periods of more than one year before filter bed scraping highly variable DEC values for SSRC were observed. Slow sand filters at locations 5 and 7 removed SSRC with 1.0–2.3 log, but at locations 6 and 8 a negative DEC was observed.

For locations 6 and 7 Coli44 was not detected in the filtrate and DEC of SSF for these susceptible bacteria at locations 5 and 8 ranged from 2.3–3.8 log for, higher than DEC for SSRC. SSF on location 5 was more effective in elimination of both indicators, most likely because of the low filtration rate (0.06–0.18 m/h) and consequently high contact time of 5–12 hrs (rates of the other locations 0.25–0.4 m/h and contact times 3–4 hrs).

Variation in elimination capacity. Waterborne outbreaks described in literature have been attributed to breakthrough of pathogens due to peak concentrations in source water coinciding with sub-optimal or inadequate treatment performance (Badenoch, 1990; Richardson *et al.*, 1991 and Craun, 1990). In Table 4.3 and Figures 4.2, 4.3 and 4.4 variation of indicator elimination in the overall treatment and the unit processes is presented by the standard deviation. The variation in elimination is considerable. The relative standard deviation of DE of the unit processes at all locations for coli44 and SSRC was 25% (n = 8) and 47% (n = 17), respectively. The variation in the elimination calculated for the overall treatment was smaller than expected from the sum of the variations of the unit processes; relative SD of 10% (n = 2) and 13% (n = 10) for coli44 and SSRC. Consequently, the low elimination in one process appears to be compensated by a higher elimination in the subsequent process(es). This phenomenon has been demonstrated before for the elimination of SSRC by CFR and RGF (Hijnen *et al.*, 1997) and for CFR/RGF and O3 (Medema, 1999). It illustrates the importance of the multiple barrier strategy applied in water treatment. Further analysis of the variation in DE values for SSRC and Coli44 revealed that the actual elimination of a number of unit processes and also of the overall treatment was positively correlated with the log transformed concentration in the influent of the process (Table 4.4), in particular for SSRC. The meaning of this observation was unclear.

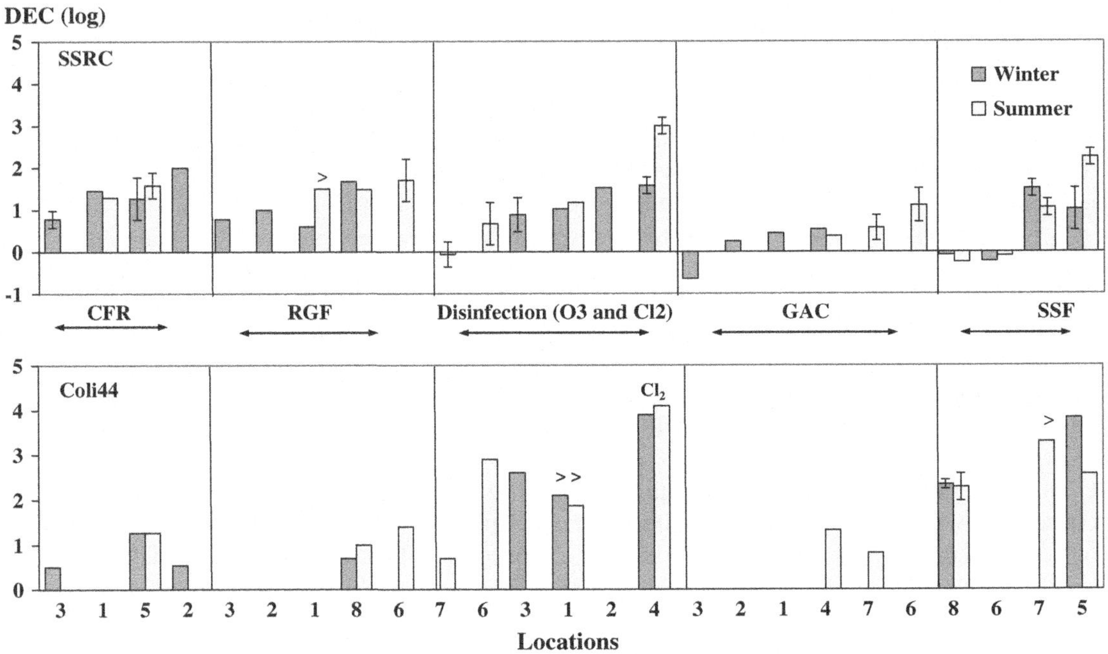

Figure 4.5 DEC values (error bars SD) of the unit-processes for SSRC (above) and Coli44 (below) arranged in increasing order for SSRC in winter and summer; locations on x-axis presented for both SSRC and Coli44; no bar means no calculation possible due to lack of data (error bars = SD)

Table 4.4 Correlation of DE value calculated from paired data of C_{in} and C_{out} with the log transferred influent concentration for a number of unit processes and total treatments

Processes	SSRC		Coli44	
	n[a]	Slope (r^2)	n[a]	Slope (r^2)
CFR	0/2	0	1/1	0.51(0.42)
RGF	1/3	1.68(0.85)	1/2	0.67(0.55)
Disinfection	2/4	1.00(0.75); 0.88(0.69)	0	–
GAC	1/3	0.62(0.72)	0	–
SSF	2/3	1.39(0.82); 1.28(0.96)	1/1	0.63(0.59)
Total Treatm.	2/5	0.87(0.88); 1.05(0.51)	1/1	1.12(0.75)

[a]*number significant linear relations ($p < 0.01$)/total number of unit processes*

The study showed that at a low-level, breakthrough of faecal micro-organisms can occur. In winter (7.5°C) Coli44 (1 CFU/100 litre) was detected in the finished water at location 3 (arrow in Figure 4.6). During this event low Coli44 and SSRC removal by CFR coincided with low Coli44 inactivation by ozonation. Hence, sub-optimal performance in one unit process can sometimes compromise the overall treatment performance. Moreover, this breakthrough-event would not have been detected by the routine (300 ml) Coli44-monitoring of finished water.

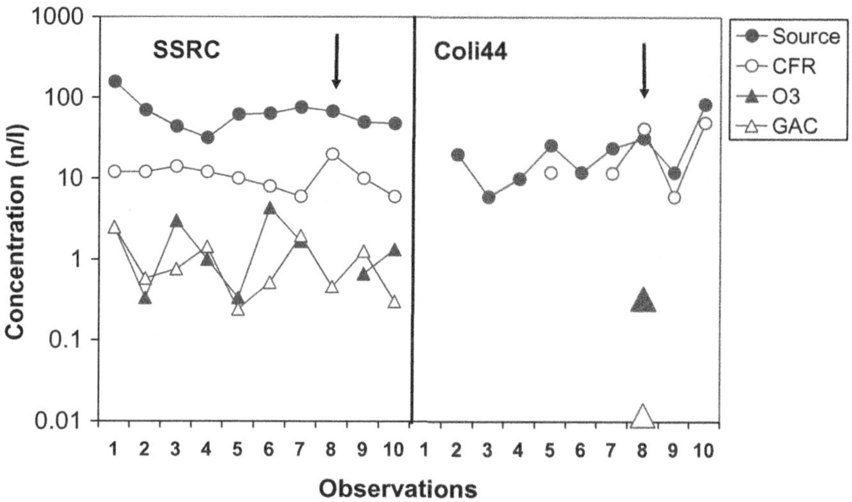

Figure 4.6 The concentration (CFU/l) of SSRC and Coli44 in the source water and in the water after the successive unit processes at location 3 monitored in the same period (arrow indicates the break-through event)

PATHOGEN REMOVAL AND QUANTITATIVE RISK ASSESSMENT

A major revision in the revised Dutch Drinking Water Decree (Anonymous, 2001 was the new risk-based regulation for pathogens in drinking water for surface water treatment. A quantitative risk analysis is required to demonstrate compliance with an annual risk of infection of 10^{-4}. This risk-level has been provisionally translated into drinking water standards for pathogens, using dose-response data for pathogens and introducing a safety factor of 10. The difference between the concentration of pathogens in source water and these provisional drinking water standards is the required treatment efficacy for pathogens.

Cryptosporidium oocyst concentrations (corrected for the recovery efficiency of the detection method) in source water of the utilities in this study were used to calculate the required elimination for this pathogen (Table 4.5). SSRC or *C. perfringens* have been proposed as process indicator for the removal of resistant pathogens such as *Cryptosporidium* and *Giardia* (Payment and Franco, 1993 and Hijnen *et al.*, 1997). The data on required removal for *Cryptosporidium* were compared to the data on SSRC removal achieved by the treatment systems (Table 4.5). The DEC of location 1 assessed for SSRC in this study was higher than the required DEC for *Cryptosporidium.* At the other locations, the obtained SSRC elimination was equal (location 2) or lower than the required DEC for *Cryptosporidium.* The difference was especially large at location 7 and 8.

DISCUSSION

Assessment of elimination capacity. The results of the presented study showed that the elimination capacity of water treatment and the unit processes for susceptible (Coli44) and persistent (SSRC) indicator bacteria can be assessed by using standard microbial methods with decreased detection limit. The decreased detection limit was realized by examining larger volumes of the water during treatment with the standard mf method and in the finished water by using the in situ sampling device, the MF sampler (Hijnen *et al.*, 2000a). The data of these limited monitoring periods showed a higher prevalence of the persistent indicator (SSRC) in the last stages of treatment than the more susceptible thermotolerant coliforms (Coli44), while in general concentrations of both organisms in the source waters were in the same order of magnitude.

Table 4.5 The required DEC of the water treatment to meet an annual risk of infection of 10^{-4} for Cryptosporidium (concentration 2.6×10^{-5}/l, Van der Kooij *et al.*, 1995) and the DEC achieved by the treatment systems as determined with SSRC

Location	DEC required (log) Cryptosporidium	DEC achieved (log) SSRC
1	2.7	3.5
2	3.1	3.1
3	3.1	1.9
4	4.8	4.0
5	4.7	4.3
6	4.7	2.9
7	3.7	1.8
8	4.2	1.2

Variation in elimination. The data also showed considerable variation in the elimination capacity, both within and between treatments. In the monitoring periods of two weeks in winter and two weeks in summer DEC of the eight treatments for SSRC ranged from 0.9–4.7 log. For Coli44 the DEC of four out of the eight treatments could be quantified and ranged from 3.0–6.6 log. For SSRC both in the source water concentration and DEC of the treatments were positively correlated with turbidity in the source waters and the turbidity reduction in treatment. This demonstrates the relationship between elimination of microbes and suspended solids. Furthermore it explains the positive correlation between source water concentrations of SSRC and DEC for these anaerobic spores. Positive correlations of the source water turbidity and turbidity reduction in treatment with levels of *Cryptosporidium* and *Giardia* in the source water and elimination of these pathogens in treatment of four drinking water facilities was observed previously (LeChevallier and Norton, 1992). At 1 log turbidity removal in conventional treatment (coagulation/sedimentation and filtration) parasite removal was approximately 0.9 log. In a larger study including 66 sites, however, such a correlation was not observed (LeChevallier *et al.*, 1991). Thus, they argued that this relationship probably does not hold for every treatment train and regulating drinking water safety for parasites with requirements for turbidity removal is not justified. Additional argument against such a regulation on turbidity removal for microbiological safety in general is that it under-estimates the overall elimination of microbes. Other processes like inactivation and predation (in slow sand filters) also contribute to micro-organisms removal.

The current study also presents variation in DEC between the different unit processes which to some degree could be related to specific process conditions. Further research is required to demonstrate the causal relationship between the hypothesized process conditions and the elimination of these process indicators. As presented before, the inactivation capacity of the full-scale ozonation and chlorination processes for susceptible micro-organisms such as Coli44 is much lower than expected on the basis of laboratory data. For the relative low DEC of ozonation at location 7 a further study indicated that ozone-dosage controlled by flow instead of the residual ozone concentration and a relatively high DOC-content (high ozone demand) in the feed water were important causes for the low performance (Hijnen *et al.*, 2001). Remarkable were the results observed for SSRC removal by GAC and slow sand filtration. DEC was highly variable and sometimes negative (higher concentration in filtrate than in influent). Multiplication of SSRC in these aerobically operated filter beds as an explanation for this observation is not plausible but can not be excluded. In previous studies (Hijnen *et al.*, 1997, 2000) negative DEC-values for GAC and slow sand filtration have been reported. On the basis of observations in the filter bed and the backwash water this phenomenon was attributed to accumulation and breakthrough in these filters with a low backwash frequency and intensity.

Finally, the collected data clearly demonstrate the benefits of the multiple barrier concept since the observed variation in elimination capacity of the overall treatment is lower than expected from the variation of the unit processes.

Increased elimination at higher concentrations. The positive correlation of the average elimination of SSRC in the eight treatments (DEC) with the source water SSRC concentration (Figure 4.1) was explained by more intensive treatment conditions due to a poor source water quality with higher turbidity and SSRC concentration. More intensive treatment conditions are partly related to process design and partly related to operation conditions. A similar correlation between the actual elimination (DE) and C_{in} was observed, however, at individual processes operated under nominal and more or less steady process conditions. Higher efficiencies of physical removal processes such as coagulation/flocculation and filtration at increased particulate concentrations have been described before (Smoluchowski, 1917; O'Melia and Ali, 1978 and Vigneswaran and Chang, 1989). Thus, the observed correlation can be due to a physical phenomenon. The observed correlation might also be related to dispersion and retardation in the process; actual DE is calculated with the actual C_{in} and C_{out} assuming that these concentrations can be paired. These observations question the validity of describing variability in elimination with DE as the parameter suggested in the present study. This statistical aspect of the collected data needs further attention.

Coli44 and SSRC as process indicators in QMRA. In a first attempt to use these data in quantitative microbial risk assessment site specific removal data of SSRC was compared with site specific information on required elimination of *Cryptosporidium* oocysts. A lack of elimination capacity was observed at a number of locations but this analysis must be considered as a first impression. The results of the current study show the potential use of both faecal indicators as process indicators to collect quantitative data on the elimination of susceptible and persistent micro-organisms. More information from comparative studies is needed for definitive conclusions on the translation of this quantitative information to the elimination of waterborne pathogens in water treatment. There are reasons to consider SSRC as a conservative process indicator for (oo)cysts removal. The anaerobic spores are smaller than *Cryptosporidium* oocysts. They are also more persistent than *Cryptosporidium*. The die-off rate of spores *C. perfringens* in river water is low (0.005 log/day; Medema *et al.*, 1997). This is probably an important cause of the low SSRC removal in filtering processes with a low filter bed cleaning frequency like SSF and GAC. Preliminary results of a slow sand filter column study indeed showed that *Cryptosporidium* oocysts are removed to a higher decree than spores of *C. perfringens* (Hijnen *et al.*, 2004a). Thus, the LVS study showed that SSRC is a useful and potential process indicator for describing the removal of *Cryptosporidium* oocysts by coagulation/filtration and disinfection processes, but most likely not for GAC and slow sand-filtration. The observed phenomenon of accumulation and delayed breakthrough in these filter beds, however, might be also a factor of concern for elimination of these persistent pathogens. In addition the use of Coli44 as a process indicator for the susceptible waterborne pathogens needs further attention.

Compliance with the standards. Finally, the findings of this study demonstrate that by additional large volume sampling for faecal indicators in the produced drinking water the level of compliance with the regulated standards for these indicators achieved with the applied treatment can be assessed. At a number of locations average indicator bacteria concentrations (Coli44 and SSRC) assessed with either small samples or large volumes were related to the number of observed positive samples (= non compliance with the standard of 0 in 100 or 300 ml samples) in the same monitoring period. These findings justified the proposed maximum average concentrations for a more than 99% compliance with the standards of the annual mandatory microbiological drinking water control program for Coli44 and SSRC at a 95% certainty level (Van der Kooij *et al.*, 1995).

CONCLUSION

With Large Volume Sampling (LVS) the concentrations of faecal indicators after treatment of the eight locations in a winter and a summer period was determined (n = 10 to 20 per location). This study shows that a high level of compliance with the drinking water standards for Coli44 and SSRC requires treatment systems with a large elimination capacity (1.6 up to 4.6 log) for these indicators. The multiple indicator concept and large volume sampling provide quantitative information about the elimination of both vegetative bacteria and resistant bacterial spores by treatment processes. It allows assessment of the contribution of unit processes in the overall treatment efficacy. In general, chemical disinfection and slow sand filtration are important barriers against Coli44. SSRC are less effectively removed by disinfection and slow sand filtration. Their ability to survive for long periods may cause accumulation and eventually breakthrough through the filters with low backwash frequency.

The efficacy of unit processes differed in the different treatment systems. In some cases, this initiated research into the cause of low efficacy of unit processes (such as the ozonation at location 7) and optimisation of the process. The study also provided information about the short-term variability of the efficacy of treatment processes. The variability in the overall treatment efficacy was found smaller than the variation in the unit processes, illustrating the robustness of multiple barrier treatment systems. Large Volume Sampling allowed the detection of low-level breakthrough of micro-organisms. Relating these breakthrough events to raw water quality or process conditions may lead to measures to prevent these risk events.

Ultimately, the data on efficacy of full-scale treatment systems against micro-organisms can be used in quantitative microbiological risk assessment of drinking water. SSRC data appear to be applicable in quantitative assessment of the risk of resistant pathogens, such as *Cryptosporidium*, for describing the efficacy of coagulation/filtration and disinfection processes, but not for slow sand of GAC filtration.

Further research includes comparative studies on the removal of faecal indicators and pathogens by unit processes and the improvement of the enumeration methods of pathogens in the source water (recovery efficiencies, specificity).

Chapter 5

Spores of sulphite-reducing clostridia (SSRC) as surrogate for verification of the inactivation capacity of full-scale ozonation for *Cryptosporidium*

Wim A.M. Hijnen[a], A.J. van der Veer[b], J. van Beveren[a] and G.J. Medema[a]

[a]*KWR Watercycle Research Institute;* [b]*Evides Water Company*

ABSTRACT

The inactivation of *C. parvum* and spores of *C. perfringens* by ozone treatment in natural water was compared in a lab-scale continuous-flow system. In addition the inactivation of the natural occurring spores of sulphite-reducing clostridia (SSRC) in this water was monitored in one of the lab-scale systems as well as in a full-scale ozonation process. The survival ratio of *C. parvum* oocysts was determined using the CD-1 neonatal mouse infectivity test and for *C. perfringens* and SSRC the survival ratio was assessed with the standard anaerobic colony count on the iron-sulphite medium.

The results of the lab-scale experiments revealed an inactivation rate constant k (Chick-Watson modelling) at 10°C for *C. parvum* of 0.14 ($SD = 0.014$; $P < 0.001$) and for *C. perfringens* of 0.25 ($SD = 0.01$; $P < 0.001$). Moreover, first results of monitoring the SSRC inactivation in full-scale ozonation processes indicated that the inactivation rate constant for these wild strains was in the same order of magnitude as determined for *C. perfringens*. Further research is needed to compare inactivation ozone kinetics for *C. perfringens* D10 and SSRC at different temperatures and in other natural waters. Results of additional lab-scale experiments with *C. perfringens* strain D10 indicated that the Ct of the gas-feed chamber should be incorporated in the design of a full-scale ozonation. Moreover, setting the Ct with the contact time was not as effective for the inactivation capacity as setting the Ct with the ozone concentration.

 Quantitative Methods to Assess Capacity of Water Treatment to Eliminate Micro-Organisms. Edited by Wim A.M. Hijnen. ISBN: 9781843393764. Published by IWA Publishing, London, UK.

INTRODUCTION

Outbreaks of waterborne diarrhoea caused by pathogenic protozoa *Cryptosporidium* and *Giardia* in the USA and the UK have increased interest in the effect of water treatment processes on these micro-organisms. These pathogens and especially *Cryptosporidium* can be present in drinking water because they are persistent and resistant to chemical disinfection processes in water treatment. Ozone is the only disinfectant, which may achieve a significant inactivation capacity during treatment against *Cryptosporidium* (Finch *et al.*, 1993). The design of ozone processes for full-scale treatment plants is based upon the results of bench-scale experiments using artificially dosed oocysts under well-defined laboratory conditions. The obtained Ct-inactivation relationship is used to design and operate full-scale ozone systems. Controlling ozone dose and contact time sets the required Ct-values. However, the hydrodynamics of the contact chambers, ozone transfer and decay may not be scaled up readily from lab to full-scale. In this light, a tool to verify the validity of the use of the lab-scale data for full-scale design would be valuable. Direct verification with *Cryptosporidium* is not appropriate. The concentration of oocysts in the incoming water is too low and dosage experiments on this scale are not suitable. There is a need for a biological surrogate to determine the achieved inactivation capacity and the variation of this capacity under full-scale conditions. A natural surrogate is preferable to the dosage of fluorescent-dyed polystyrene microspheres as proposed by Mariñ as *et al.* (1999). Spores of sulphite-reducing clostridia (SSRC) have been proposed as a surrogate for the assessment of the effect of water treatment processes on persistent protozoa (Payment *et al.*, 1993 and Hijnen *et al.*, 1997). One of the arguments was their relatively high resistance against ozone observed under full-scale conditions (Hijnen *et al.*, 2000b) which can easily be determined using a large volume sampling apparatus (Hijnen *et al.*, 2000a). The objective of this study was to verify the use of SSRC as a surrogate for the estimation of the inactivation capacity of a full-scale ozonation for *Cryptosporidium*.

EXPERIMENTAL WORK ON OZONE DISINFECTION

The experimental set up. The inactivation of *Cryptosporidium parvum, Clostridium perfringens* and the natural SSRC by ozonation was determined under lab-scale conditions and compared with the inactivation of the natural SSRC in full-scale ozonation. To eliminate the effect of the water quality on the inactivation kinetics (Haas *et al.*, 1996 and Oppenheimer *et al.*, 2000) all experiments were carried in the influent of the full-scale ozone process operated in a demonstration plant (100 m^3/h) of Water Company Europoort (WBE-water). This is river Meuse water, pre-treated with impoundment reservoirs, microstraining, coagulation and sedimentation and rapid granular filtration. The quality of this WBE-water shows no large fluctuations for most of the relevant parameters (Table 5.1) and the concentration of SSRC in this water range from 1 up 10 CFU per litre. For the lab-scale experiments the water was transported in large stainless steel vessels (30, 200 and 700 litre).

Table 5.1 The range of values of the most relevant water quality parameters of the WBE-water

Parameter	Average (10- and 90-percentile)
Temperature (°C)	13.2 (3.5–21.9)[a]
Turbidity (Ftu)	0.08 (0.012–0.15)
Fe (mg/l)	0.018 (0.05–0.045)
NH_4^+ (mg/l)	0.02 (<0.01–0.04)
PH	7.4 (7.27–7.54)
DOC (mg/l)	2.05 (1.75–2.36)
Uve (m^{-1})	5.3 (4.6–6.3)
Alkalinity (mmol $CaCO_3$)	1.57 (1.53–1.62)

[a]*Minimum and maximum temperature, respectively*

The ozone systems. Laboratory experiments with *C. parvum* and *C. perfringens* were carried in a three-stage continuous-flow bench-scale system of Montgomery Watson (MW) in the US (Figure 5.1 from Oppenheimer *et al.*, 2000). The experiments with *C. perfringens* and natural SSRC were performed in the Kiwa continuous-flow system (Figure 5.2), a lab-scale duplicate of the full-scale WBE ozonation process. All technical information of these three systems is summarised in Table 5.2.

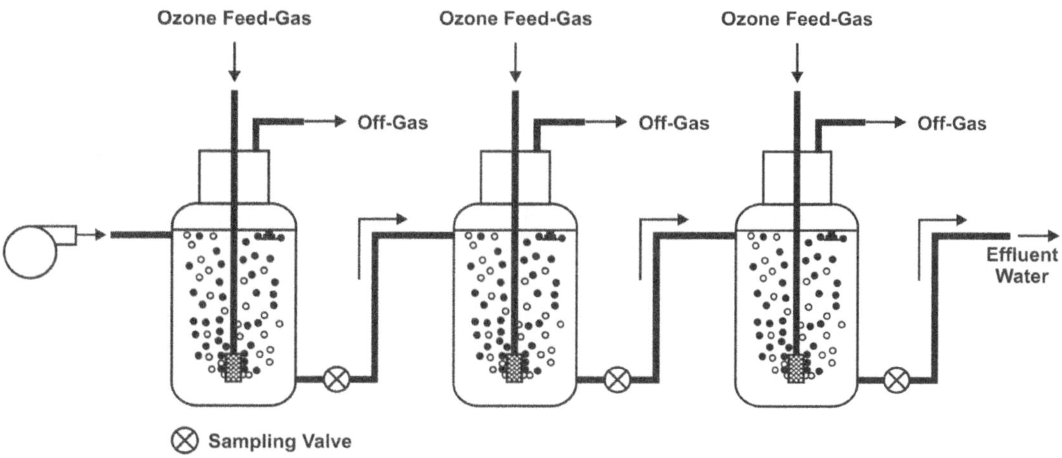

Figure 5.1 Schematic of the MW-system (from Oppenheimer *et al.*, 2000)

Figure 5.2 Kiwa-system used for the ozone experiments with spores of *C. perfringens* and SSRC

Micro-organisms and microbiological analysis. For the experiments with *Cryptosporidium parvum* the Harley Moon-isolate was produced in neonatal Holstein calves and used not more than 2 month after isolation, as described in Oppenheimer *et al.* (2000). For *C. perfringens* an isolate from a patient suffering from diarrhoea caused by food infection was used (strain D10).

Table 5.2 Technical characteristics of the applied ozone systems

Characteristics	MW-system	Kiwa-system	WBE-system
Vol. chambers (l)	0.43[a]	1.4[a]; 2.6[b]	4.100[a]; 4.100[b]
Water flow (l/h)	6.6	42	100.000
Gas flow (l/h)	3.6	12	26.000
T_{10}/HRT (min./min.)	0.35	0.8	0.72

[a]Gas-feed Counter current Chambers (GCC); [b]Reactive Flow Chambers (RFC)

From a frozen stock culture (−70°C) in peptone/glycerol solution D10 was inoculated anaerobically on Perfringens-agar-base plates (PAB Oxoid CM587). After 5 days of incubation at 37°C the produced biomass was suspended with 2 ml of sterile water in a 200 ml glass container with 150 ml sterile drinking water. The concentration of D10 in the two stock solutions used in the experiments was 2×10^6 and 10^8 per ml, respectively. Pasteurized (30 minutes at 70°C) and non-pasteurized samples of this solution yielded similar concentrations, indicating a high percentage of sporulation. After incubation for 20–25 days at 37°C these containers were stored at 3–5°C for 70–250 days. Before the inoculation of the WBE-water this D10 stock solution was filtered through an 8 μm sterilized membrane filter (Schleicher & Schüll AE99) to remove suspended solids and large aggregates of spores.

The water samples before and after ozonation were collected in sterile bottles with sterilized thiosulphate solution as a quenching agent. 10 ml/l of 1% solution and 2 ml/l of a 30 g/l thiosulphate solution were added to the samples of the MW-system and of the Kiwa-system, respectively.

The inactivation of the oocysts of *Cryptosporidium parvum* was determined with the CD-1 neonatal mouse infectivity test (Finch *et al.*, 1993 and Oppenheimer *et al.*, 2000). *C. perfringens* D10 and SSRC concentrations in water were determined on sulphite-iron agar by the method previously described (Hijnen *et al.*, 1997). 1 or 0.1 ml samples with high concentrations of D10 were inoculated directly in the liquefied medium, with or without further dilution in 9 ml of sterile drinking water. The standard membrane filtration method was used for sample volumes from 10 ml up to 10 litres and for SSRC in large volumes up to 56 litre the MF-sampling technique (Hijnen *et al.*, 2000a) was used.

Inactivation kinetics. Disinfection is the inactivation of micro-organisms or the reduction of the concentration viable micro-organisms N due to the exposure to a concentration disinfectant C during a specific contact time t. The inactivation kinetic is most commonly described by the first order disinfection model of Chick-Watson (1908):

$$\frac{dN}{dt} = -kC^nN$$

where k is the inactivation rate constant and n is the dilution constant. Based on this model the linear relationship between the log inactivation of N and the Ct-value ((mg/l)*min) is described by:

$$^{10}\log\left(\frac{N_t}{N}\right) = -k * C^n t$$

N_t is the microbial concentration after contact time t.

Ozone concentrations and CT-calculation. Ozone in water was determined by the Indigo method (Bader and Hoigne, 1982). In the Kiwa-system ozone was measured by the potentiometric method with an Orbisphere Indication instrument (model 26505) and sensor (model 2301) calibrated by the Indigo method. The Ct-value ($Ct_{calc.}$) was calculated from the average ozone concentration C (mg/l) and the contact time t (minutes) in each of the successive chambers (including the gas-feed chamber). The three in-series chambers of the MW-system can be considered as continuously stirred tank reactors (CSTR) for which the average ozone concentration in the chambers is the concentration monitored in the outlet of each chamber ($C_{O3,out}$). $Ct_{calc.}$-values of this system were derived from the monitored inactivation values and the ozone concentrations by a model specially developed by Montgomery-Watson for this system where the hydraulic characteristics were taken into account (Oppenheimer *et al.*, 2000). For the plug-flow systems of Kiwa and WBE the average concentrations in the chambers were calculated from the in- and outcoming concentrations ($C_{O3,in}$ + $C_{O3,out}$)/2. To consider the influence of the contact time distribution on the disinfection kinetics in the full-scale ozonation t10 was used which is defined as the contact time in which 90% of the incoming water volume remains in the chamber (Von Huben, 1991). t10-values of the different systems were verified with tracer tests (I- or Cl-) and presented as ratio of the average hydraulic retention time (HRT) in Table 5.2.

THE RESISTANCE OF OOCYSTS AND SPORES TO OZONE

Inactivation of C. parvum and C. perfringens. The inactivation of *C. parvum* and *C. perfringens* by ozone in the MW-system was determined at 10°C in separately transported batches of the WBE-water. The values of the water quality parameters were in the range presented in Table 5.1. Adjusting the ozone dosage (constant contact time) set the Ct of the ozone process. From the linear fit of the relation between the log inactivation and Ct-value presented in Figure 5.3 the inactivation rate constant k for both micro-organisms was calculated.

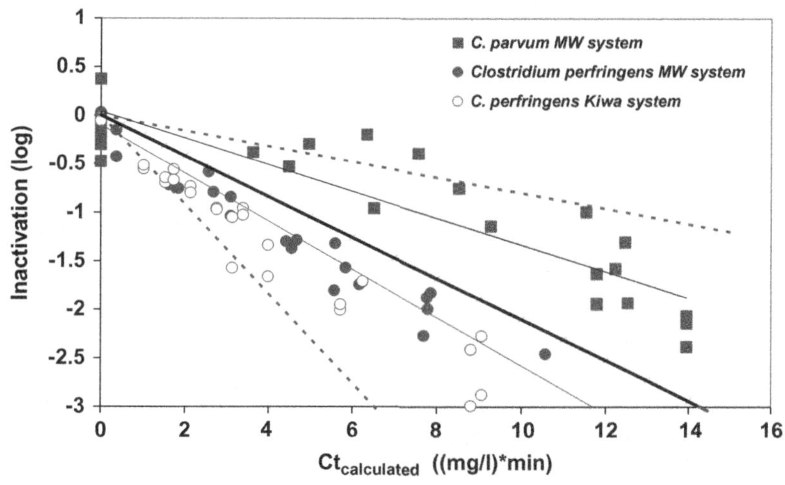

Figure 5.3 Linear relationship between the log inactivation and the $Ct_{calc.}$-value for *C. parvum* and *C. perfringens* separately determined in WBE-water at 10°C in the MW-system and for *C. perfringens* as well as the Kiwa-system, respectively (dashed lines and bold line: the inactivation data for *C. parvum* in natural waters from Oppenheimer *et al.*, 2000 based on the minimum, maximum and median inactivation rate constant at 10°C)

The k-value for *C. parvum* was 0.14 (SD = 0.014; $r^2 = 0.81$; $P < 0.001$). The k-value determined for *C. parvum* in 16 different types of natural water (including the WBE-water) for a temperature of 10°C determined by Oppenheimer *et al.* (2000) ranged from 0.08 up to 0.46 and the median value was 0.21. Based on the water quality data no explanation was found for the relatively low k-value determined for the WBE-water. From the comparison of the disinfection data of both micro-organisms (Figure 5.3) it was concluded that the inactivation kinetics of *C. perfringens* and *C. parvum* were in the same order of magnitude. The k-value for *C. perfringens* was 0.25 (SD = 0.01; $r^2 = 0.96$; $P < 0.001$).

To verify the reproducibility of the inactivation kinetics for *C. perfringens* in WBE-water the experiments carried out in the US were repeated in the Kiwa-system. The results revealed that for the involved range of $Ct_{calc.}$-values the inactivation kinetics for both systems was similar. The k-value for the linear inactivation kinetics determined in the Kiwa-system was 0.27 (SD = 0.016; $r^2 = 0.93$; $P < 0.001$).

EFFECT OF PROCESS CONDITIONS

Inactivation in the gas-feed chamber. In one of the experiments with the Kiwa-system the log inactivation of *C. perfringens* in the Gas-feed Counter current Chamber (GCC) was determined separately from the overall log inactivation in the total system. The results revealed that a proportional and significant part of the overall log inactivation was obtained in the GCC (Figure 5.4). This indicates that also this chamber should be incorporated in the design of an ozone process.

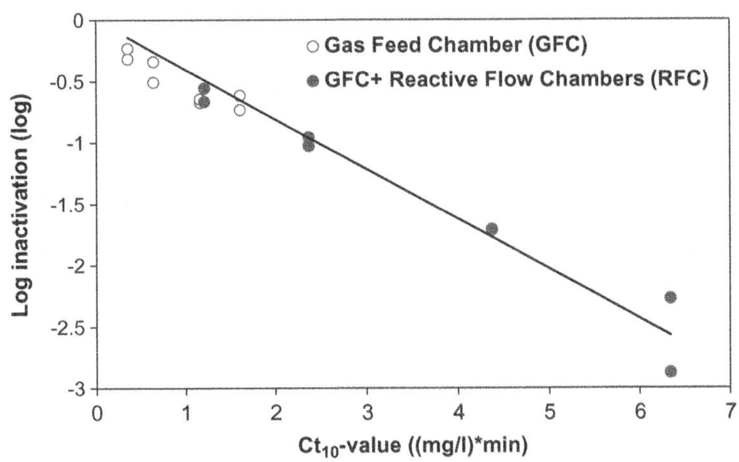

Figure 5.4 The log inactivation of *C. perfringens* strain D10 in the gas-feed chamber (GCC) and in the total Kiwa-system (GCC + RFC) as function of the Ct_{10}

Variation of the contact time. In the Kiwa-system the inactivation of *C. perfringens* strain D10 in WBE-water was measured at 10°C by varying the Ct_{10}-values of the ozone process with the contact time at a constant ozone dosage of 1.2 mg/l. Unlike the former results where Ct was set with the ozone dosage, this experiment shows that the log inactivation did not increase with the increasing Ct_{10}-values (Figure 5.5).

From these results it was concluded that setting Ct-value with the contact time under the described conditions with low ozone concentrations in the reactive flow chambers (measured by the indigo-method and presented in Figure 5.5) did not affect the inactivation capacity of the ozone process. Consequently, setting the Ct-value and inactivation capacity with the ozone dosage seems to be more effective.

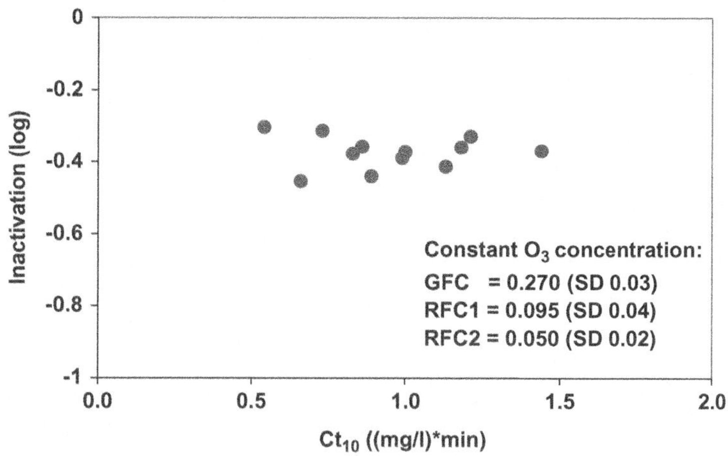

Figure 5.5 The relationship between Ct_{10} and the log inactivation of *C. perfringens* strain D10 under the condition of a constant ozone dosage of 1.2 mg/l and varying contact times of 4 up to 20 minutes in the Kiwa-system

CULTURED AND ENVIRONMENTAL SPORES

Inactivation of C. perfringens and emvironmental SSRC. A previously described study revealed that under full-scale conditions at 5 drinking water production plants in the Netherlands the log inactivation of ozonation for SSRC easily could be assessed by large volume sampling (Hijnen *et al.*, 2000b). The processes were operated at a Ct-range of 1 up to 4 ((mg/l)*min) and the average observed SSRC log inactivation in winter (3–10°C) and summer (10–20°C) was 0.7 (minimum and maximum log inactivation ranged from 0 up to 1.2 log). In the period March and April 1998 at water temperatures of 8–11°C the SSRC inactivation in the full-scale WBE system was monitored intensively by sampling large volumes (35–50 litre; 13 observations) with the MF-sampling technique. The average log inactivation was 0.8 log (SD = 0.2; n = 13) and is plotted against the average Ct_{10}-value (n = 20) in this period of 1.5 (SD = 0.34) (mg/l)*min (Figure 5.6; ●). In the same figure the average SSRC log inactivation mentioned before for a number of full-scale processes (n = 5) was plotted against the applied average Ct-values (o). For both data sets a wide range of Ct_{10}- and log inactivation values was observed presented by an error bar in the Figure (both parameters were not monitored simultaneously). A comparison of these results with the inactivation kinetics of *C. perfringens* in the WBE-water at 10°C show that the susceptibility to ozone of envionmental SSRC under full-scale conditions and *C. perfringens* under lab-scale conditions was in the same order of magnitude (Figure 5.6).

Inactivation of environmental SSRC under full-scale and lab-scale conditions. In the full-scale WBE-ozone process a SSRC log inactivation of 1.2 log (SD = 0.13) was observed at a water temperature of 21°C and a Ct_{10}-value of 1.0 (Figure 5.6; ■). The inactivation of the environmental SSRC in WBE-water was determined simultaneously in the lab-scale Kiwa-system, a small copy of the full-scale WBE-system (□). The results indicate that the inactivation kinetic under full- and lab-scale conditions was similar. The relationship between the log inactivation and the Ct_{10}-value in both systems was described by the same linear fit (Figure 5.6). Moreover, comparison of these results with the results obtained at 10°C indicate that a temperature shift of 10°C will double the log inactivation for SSRC by ozone. Oppenheimer *et al.* (2000) determined a larger effect of the temperature on the log inactivation rate of *Cryptosporidium* by ozone. This value increases by a factor of 4.5 for every 10°C increase in temperature.

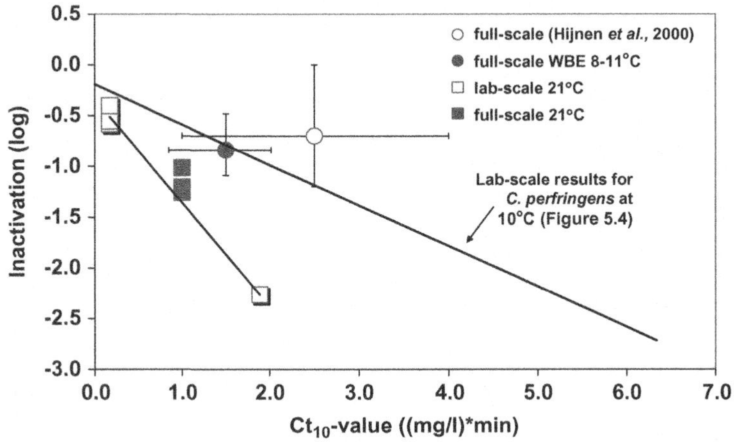

Figure 5.6 The relationship between the log inactivation of environmental SSRC and *C. perfringens* in the WBE-water and the Ct_{10} of the full-scale and lab-scale ozone processes (error bars indicate the range of log inactivation and Ct_{10})

CONCLUSION

This study revealed that SSRC is an appropriate tool to study, control and optimise disinfection processes with ozone under full-scale conditions for the inactivation of *Cryptosporidium*. The inactivation rate constant k for *Cryptosporidium* and SSRC determined at 10°C under lab-scale and full-scale conditions in the WBE-water, respectively were in the same order of magnitude. First results show a significant difference in the effect of temperature on the inactivation kinetics for both micro-organisms. Further research is needed before SSRC can be used as a surrogate for quantitative assessment of the inactivation capacity of full-scale ozonation processes for *Cryptosporidium*.

Chapter 6

Inactivation credit of UV radiation for viruses, bacteria and protozoan (oo)cysts in water: a review

Wim A.M. Hijnen, E.F. Beerendonk and G.J. Medema

KWR Watercycle Research Institute

ABSTRACT

UV disinfection technology is of growing interest in the water industry since it was demonstrated that UV radiation is very effective against (oo)cysts of *Cryptosporidium* and *Giardia*, two pathogenic micro-organisms of major importance for the safety of drinking water. Quantitative Microbial Risk Assessment, the new concept for microbial safety of drinking water and waste water, requires quantitative data of the inactivation or removal of pathogenic micro-organisms by water treatment processes. The objective of this study was to review the literature on UV disinfection and extract quantitative information about the relation between the inactivation of micro-organisms and the applied UV fluence. The quality of the available studies was evaluated and only high-quality studies were incorporated in the analysis of the inactivation kinetics.

The results show that UV is effective against all waterborne pathogens. The inactivation of micro-organisms by UV could be described with first-order kinetics using fluence-inactivation data from laboratory studies in collimated beam tests. No inactivation at low fluences (offset) and/or no further increase of inactivation at higher fluences (tailing) was observed for some micro-organisms. Where observed, these were included in the description of the inactivation kinetics, even though the cause of tailing is still a matter of debate. The parameters that were used to describe inactivation are the inactivation rate constant k (cm^2/mJ), the maximum inactivation demonstrated and (only for bacterial spores and *Acanthamoeba*) the offset value. These parameters were the basis for the calculation of the Microbial Inactivation Credit (MIC = 'log-credits') that can be assigned to a certain UV fluence. The most UV resistant organisms are viruses, specifically Adenoviruses, and bacterial

 Quantitative Methods to Assess Capacity of Water Treatment to Eliminate Micro-Organisms. Edited by Wim A.M. Hijnen. ISBN: 9781843393764. Published by IWA Publishing, London, UK.

spores. The protozoon *Acanthamoeba* is also highly UV resistant. Bacteria and (oo)cysts of *Cryptosporidium* and *Giardia* are more susceptible with a fluence requirement of $<20\,mJ/cm^2$ for a MIC of 3 log.

Several studies have reported an increased UV resistance of environmental bacteria and bacterial spores, compared to lab-grown strains. This means that higher UV fluences are required to obtain the same level of inactivation. Hence, for bacteria and spores, a correction factor of 2 and 4 was included in the MIC calculation, respectively, whereas some wastewater studies suggest that a correction of a factor of 7 is needed under these conditions. For phages and viruses this phenomenon appears to be of little significance and for protozoan (oo)cysts this aspect needs further investigation. Correction of the required fluence for DNA-repair is considered unnecessary under the conditions of drinking water practice (no fotorepair, dark repair insignificant, esp. at higher ($60\,mJ/cm^2$) fluences) and probably also wastewater practice (fotorepair limited by light absorption). To enable accurate assessment of the effective fluence in continuous flow UV systems in water treatment practice, biodosimetry is still essential, although the use of Computational Fluid Dynamics (CFD) improves the description of reactor hydraulics and fluence distribution. For UV systems that are primarily dedicated to inactivate the more sensitive pathogens (*Cryptosporidium*, *Giardia* and pathogenic bacteria), additional model organisms are needed to serve as biodosimeter.

INTRODUCTION

The first application of UV irradiation in drinking water as disinfection process was in 1910 in Marseille (Henry *et al.*, 1910), after the development of the mercury vapour lamp and the quartz tube and establishing the germicidal effect of UV irradiation. According to Wolfe (1990) and Hoyer (2004) general application was hampered because of high costs, poor equipment reliability, maintenance problems and the advent of chlorination (cheaper, more reliable and potential to measure disinfectant residual). Due to the increased information on the production of hazardous oxidation by-products during chlorination and ozonation, UV irradiation gained more attention; low pressure UV produces almost no by-products. Also, unlike chemical disinfectants, the biological stability of the water is not affected by low pressure lamps. In Europe, UV has been widely applied for drinking water disinfection since the 1980's, for the control of incidental contamination of vulnerable groundwater and for reduction of Heterotrophic Plate Counts (Kruithof *et al.*, 1992). The breakthrough of UV applicability as a primary disinfection process in the US and Europe came after the discovery of the high efficacy of UV irradiation against *Cryptosporidium* (Clancy *et al.*, 1998) and *Giardia*. Chemical disinfection with chlorine is not effective against these pathogens and ozone applied at low CT-values to limit formation of bromate has relatively little effect on the infectivity of the protozoan (oo)cysts. In contrast, infectivity of these pathogens is significantly reduced by UV fluences that can readily be applied in drinking water treatment. UV is now regarded as being broadly effective against all pathogens, bacteria, protozoa and viruses, that can be transmitted through drinking water.

The introduction of the Quantitative Microbiological Risk Assessment (QMRA) to define the microbiological safety of drinking water (Haas, 1983; Regli *et al.*, 1991; Teunis *et al.*, 1997 and Medema *et al.*, 2003) is a development of growing interest. Besides knowledge about the presence of pathogenic micro-organisms in the source water, QMRA requires quantitative knowledge about the capacity of water treatment processes, including UV disinfection, to eliminate (remove or inactivate) pathogenic micro-organisms.

The aim of this study was to evaluate available literature data and create a well-defined database which enables calculation of the Microbial Inactivation Credit (MIC) of UV disinfection for viruses, bacteria and protozoan (oo)cysts in water. Most studies are lab-scale based. For full-scale chemical disinfection processes with chlorine and ozone it was demonstrated that the MIC for *E. coli* was lower than expected from the applied CT-values and the known dose-response curves determined under laboratory conditions (Hijnen *et al.*, 2000a, 2004a). Therefore literature was evaluated to verify the influence of process conditions on the MIC of full-scale UV-disinfection processes. The results of this review were also used to identify further research needs.

LITERATURE REVIEW METHOD

Selection of the reviewed literature. Literature on the inactivation of viruses, bacteriophages, bacteria, bacterial spores and protozoan (oo)cysts by UV irradiation in water was collected and evaluated on technical and microbiological aspects. In the review, only those studies were used where inactivation was assessed using generally accepted microbial culturing (solid media or tissues) or animal infectivity methods. Similarly, only studies were evaluated in which assessment of the UV fluence was clearly described and based on either UV sensor measurements, fluence calculations and/or Reduction Equivalent Fluence (REF) assessment with (bio)dosimetry. The description and quality of the microbiological data, quality assurance and reproducibility of the experimental data, as well as the availability and quality of the technical and experimental conditions were reviewed. Only those studies where process and experimental conditions were well documented were used.

Experimental conditions. The inactivation of micro-organisms by UV irradiation has been studied under different experimental conditions. Many studies used a collimated beam apparatus (CB-tests) under bench-scale and well-defined laboratory conditions. A volume of inoculated water is irradiated during varying periods of time under a lamp emitting UV light. Other studies used continuous flow systems (CF-systems) in a laboratory, in a pilot- or demonstration plant or under full-scale conditions where the water passes a reactor with one or more UV lamps and UV irradiation and contact time vary over the reactor.

Drinking water and wastewater studies. UV disinfection is applied in the drinking water industry and for disinfection of treated wastewater. Studies from both applications were reviewed and identified as such. Papers on wastewater studies (WWS) usually describe disinfection of secondary effluent with or without an additional pre-treatment. In the drinking water studies (DWS), water is generally of much lower turbidity and higher UV transmission than in waste water. When relevant, the influence of these parameters is discussed.

UV fluence data. UV fluence cannot be measured directly, so it has to be inferred from monitoring the UV irradiance with a UV sensor and the time that the micro-organisms are exposed to UV. For the Collimated Beam experiments, the average UV irradiance and contact time are well-characterised and have small confidence intervals. For continuous flow systems the average UV fluence can be calculated from the same parameters. However, the confidence intervals are much larger, due to the much larger variation in contact time and in UV irradiation at different points in the reactor, compared to the Collimated Beam experiments. By modelling the hydraulic retention time in the UV reactor using Computational Fluid Dynamics (CFD) accuracy of these calculations has increased the last few years. Alternatively, the Reduction Equivalent Fluence of CF-systems can be determined with biodosimetry (Qualls and Johnson, 1983a; Sommer *et al.*, 1999; Österreichisches normungsinstitut, 1999; Hoyer, 2004 and USEPA, 2003). Biodosimetry is performed by challenging the UV reactor with a micro-organism with calibrated UV inactivation kinetics (biodosimeter) assessed with CB-tests in the test water. With the measured inactivation of the biodosimeter and the calibration curve, the Reduction Equivalent Fluence or REF (mJ/cm^2) can be calculated.

Inactivation kinetics of the micro-organisms. The inactivation kinetics of a large number of pathogenic micro-organisms and indicator micro-organisms that are significant to the microbial safety of water have been calculated from studies where UV fluence has been determined under optimal conditions: CB-tests with 'drinking water' (low turbidity and high UV transmission).

Inactivation by UV is based on the damage caused to the nucleic acids (DNA/RNA) of the cell or virus. Primarily the formation of pyrimidine dimers, but also of other photoproducts of nucleic acids and nucleic acid lesions (von Sonntag *et al.*, 2004), inhibit replication and transcription and hence, prevent the cell or virus from multiplying. The UV absorbance of DNA peaks around 260 nm; at lower and higher wavelengths the absorbance decreases. Below 230 nm the absorbance increases again. Most studies used low-pressure mercury lamps with a major wavelength output (85%) at 254 nm (monochromatic (MC) UV radiation) but for some micro-organisms the UV inactivation was (also) determined with polychromatic (PC) UV radiation from medium pressure lamps.

The UV sensitivity of the selected micro-organisms is described by the parameters of the inactivation kinetics. Inactivation is defined as the reduction of the concentration of culturable micro-organisms N due to the exposure to a concentration disinfectant C during a specific contact time t. The inactivation kinetic for chemical disinfectants is most commonly described by the first-order disinfection model of Chick-Watson (1908) and the same model can be applied for UV disinfection. The inactivation of micro-organisms is usually described by the log inactivation of N. Based on the first-order model the linear relationship between log inactivation and the UV dose or fluence is described by:

$$^{10}\log\left(\frac{N_t}{N}\right) = -k * Fluence \tag{1}$$

N_t is the microbial concentration after contact time t. Fluence is the product of the UV fluence rate (mW/cm^2) and the exposure t ($mWs/cm^2 = mJ/cm^2$). In the literature, two main deviations from first-order UV disinfection kinetics have been observed. Some authors (Knudson, 1985; Hoyer, 1998; Sommer *et al.*, 1998 and Mamane-Gravetz and Linden, 2005a) observed no inactivation of bacteria or bacterial spores at low UV fluences followed by a normal log-linear relationship at higher UV fluences. This can be described by a shoulder model and is presented by the following equation

$$DI = -k * Fluence - b \tag{2}$$

where DI is the decimal inactivation $^{10}\log(N_t/N)$, b is the y-intercept, a negative value since the curve is crossing the fluence-axis at the UV fluence where log-linear relationship starts (offset). The second deviation from the linear kinetics is no further increase in inactivation at high fluences, called tailing. Tailing is excluded in the k-value calculation in this review, by excluding the inactivation data at higher fluences in studies where tailing was observed (from the plots of inactivation versus fluence).

Influence of process conditions. The inactivation kinetics can be used to determine the disinfection efficacy or Micro-organism Inactivation Credit (MIC; log) of full-scale UV systems and to assess the fluence requirement to obtain a certain MIC. However, for translation of CB-results to full-scale UV systems it is essential to know the effect of process conditions on the efficiency of the radiation process. In contrast to oxidative disinfection processes with chemicals like chlorine and ozone, the efficacy of UV disinfection is not affected by conditions like temperature, pH (Severin *et al.*, 1983 and USEPA, 2003) and reactive organic matter. UV absorbance by organic and inorganic matter is included in the UV fluence calculation. But the following factors may affect the efficiency of UV disinfection at full-scale:

- Factors related to the micro-organisms: physiological state (pre-culturing, growth phase), strain diversity, repair mechanisms and particle association;
- Factors related to the fluence assessment: fluence-distribution due to the distribution of the hydraulic retention time, adsorption, reflection and refraction of UV light through the water and lamp intensity (aging and fouling).

Several studies have addressed these aspects and are discussed to determine whether adaptation of the required fluence for a certain MIC, calculated by applying the UV sensitivity data to full-scale UV systems, is required and, if possible, to quantify to what extent.

UV SENSITIVITY OF MICRO-ORGANISMS

The UV sensitivity of the micro-organisms is described with the inactivation rate constant k (cm^2/mJ). A UV sensitive micro-organism has a high k-value and requires a low fluence for inactivation according to equation 1.

Viruses and bacteriophages. The number of studies where UV sensitivity of specific pathogenic viruses and bacteriophages is determined under well-defined laboratory conditions with collimated beam apparatus (CB-tests) ranged from 1 to 6. The total number of data per virus or bacteriophage ranged from 3 up to 109 (Table 6.1 and 6.2). The calculated k-value (no shoulder; intercept = 0) showed a narrow 95% confidence-interval (CI) and a high goodness-of-fit (13 out of 18 $r^2 > 0.85$). The six authors describing inactivation of seeded Poliovirus type 1 yielded a total of 61 data points presented in Figure 6.1. The inactivation rate constant k calculated for a UV fluence range of 5 to 50 mJ/cm^2 was 0.135 (95%-CI = 0.007; $r^2 = 0.79$). Due to the observed tailing by Sommer *et al.* (1989) and Maier *et al.* (1995), MIC_{max} (the maximum observed MIC) is set at 5.4 log (≥50 mJ/cm^2).

Table 6.1 UV sensitivity of viruses for monochromatic (MC) and polychromatic (PC) UV radiation determined with collimated beam tests

	Studies (n)	UV fluence (mJ/cm^2)	UV	k[a] (±95%CI; r^2)	MIC_{max} (Log)
Poliovirus type 1	6 (61)[b,c,d,e,f,g]	5–50	MC	0.135 (0.007; 0.79)	5.4
Adenovirus ST2,15,40,41	5 (98)[g,h,i,j,k]	8–306	MC	0.024 (0.001; 0.87)	6.4
Adenovirus ST40	1 (29)[i]	8–184	MC	0.018 (0.004; 0.88)	3.0
Adenovirus ST2,41	1 (18)[k]	30–90	PC	0.040 (0.003; 0.77)	4.3
Rotavirus SA-11	5 (55)[b,d,e,k,l]	5–50	MC	0.102 (0.006; 0.78)	4.1
Rotavirus SA-11	1 (11)[k]	5–30	PC	0.154 (0.011; 0.92)	4.6
Calicivirus feline, canine	3 (29)[i,m,n]	4–49	MC	0.106 (0.010; 0.67)	5.5
Calicivirus bovine	1 (20)[k]	4–33	MC	0.190 (0.008; 0.96)	5.7
Calicivirus bovine	1 (20)[k]	2–15	PC	0.293 (0.010; 0.97)	5.9
Hepatitis A	3 (13)[e,l,o]	5–28	MC	0.181 (0.028; 0.70)	5.4
Coxsackie virus B5	2 (12)[h,l]	5–40	MC	0.119 (0.006; 0.97)	4.8

[a]*linear regression, intercept = 0;* [b]*Chang* et al., *1985;* [c]*Harris* et al., *1987;* [d]*Sommer* et al., *1989;* [e]*Wilson* et al., *1992;* [f]*Maier* et al., *1995;* [g]*Meng and Gerba, 1996;* [h]*Gerba* et al., *2002;* [i]*Thurston-Enriquez* et al., *2003;* [j]*Thompson* et al., *2003;* [k]*Malley* et al., *2004;* [l]*Battigelli* et al., *1993;* [m]*De Roda Husman* et al., *2003;* [n]*Duizer* et al., *2004;* [o]*Wiedenmann* et al., *1993*

Table 6.2 UV sensitivity of bacteriophages for monochromatic (MC) and polychromatic (PC) UV radiation determined with collimated beam tests

	Studies (n)	UV fluence (mJ/cm^2)	UV	k[a] (±95%CI; r^2)	MIC_{max} (Log)
MS2-phages	5 (109)[b,c,d,e,f,g]	5–139	MC	0.055 (0.002; 0.93)	4.9
MS2-phages	1 (11)[f]	12–46	PC	0.122 (0.009; 0.92)	5.3
φX174	4 (30)[e,h,i,j]	2–12	MC	0.396 (0.025; 0.85)	4.0
PRD1	1 (4)[d]	9–35	MC	0.128 (0.014; 0.98)	3.8
B40-8	1 (14)[k]	1–39	MC	0.140 (0.010; 0.96)	5.6
T7	1 (3)[k]	5–20	MC	0.232 (0.080; 0.90)	4.6
Qβ	1 (5)[k]	10–50	MC	0.084 (0.003; 0.99)	4.2

[a]*linear regression, intercept = 0;* [b]*Havelaar* et al., *1990;* [c]*Wilson* et al., *1992;* [d]*Meng and Gerba, 1996;* [e]*Sommer* et al., *1998;* [f]*Malley* et al., *2004;* [g]*Mamane-Gravetz* et al., *2005b;* [h]*Battigelli* et al., *1993;* [i]*Oppenheimer* et al., *1993;* [j]*Sommer* et al., *2001;* [k]*Clancy* et al., *2004*

In Figure 6.1 the fluence-response curves for Adenoviruses serotypes (ST)2, 15, 40 and 41 are also presented. These data demonstrate that the UV sensitivity of these serotypes for monochromatic UV radiation shows small differences. Adenovirus is the most persistent virus type presented in Table 6.1. This conclusion is supported by the recently published study of Nwachuku *et al.* (2005) who found k-values for serotypes 1 and 6 in the same order of magnitude as for the types presented in Table 6.1. Thurston-Enriquez *et al.* (2003) found the lowest k-value of 0.018 cm^2/mJ with Adenovirus ST40. Malley *et al.* (2004) determined the UV sensitivity of Adenovirus ST2 and ST41 for polychromatic (PC) UV radiation (medium pressure (MP) lamps). Up to a UV fluence of 90 mJ/cm^2, the UV sensitivity was a factor of 1.7 higher than that observed for monochromatic (MC) UV radiation (low pressure lamps) (Table 6.1), but above this fluence they observed tailing (Figure 6.1). By using bandpass filters they distinguished the germicidal effect of different wavelengths in the polychromatic UV light at fluence-ranges up to 90 mJ/cm^2 and showed that at wavelengths of 220 and 228 nm UV was significantly (a factor of 5–7) more effective in inactivating Adenovirus ST2 than UV light with a wavelength of 254 nm.

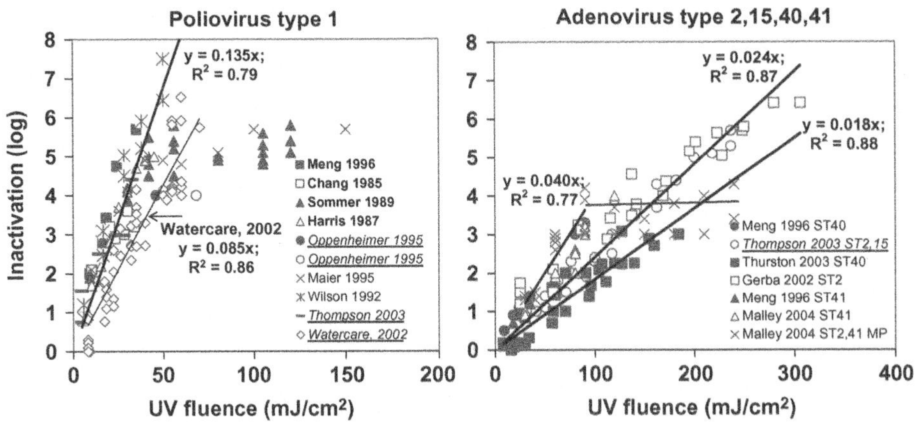

Figure 6.1 UV fluence-response curves for Poliovirus and Adenovirus (regular font DWS, *italic: WWS*)

The fluence-response data and lines for Rotavirus type SA11 and for three types of Caliciviruses are presented in Figure 6.2. Again Sommer *et al.* (1989) showed no further increase in inactivation at fluences above 50 mJ/cm^2 and Malley *et al.* (2004) showed that monochromatic UV radiation was less efficient than polychromatic UV radiation for inactivation of rotavirus SA-11.

Using MP-lamps, the k-values were 1.7 times higher (Table 6.1). Caliciviruses from different non-human hosts (feline, canine and bovine) showed highest UV sensitivity for the bovine type (Malley *et al.*, 2004) and the same study showed a 1.5 times higher inactivation with polychromatic UV radiation compared to monochromatic UV. The k-value for inactivation of feline Calicivirus (0.106 cm^2/mJ) was in the same order of magnitude as that observed for Rotavirus, Poliovirus and Coxsackie virus B5 (Table 6.1). Hepatitis A virus was more sensitive to UV radiation.

Noroviruses are part of the human caliciviruses and are not culturable. Using RT-PCR, Watercare (2002) determined that environmental Norovirus was less effectively inactivated by UV than the other viruses as determined with culture assays. In their wastewater treatment plant, ten samples before and 10 samples after UV were analysed; seven samples before UV and one sample after UV were positive for Norovirus. The mean fluence was 23 mJ/cm^2. From these (presence/absence) data, a 0.8 log inactivation was estimated for a UV fluence of 20 mJ/cm^2. At higher fluences (40 and 70 mJ/cm^2) all samples were negative. However, it is uncertain to which degree inactivation assessed with RT-PCR is representative for inactivation assessed with infectivity assays.

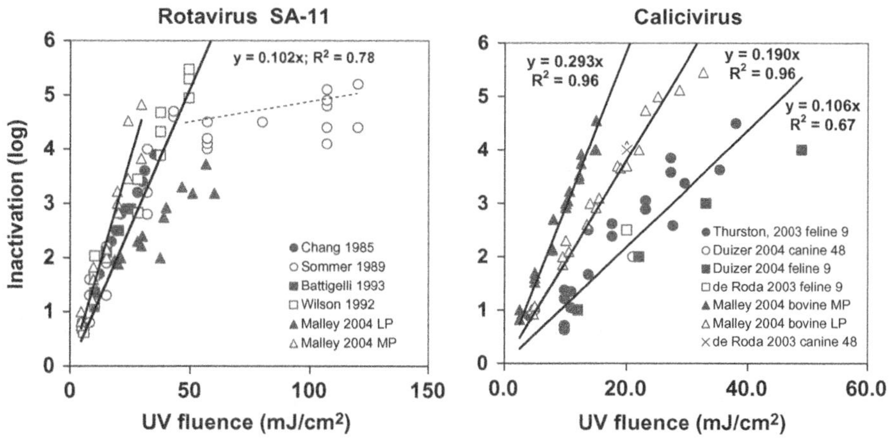

Figure 6.2 UV fluence-response curves for rotavirus and calicivirus

UV sensitivity of bacteriophages used or proposed as model organisms for the assessment of the REF of a UV system on full-scale is also presented in Table 6.2. MS2 phage is the most persistent of the tested phages with a k-value of 0.055 cm^2/mJ. k-Values of the other bacteriophages ranged from 0.128 cm^2/mJ for PRD1 up to 0.396 cm^2/mJ for PhiX174. The fact that the k-values of these micro-organisms are in the same order of magnitude as observed for most pathogenic viruses (compare Table 6.1 and 6.2) supports the use of these micro-organisms as surrogates for virus inactivation by UV. Only Adenoviruses are more resistant to UV.

Recently, Mamane-Gravetz *et al.* (2005b) demonstrated that MS2 is more (three times) sensitive to low wavelengths near 214 nm emitted by MP-lamps compared to 254 nm output of LP-lamps, an observation in line with those of Malley *et al.* (2004) for Adenoviruses as described before. The k-values determined by Mamane-Gravetz *et al.* (2005b) at wavelengths of 254 and 214 nm were 0.055 and 0.161 cm^2/mJ, respectively. These values are in the same order of magnitude as the k-values calculated in this study for LP- and MP-lamps, respectively (Table 6.2). The UV fluence of the MP-lamp in the Malley study was calculated based on the average irradiance measured by a UV-sensor and weighted by a germicidal factor at each wavelength (based on the DNA absorbance, relative to 254 nm). Thus, the fluence of LP- and MP-lamps was compensated for the wavelengths emitted by these lamps. Malley *et al.* (2004) argued that this weighting may have been biased for MP-lamps. On the other hand their results may indicate a higher inactivation efficiency of MP-lamps compared to LP-lamps, a conclusion supported by the observations of Mamane-Gravetz *et al.* (2005b).

Bacteria and bacterial spores. Bacteria (vegetative cells) are significantly more susceptible to UV radiation than viruses and therefore less extensively studied. In Figure 6.3, fluence-response curves for some selected pathogenic bacteria are presented. With the exception of *E. coli* in five studies, only one or two studies were found for individual pathogenic bacteria. Wilson *et al.* (1992) tested the UV sensitivity of seven of the ten bacterial species presented in Table 6.3.

The number of data points ranged from 4 up to 41 for *E. coli*. The k-values varied from 0.312 cm^2/mJ for *Streptococcus faecalis* to 1.341 cm^2/mJ for *Vibrio cholerae* (both Wilson *et al.*, 1992). Linear regression analysis showed low variation (95% confidence interval) and high goodness-of-fit (r^2). Sensitivity of *Legionella pneumophila* published in literature was highly variable. From the data of Antopol *et al.* (1979) and Wilson *et al.* (1992) a k-value of 1.079 and 0.400, respectively, was determined (Table 6.3). Knudson (1985) published a higher sensitivity (k = 1.916) and the k-value presented by Oguma *et al.* (2004) was 0.62. The latter author also demonstrated that the sensitivity of both *L. pneumophila* and *E. coli* to monochromatic and polychromatic was similar (Table 6.3).

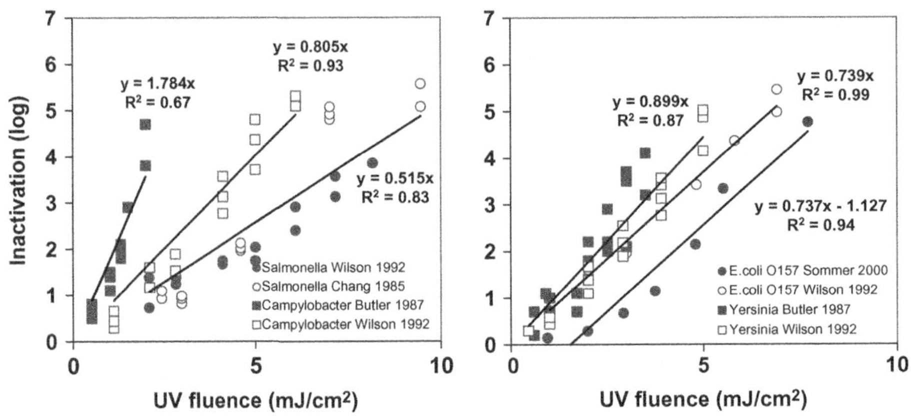

Figure 6.3 UV fluence-response curves for pathogenic bacteria

Table 6.3 UV sensitivity of bacteria and bacterial spores for monochromatic (MC) and polychromatic (PC) UV radiation determined with collimated beam tests

	Studies (data)	**Fluence (mJ/cm²)**	**UV**	**k (±95%CI; r^2)**	**Offset[a] (mJ/cm²)**	**MIC_{max} (Log)**
Salmonella typhi	2 (26)[b,c]	2–10	MC	0.515 (0.047; 0.83)	0	5.6
Campylobacter jejuni	2 (27)[c,d]	0.5–6	MC	0.880 (0.124; 0.65)	0	5.3
Yersinia enterocolitica	2 (34)[c,d]	0.6–5	MC	0.889 (0.060; 0.87)	0	5.0
Shigella dysenteriae	1 (9)[c]	1–5	MC	1.308 (0.087; 0.95)	0	5.9
Shigella sonnei	1 (9)[b]	3–8	MC	0.468 (0.053; 0.89)	0	4.7
Vibrio cholerae	1 (10)[c]	0.6–4	MC	1.341 (0.113; 0.94)	0	5.8
Legionella pneumophila	1(15)[c]	1–12	MC	0.400 (0.040; 0.92)	0	4.4
Legionella pneumophila	1 (4)[e]	0.5–3	MC	1.079 (0.077; 0.99)	0	3.0
Escherichia coli O157	2 (16)[c,f]	1–7	MC	0.642 (0.082; 0.85)	0	5.5
Escherichia coli	6 (41)[b,d,g,h,i,j]	1–15	MC	0.506 (0.049; 0.71)	0	6.0
Escherichia coli	1 (23)[k]	1.5–9	PC	0.539 (0.070; 0.64)	0	5.2
Streptococcus faecalis	2 (19)[b,g]	2.5–16	MC	0.312 (0.032; 0.85)	0	4.6
Bacillus subtilis	4 (30)[b,h,l,m]	5–78	MC	0.059 (0.007 0.91)	12.3	4.0
C. perfringens	1 (9)[m]	48–64	PC	0.060 (0.027; 0.81)	18	3.0

[a]*Offset is threshold-value > 0: linear regression with intercept ≠ 0;* [b]*Chang* et al., *1985;* [c]*Wilson* et al., *1992;* [d]*Butler* et al., *1987;* [e]*Antopol* et al., *1979;* [f]*Sommer* et al., *2000a;* [g]*Harris* et al., *1987;* [h,i]*Sommer* et al., *1989, 1996a;* [j]*Zimmer* et al., *2002;* [k]*Oguma* et al., *2002;* [l]*Sommer* et al., *1998;* [m]*Hijnen* et al., *2004b (continuous flow system)*

Aerobic spores of *Bacillus subtilis* and anaerobic spores of *Clostridium* perfringens are clearly less sensitive to UV than the vegetative bacterial cells (Table 6.3) and also most of the viruses and phages (Table 6.1 and 6.2). The data on the UV sensitivity of *C. perfringens* were derived from a continuous flow-system with medium pressure lamps, in which the UV fluences were determined with biodosimetry (REF) with UV calibrated spores of *B. subtilis*.

Pathogenic protozoa. Interest in UV as a disinfection process for water has increased after Clancy *et al.* (1998) showed that *Cryptosporidium parvum* oocysts were highly susceptible to UV when the effect on the infectivity was assessed with the neonatal mouse model. Since then, Clancy and several other authors have studied inactivation of *Cryptosporidium parvum* and *Giardia* muris by UV radiation (Table 6.4). Figures 6.4 and 6.5 show substantial inactivation of (oo)cysts of both protozoa at low UV fluences ($<20\,mJ/cm^2$) by LP- and MP-lamps. Recently, Johnson *et al.* (2005) demonstrated a similar UV sensitivity for *C. hominis* oocysts which predominates in human cryptosporidiosis infections. Based on the regression analysis of these fluence-response data, efficacy of LP- and MP-lamps for oocyst inactivation is in the same order of magnitude (Table 6.4). Comparison of these k-values with the k-values from Table 6.1 and 6.2 show that these protozoa are more sensitive to UV than viruses, but less sensitive compared to most bacteria. The regression analysis of the accumulated data shows a low goodness-of-fit (r^2 = 0.37; 0.49 and 0.81) and positive intercept values. Furthermore, Craik *et al.* (2000, 2001) observed considerable tailing for a number of inactivation data at high UV fluences (Figure 6.4c). Qian *et al.* (2004) described the protozoan data with a statistical method (Bayesian meta-analysis) which resulted in the UV fluence requirement curves presented in Figure 6.4 and 6.5 (USEPA, 2003). These curves were calculated for an inactivation requirement of up to 3 log and could be described by a log-log relationship (Log inactivation of *Giardia* = 1.2085 LN UV fluence + 0.0715; r^2 = 0.99; Log inactivation of *Cryptosporidium* = 1.2344 LN UV fluence − 0.1283; r^2 = 0.99).

Table 6.4 UV sensitivity of protozoa and Acanthamoeba spp. for monochromatic (MC) and polychromatic (PC) UV radiation determined with collimated beam tests

	Studies (data)	k (±95% CI; r^2)	Range (mJ/cm^2)	Intercept (95%)	MIC_{max}
C. parvum	6 (38)[a,b,c,d,e,f]	0.243 (0.08; 0.49)	0.5–6.1; PC	1.502 (0.538)	3.0
C. parvum	4 (65)[a,c,f,g]	0.225 (0.07; 0.37)	0.9–13.1; MC	1.087 (0.403)	3.0
Giardia muris	1 (4)[h]	0.122 (0.178; 0.81)	1.5–11; MC	1.303 (1.280)	2.4
Giardia lamblia	1 (2)[i]	nd	0.05–1.5; MC	nd	2.5
Acanthamoeba pp.	1 (16)[j]	0.021 (0.004; 0.94)	43–172; MC	0.499 (0.449)	4.5

[a,b]*Clancy* et al., *2001, 2002;* [c]*Craik* et al., *2001;* [d]*Shin* et al., *2001;* [e]*Morita* et al., *2002;* [f]*Rochelle* et al., *2004;* [g]*Bolton* et al., *1998;* [h]*Craik* et al., *2000;* [i]*Linden* et al., *2002 ;* [j]*Maya* et al., *2003*

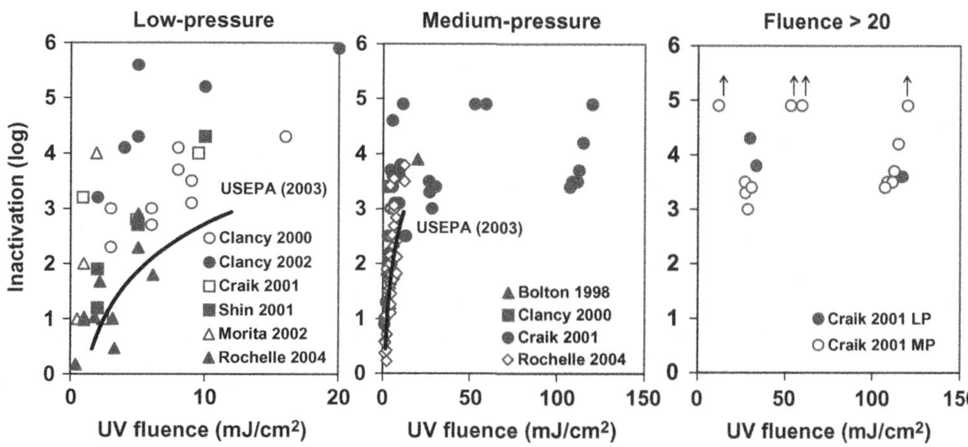

Figure 6.4 UV fluence-response curves for *Cryptosporidium parvum* (multiple strains) and LP- and MP-lamps and tailing (Fluence > 20) in the inactivation data observed by Craik *et al.*, 2001 for fluences above 20 mJ/cm^2

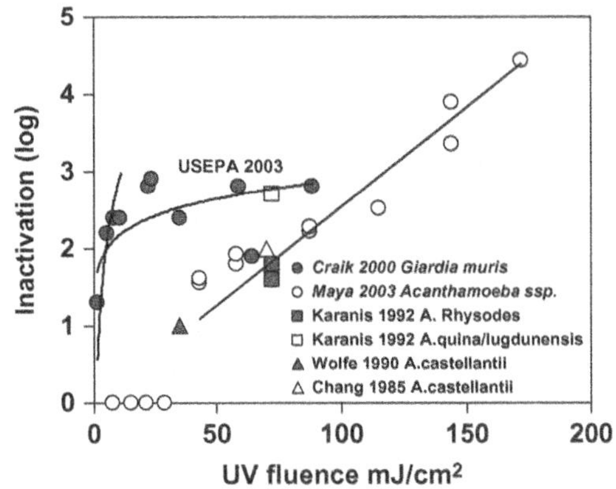

Figure 6.5 UV fluence-response curves for Giardia muris and Acanthamoeba

One study described the inactivation of *Acanthamoeba* spp. by UV in CB-tests using CD1 neonatal mouse model test to measure infectivity (Maya *et al.*, 2003). Just as observed for *B. subtilis*, an offset UV fluence is required for this organism to see an effect on infectivity of these pathogens. This offset value of 30 mJ/cm^2 as well as the low k-value of 0.021 calculated from the successive log-linear relationship, show that the sensitivity of this micro-organism and of the most resistant virus type Adenovirus to UV are in the same order of magnitude.

INFLUENCE OF PROCESS CONDITIONS

The k-values summarized in Tables 6.1, 6.2 and 6.3 can be used to determine the scale of new full-scale UV treatment processes or to calculate the inactivation efficiency of operational UV systems. Translation of UV sensitivity assessed with CB-tests and seeded micro-organisms to the efficiency of UV disinfection under full-scale conditions, however, is influenced by factors related to the micro-organisms and by factors related to the fluence assessment. This is similar to the translation of lab-scale tests for chemical disinfection to full-scale, as illustrated for ozone by Smeets *et al.* (2005, 2006). The literature on the influence of these factors has been reviewed. For pathogenic micro-organisms two studies were found (Figure 6.6) that investigated the inactivation of environmental pathogenic micro-organisms under different conditions. For indicators, more studies were available. Most of the evaluated data came from wastewater studies and to a lesser extent from drinking water studies. Inactivation of seeded or environmental indicator micro-organisms (coliforms, enterococci, clostridia spores, FRNA-phages, *Bacillus* spores) has been determined in either CB-apparatus or CF-systems. Inactivation data are presented in Figure 6.7–6.11, where CB-test results are separated from results from CF-systems. The findings are reviewed in the two following paragraphs.

Micro-organism related factors. The UV sensitivity of seeded and environmental micro-organisms is compared in some studies under identical conditions or by comparing results from a study with environmental organisms with the overall data for seeded organisms tested in CB-apparatus or CF-systems (DWS, Figure 6.6–6.9).

Environmental *Salmonella*, faecal coliforms and enterococci in CB-tests in wastewater (Maya *et al.*, 2003) were more resistant to UV light than the seeded micro-organisms of the same species (DWS, Figure 6.6, 6.8 and 6.9). A higher UV resistance of environmental spores compared to seeded spores (which were surviving isolates from the environmental spores) was also observed for *Bacillus* spp. (Mamane-Gravetz *et al.*, 2005a; CB-tests) and sulphite-reducing clostridia SSRC (Hijnen *et al.*, 2004b; CF-system), both DWS (Figure 6.10). A higher resistance to UV of environmental bacteria is also demonstrated by the inactivation data of thermotolerant coliforms in the WWS studies of Watercare (2002) and Gehr and Nicell (1996) as shown in

Figure 6.8. Based on the difference in k-value between the Watercare data and the DWS data assessed in CB tests, the k-value decreased a factor of seven (from 0.506 to 0.066; Figure 6.8).

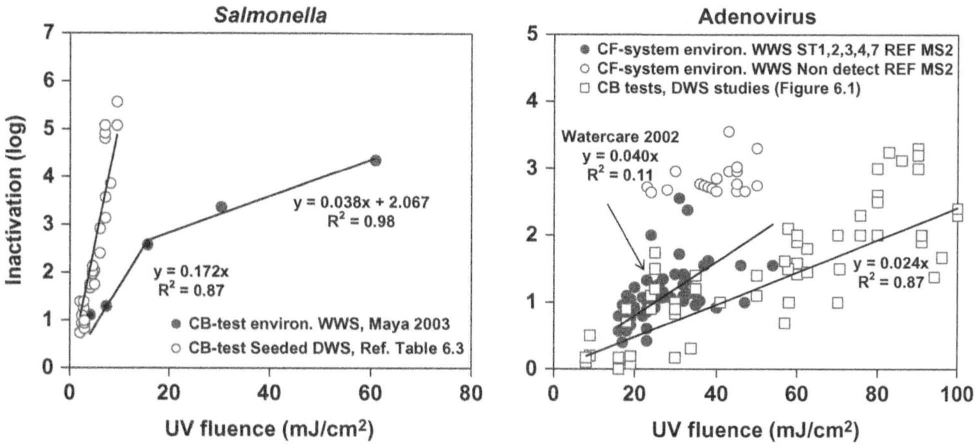

Figure 6.6 Comparison of UV fluence-response curves for seeded and environmental Salmonella and Adenoviruses

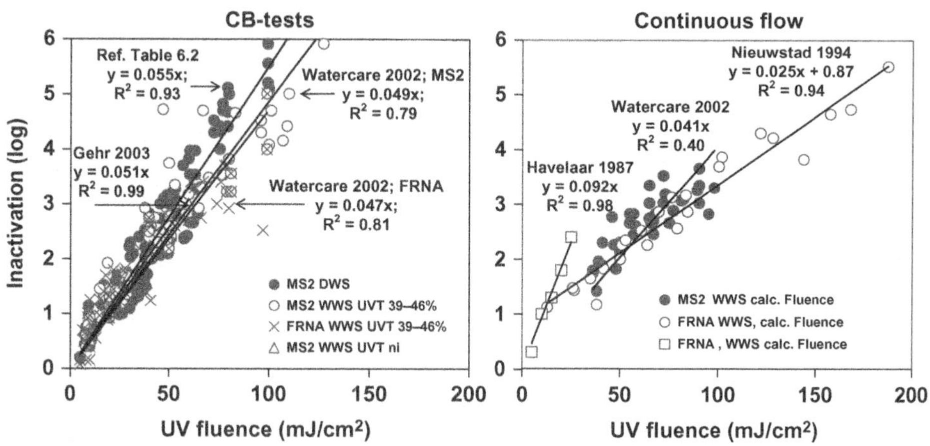

Figure 6.7 UV fluence-response curves for seeded MS2 FRNA phages and environmental FRNA phages determined under different conditions (ni = no information)

In a study with a CF-system (LP-lamps) operated at a fluence of 25 mJ/cm^2 complete inactivation of environmental Poliovirus (type 1) in a chalk well water was observed with an estimated inactivation of more than 2.3 log (Slade *et al.*, 1986; 21 samples of 0.15–1 m^3 tested over a period of one year). This might indicate little or no difference in sensitivity between environmental and lab-cultured Polioviruses; the latter are inactivated with 3.4 log at this fluence calculated from the k-value of Table 6.1. In the WWS-study of Watercare (Watercare, 2002; Simpson *et al.*, 2003 and Jacangelo *et al.*, 2004) the UV sensitivity of environmental F-specific RNA phages (FRNA) was comparable to the UV sensitivity of seeded MS2 phages tested under similar conditions (CB-tests; Figure 6.7). A k-value of 0.049 cm^2/mJ was calculated for the seeded MS2-phages while for environmental FRNA tested under similar conditions in CB-tests resulted in a k-value of 0.047 cm^2/mJ (Figure 6.7). In contrast, a lower

UV resistance of environmental Adenoviruses was observed in a CF-system (Watercare study) compared to seeded Adenoviruses tested in DWS-studies (CB-tests, Figure 6.6). Because 35% of the observations in this study yielded a higher inactivation than could be detected (>2.7 up to >3.3 log at a fluence range of 23 up to 50 mJ/cm^2; Figure 6.6), the difference is even larger; these data have not been used in the k-value calculation presented in Figure 6.6. Predominant Adenovirus types in the Watercare study were serotypes 1, 2, 3, 4 and 7, with less commonly serotypes 5, 8, 11, 13, 15, 19, 25 and 29. The higher susceptibility of the environmental Adenovirus in this study could be the result of the absence of Adenovirus, type 40, the most persistent serotype.

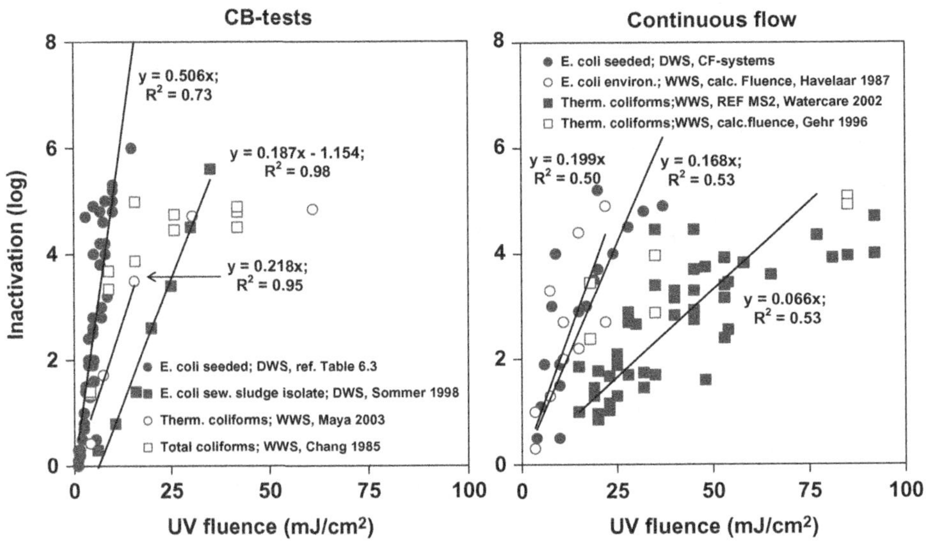

Figure 6.8 UV fluence-response curves for seeded and environmental coliforms (E. coli, thermotolerant coliforms and total coliforms) determined under different conditions

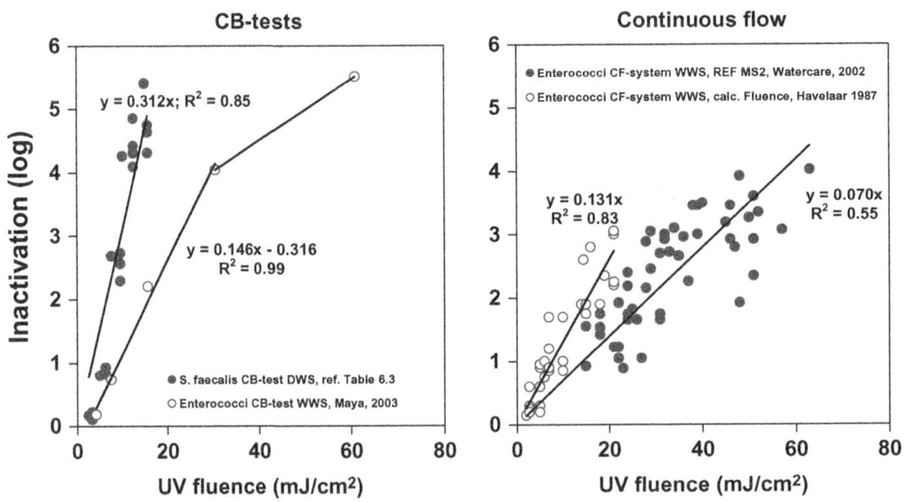

Figure 6.9 UV fluence-response curves for seeded and environmental enterococci determined under different conditions

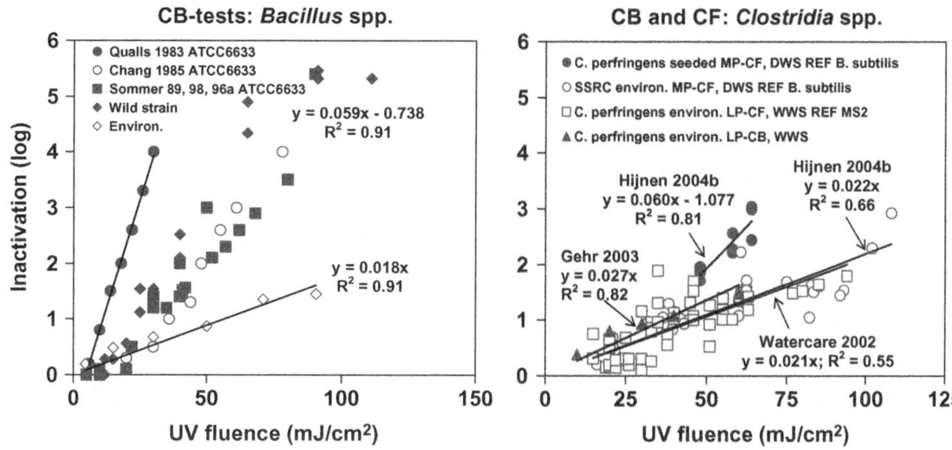

Figure 6.10 UV fluence-response curves for seeded and environmental bacterial spores determined under different conditions

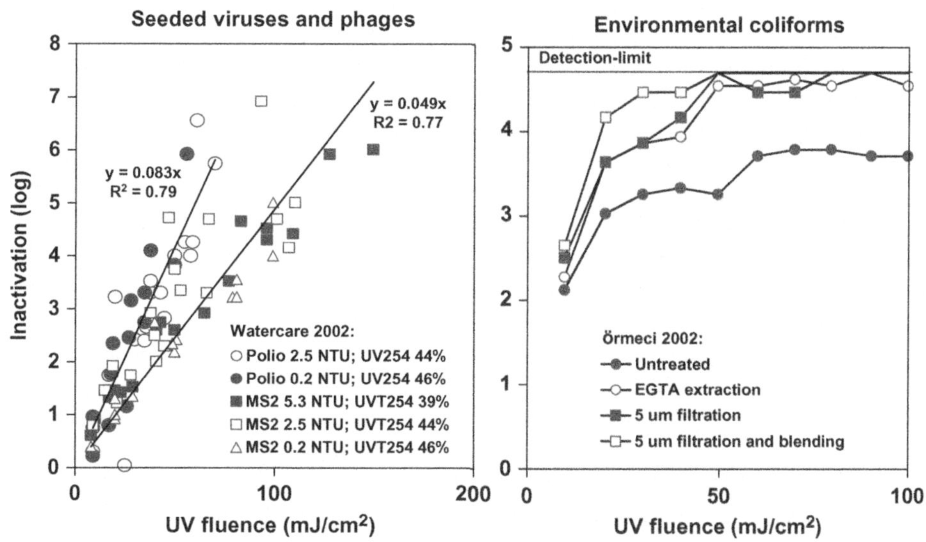

Figure 6.11 Effect of water quality and association with particles on UV fluence-response curves for viruses and phages and for environmental coliforms, respectively

Overall, increased UV resistance of environmental micro-organisms was more explicit for the bacterial spores and for vegetative bacteria, and was of less significance for FRNA-phages and viruses. To the authors knowledge no data have been published on the UV sensitivity of environmental (oo)cysts of *Cryptosporidium* and *Giardia*. The observed difference in UV sensitivity for bacteria and spores may be attributable to the physiological state of the micro-organisms, strain diversity, DNA-repair mechanisms and particle association. These factors are discussed in more detail below.

a) **Physiological state.** The physiological state of micro-organisms affects the sensitivity to environmental stress factors such as UV radiation. Martiny *et al.* (1990), Mofidi *et al.* (2002) and Malley *et al.* (2004) showed that UV sensitivity was related to the growth-phase of the bacteria with the highest sensitivity in the active growth phase and lower sensitivity in the stationary phase.

b) **Strain variation.** Different strains of one species may have different UV sensitivity, as demonstrated for *E. coli* by Sommer *et al.*, 1998; Sommer *et al.*, 2000a (Figure 6.8) and Malley *et al.* (2004). UV sensitivity of different *E. coli* strains in these studies varied by a factor of 5.8 and 3.7, respectively. The latter study demonstrated a higher sensitivity of *E. coli* O157:H7 compared to non-pathogenic/toxic strains. In contrast, Clancy *et al.* (2002) and Rochelle *et al.* (2004) showed that the high inactivation efficiency of UV radiation for *Cryptosporidium* was observed in multiple strains of *C. parvum*. The similar UV sensitivity observed for *C. hominis* (Johnson *et al.*, 2005) suggests that this high sensitivity of *Cryptosporidium* oocysts is common for all sub-species.

c) **Repair.** Exposure to UV results in damage to the nucleic acids of the cell. Although also other components of the cell may be damaged by UV, micro-organisms may still retain metabolic functions such as enzyme activity. Over time, organisms have developed mechanisms to repair DNA damage as a result of exposure to UV from the sun. The mechanisms of repair are comprehensively described in von Sonntag *et al.* (2004). Two types of repair have been described: dark-repair and photo-reactivation.

Dark-repair does not require light and has been demonstrated in almost all bacteria. Spores have no active metabolism, but repair starts upon germination. Viruses have no metabolism so cannot repair damage to their genome themselves. However, several viruses have been shown to use the repair enzymes of the host cell. This is suggested as the cause of the high resistance of Adenovirus, a double-stranded DNA virus, which can use the host cell's repair mechanism, while RNA-viruses may not. Some viruses even carry the genes for repair-enzymes (Lytle, 1971; refs. von Sonntag), but this is not the case for viruses that are transmitted via water.

Photo-reactivation occurs in conditions of prolonged exposure to (visible) light and is specifically targeting pyrimidine-dimers. For bacteria several CB studies demonstrated repair after light exposure. The significance of this phenomenon to the required fluence to achieve a certain inactivation can be deduced from the influence of repair on the inactivation kinetics (dose-response curves), but also from the occurrence of these repair mechanisms under conditions of disinfection practice.

Most photo-reactivation studies with CB tests used low fluences and optimal conditions for light exposure for repair (thin layer of fluid). The results show that under these conditions fluence requirement increases with increasing fluence (lower k-values). Quantitative data showed a 2.8–4.6 higher UV fluence requirement for 1–3 log inactivation of *Legionella pneumophila* (Knudson, 1985); based on these data k-value decreased a factor of 3.2. Oguma *et al.* (2004) observed a comparable log-repair at an initial inactivation of 3 log after UV disinfection with LP- and MP-lamps and complete photo-reactivation. For *E. coli* Bernhardt (1994) showed an increased offset value and decreased k-value. For several bacteria spp. (*E. coli*, *Yersinia enterocolitica*, *Salmonella typhi* and *Vibrio cholera*) he calculated an increased fluence requirement for a 4 log inactivation ranging from a factor of 1.8 up to 4.2 to account for complete photo-reactivation. For *E. coli* similar enhancement of fluence requirement was observed by Hoyer (1998; 3.5 times for 4 log inactivation). Sommer *et al.* (2000a) showed a decrease in k-value after photo-reactivation and also demonstrated that dark-repair is of less importance for *E. coli*. The latter observation was confirmed by Zimmer *et al.* (2002), who demonstrated that photo-reactivation of *E. coli* did not occur after MP-lamps, an observation also supported by Oguma *et al.* (2002, 2004).

Morita *et al.* (2002) demonstrated photo-reactivation and dark-repair of DNA in *Cryptosporidium parvum* with the endonuclease-sensitivity site assay. The animal infectivity, however, was not restored. Furthermore, they concluded that UV radiated oocysts are able to excyst but have lost their ability to infect host cells. Similar observations were reported by Shin *et al.* (2001) and Zimmer *et al.* (2003). Craik *et al.* (2000) and Linden *et al.* (2002) came to the same conclusion for *Giardia muris* and *Giardia lamblia* cysts, respectively. Belosevic *et al.* (2001), however, showed the ability of DNA-repair by some *Giardia* spp. after UV radiation with MP-lamps. This was also presented by Kruithof *et al.* (2005); in vivo reactivation (dark-repair) of *G. muris* cysts occurred at fluence values as low as 25 mJ/cm^2, but not at 60 mJ/cm^2, after prolonged time of incubation (3, 14 and 20% reactivation after 10, 20 and 30 days, respectively). Also DNA-repair of *G. lamblia* cysts after exposure to monochromatic UV irradiation is recently reported (Shin *et al.*, 2005). An extensive study was presented on repair in *C. parvum* oocysts by Rochelle *et al.* in 2004. Identification of possible DNA repair genes in *C. parvum* showed that the oocysts contain all of the major genetic components of the nucleotide excision repair complex. Nevertheless, inactivation displayed by oocysts immediately after UV exposure or displayed by oocysts after UV exposure followed by various repair conditions were generally in the same order of magnitude. This suggests that repair of UV induced damage in *C. parvum* after UV exposure in drinking water is not likely to occur.

d) **Particle association.** Higher resistance of particle associated faecal bacteria has been observed in secondary effluents. This phenomenon was demonstrated by Qualls *et al.* (1983b) and Havelaar *et al.* (1987) for thermotolerant coliforms and the enterococci, respectively. A lower inactivation rate was observed in non-filtered effluent of sewage water plants compared to filtered samples (pore size 8 µm). Recently, Örmeci and Linden (2002) applied different techniques (extraction with EGTA, filtration of 5 µm filters with or without homogenization by blending) to separate particle- and non-particle associated coliforms and showed an increased resistance of environmental coliforms associated with particles to UV (Figure 6.11). Aggregates of *B. subtilis* spores were artificially made with clay in a Jar Test apparatus by Mamane-Gravetz and Linden (2004) and caused a reduction in inactivation efficiency. The k-value decreased from 0.0617 for the suspended spore-clay solution to 0.0579 cm^2/mJ for the aggregated spore-clay suspension. The same authors published a new study in 2005, in which they found evidence for a correlation of hydrophobicity of spores with aggregation. Aggregation may be a cause of tailing (no further increase of inactivation at higher fluence) observed in the kinetics. The k-values of isolated environmental *Bacillus* strains in the tailing phase of the kinetics were similar to the k-values of the original and natural *Bacillus* spore population. This indicates that a shielding effect of aggregation or particle association is a significant factor in the low susceptibility of environmental *Bacillus* spores, and also observed for the environmental clostridia spores as presented in Figure 6.10.

Fluence related factors. Variability in fluence may be caused by water quality (adsorption, reflection and refraction) and the distribution of the hydraulic retention time in continuous flow systems.

a) **Water quality.** The presence of UV absorbing organic and inorganic compounds in water will reduce the UV fluence but results showed that fluence values can easily be corrected for the UV transmission of the water. Havelaar *et al.* (1990) placed 0.22 µm membrane filtered secondary effluent (UV transmission of 40–60%) in the UV pathway of the collimated beam apparatus and showed no decrease of the inactivation rate constant for MS2 phages after correction of the fluence for the transmittance. In studies with CF-systems both Schoenen *et al.* (1995) and Sommer *et al.* (1997) showed that the inactivation efficiency assessed at similar sensor readings was more sensitive for change in lamp intensity than for change in the water transmittance.

The influence of water quality on the efficiency of UV disinfection can be demonstrated by comparing results of drinking water studies (DWS) with wastewater studies (WWS) conducted under similar conditions. CB-tests with seeded Polioviruses in secondary effluent with high turbidity and low

UV transmission were published by Oppenheimer *et al.* (1995), Watercare (2002) and Thompson *et al.* (2003) (Figure 6.1). In the first two studies a lower inactivation was observed in wastewater than in drinking water studies. In the Watercare study, the inactivation rate constant k of seeded Poliovirus in filtered secondary effluent was a factor 1.6 lower than the k-value in calculated from the drinking water studies (Figure 6.1). The same study (Watercare, 2002) observed a slightly lower inactivation of MS2 in secondary effluent compared to the inactivation of MS2 in drinking water at the same fluence (Figure 6.7). CB-tests were conducted with secondary effluent with high and low turbidity (0.2–2.5 NTU and UVT of 40–68%). They demonstrated that there was no impact of turbidity on the inactivation of seeded Polioviruses and MS2 phages (Figure 6.11). In addition, data from the WWS-study of Thompson *et al.* (2003) showed no decreased inactivation of seeded Poliovirus and Adenovirus compared to inactivation of these organisms tested in DWS-studies (Figure 6.1).

b) **Fluence determination in CF-systems.** In the literature, only few drinking water studies have been published where fluence-response curves were determined with continuous flow systems (CF-systems). Results from the studies of Martini *et al.* (1990), Schoenen *et al.* (1991) and Bernhardt *et al.* (1992, 1994) showed lower inactivation rate constants for *E. coli* (Figure 6.8) when compared to the k-value determined from CB-tests. The fluence in the CF-systems in these studies were based on information of the supplier of the UV equipment or on actinometry.

Information about the precision of fluence calculations can be obtained with biodosimetry. Spores of *Bacillus subtilis* and MS2 phages are used as model organisms in biodosimetry assays to assess the Reduction Equivalent Fluence (REF) of CF-systems. Sommer *et al.* (2000b) determined the REF with UV_{254} calibrated spores of *B. subtilis* of more than 30 commercial available CF-systems and presented the results of six systems. In one system, the REF was equal to the UV fluence calculated according to the supplier's instructions, four systems showed that that REF was 19–38% lower than the calculated fluence and in one system REF was 33% higher than the calculated fluence (Table 6.5). The average ratio of the REF to the calculated fluence was 0.83 with a relatively high standard deviation of 0.25 (Table 6.5). The overestimation of the effective fluence (REF) by fluence calculation was supported by data presented by Hijnen *et al.* (2004b). In conclusion, calculated fluence data in CF-systems frequently do not match those obtained by biodosimetry. Biodosimetry is essential to determine the efficacy of CF-systems (DVGW, 1997; Sommer *et al.*, 2000b and USEPA, 2003). The introduction of Computational Fluid Dynamics for fluence calculations (no data presented) is improving the quality and precision of fluence calculations.

Table 6.5 Calculated UV fluence versus fluence assessed with biodosimetry (REF)

System	Study	Model organism	n	Ratio $UV_{calculated}$/REF (±SD)
A[a]	DWS	*B. subtilis*	3	1.33 (0.07)
B[a]	DWS	*B. subtilis*	3	0.81 (0.02)
C[a]	DWS	*B. subtilis*	3	0.79 (0.10)
D[a]	DWS	*B. subtilis*	3	1.00 (0.28)
E[a]	DWS	*B. subtilis*	3	0.73 (0.19)
F[a]	DWS	*B. subtilis*	3	0.62 (0.14)
G[b]	DWS	*B. subtilis*	9	0.59 (0.03)
H[c]	WWS	*MS2 phage*	37	0.83 (0.25)

[a]*Sommer* et al., *2000b;* [b]*Hijnen* et al., *2004b;* [c]*Watercare, 2002*

c) **Reflection.** Reflection caused by the construction materials of the UV reactor will have an influence on the inactivation efficiency determined by biodosimetry (Sommer *et al.*, 1996). This factor is of greater influence in single lamp systems than in multiple lamp systems because of the higher surface-volume ratio.

GENERAL DISCUSSION

Kinetics of UV inactivation. Most of the inactivation data can be adequately described with the first-order disinfection model, at least for a certain fluence range. An offset UV fluence before inactivation starts, i.e., a shoulder model, is observed for *Bacillus* spores and *Acanthamoeba* spp. The simple inactivation model, where the shoulder is given as an offset of the first-order model, is used in this study.

Another deviation from first-order kinetics is the reduction of inactivation rate at higher UV fluences (tailing). This is observed in several drinking water studies with CB-tests (Polioviruses, rotaviruses, *E. coli*, *C. parvum* and *G. muris*) and also for environmental bacteriophages and bacteria in wastewater studies in CF-systems. Tailing normally starts after at least 99% of the initial available micro-organisms are inactivated and is observed to a larger extent in the more UV susceptible micro-organisms. For the most resistant organisms (Adenoviruses, MS2 phages, bacterial spores and *Acanthamoeba* spp.) tailing was not observed. The cause of tailing is still under debate. Several causes have been hypothesised, such as experimental bias, hydraulics, aggregation of micro-organisms or a resistant subpopulation, but no conclusive evidence is available for any of these. For micro-organisms where tailing is observed, we have used the first-order model only for the fluence range that yielded a linear relation with the inactivation in the experiments. Because of the observed tailing, extrapolation of this inactivation rate to higher fluences is yielding uncertain results. For use in QMRA, the higher fluences can be assumed to yield (at least) the same inactivation credits as the highest fluence in the linear relation.

Significance for water disinfection. This study provides an extensive overview of the efficacy of UV disinfection for viruses, bacteria and bacterial spores and protozoan (oo)cysts, obtained from the reviewed literature. The k-values that were calculated from the reviewed studies can be used in QMRA and treatment design to determine the efficacy of a UV fluence in the inactivation of the range of reported bacterial and viral pathogens and indicator organisms. For *Cryptosporidium* and *Giardia*, the logarithmic functions given in USEPA (2003) were used for calculating the inactivation efficacy.

In the group of pathogenic micro-organisms, viruses are generally more resistant than *Cryptosporidium*, *Giardia* and the bacterial pathogens. Adenovirus 40 is the most UV resistant waterborne pathogen known. *Acanthamoeba* is also very resistant. Bacterial spores, esp. environmental spores of *Clostridium* are also resistant to UV, with k-values that are comparable to the Adenoviruses (Figure 6.10).

Correction of the required fluence: micro-organism related factors. Based on the increased UV resistance observed for environmental *Salmonella*, enterococci, thermotolerant coliforms, FRNA phages and spores of sulphite-reducing clostridia, correction of the fluence requirement for inactivation of bacteria and bacterial spores from the environment seems appropriate. The evaluated studies suggest a two times increased fluence requirement for bacteria and four times for bacterial spores in drinking water. For wastewater this is most likely not enough and based on Figure 6.8 a factor of seven seems more appropriate. The results of environmental polioviruses (Slade *et al.*, 1986) and FRNA phages and Adenoviruses (Watercare, 2002) indicate that such a correction is not needed for phages and viruses. However, further research is needed to support these findings. Similarly, studies on the increased resistance of environmental protozoan (oo)cysts to UV are appropriate.

Data from the evaluated studies indicated that photo-reactivation can result in a significant increase of the required fluence for bacteria to achieve the same level of inactivation as without photo-reactivation. Dark-repair does not seem to be very significant for the UV disinfection practice for most pathogens. Though for *Giardia* dark-repair was observed in two studies at lower (5–25 mJ/cm^2) fluences, but not at higher fluences (60 mJ/cm^2). Consequently, correction of the required fluence of full-scale UV disinfection because of photo-reactivation of bacteria and in the case of *Giardia*, also because of dark-repair at low fluences, could be necessary. For viruses, it is assumed that repair is included in the available fluence-response curves, as suggested for the double-stranded DNA Adenoviruses. After UV disinfection of drinking water photo-reactivation is not likely to occur but in the case of wastewater disinfection light exposure is likely. Translation of the presented photo-reactivation data to full-scale conditions, however, is not straight forward. These data have been observed under conditions favouring the induction of photo-repair (low fluence values, thin layer with optimal conditions for reactivation). The conditions in wastewater practice will be less favourable for exposure to light and hence for photo-reactivation to occur. Furthermore, the applied UV fluences in practice are usually higher than applied in the reviewed studies. Lindenauer and Darby (1994) and Gehr and Nicell (1996) showed a decrease in repair at higher fluences due to tailing in the inactivation kinetics. The former author also hypothesized that extended DNA damage at higher fluence values will reduce the potential for photo-repair. From their wastewater study and that of Whitby and Palmateer (1993), Gehr and Nicell (1996) suggested that in practice the overall impact of photo-repair might be negligible, because of the limited exposure to light and therefore limited induction of photo-repair. In conclusion this needs further verification, but we assume that the necessity for a fluence correction as a result of photo-reactivation in UV disinfection practice is less than suggested by the experimental data in the photo-reactivation studies.

Correction of the required fluence: fluence related factors. Most of the studies that have been reviewed have been executed under well-controlled laboratory conditions in which UV fluence was assessed with sensors and seeded micro-organisms. Information about the efficacy of UV systems under full-scale conditions was limited and those which have been evaluated, generally showed lower inactivation efficiency than in the laboratory. This reduced efficiency may be caused by factors related to the micro-organisms as described previously, but also by imperfections in the calculation of the fluence to which the micro-organisms are exposed in full-scale UV systems. The latter can be largely overcome by applying biodosimetry to full-scale UV systems to determine the Reduction Equivalent Fluence (REF). This is already enforced for the application of UV systems in drinking water practice in Austria (Österreichisches normungsinstitut, 1999). In Germany, a similar protocol is used as guideline (DVGW, 1997), and in the USA, the draft EPA Ultraviolet Disinfection Guidance Manual also appoints credits for inactivation of *Cryptosporidium* on the basis of biodosimetry (USEPA, 2003). Commonly used biodosimeters are spores of *B. subtilis* or MS2 phages. Cabaj *et al.* (1996), however, demonstrated that the REF decreases with increased broadening of the fluence distribution and increased inactivation rate constant of the used model organism. Consequently, susceptible model organisms (high k-value) are more sensitive to a broad fluence distribution, which will enlarge the gap between the REF and the arithmetic mean fluence. MS2 phages and spores of *B. subtilis* are less sensitive to UV than most other pathogenic micro-organisms (Table 6.1, 6.2, 6.3 and 6.4). The EPA manual (USEPA, 2003) introduced a REF bias based on effects of fluence distribution and inactivation rate constants to account for the difference in sensitivity between model organism and target pathogens. Another approach is the use of alternative model organisms. *E. coli* is suggested and also this review indicates that it can be used as model for the more susceptible bacteria and also *Cryptosporidium* and *Giardia*. More recently, Clancy *et al.* (2004) suggested two potential bacteriophages Qβ and T7 as model organisms. The use of T7 as UV dosimeter was previously proposed by Rontó *et al.* (1992). The k-values of these organisms (Table 6.2) are more in the range of the k-values calculated for the more sensitive pathogens. For MP-systems, the germicidal fluence is usually obtained with the DNA absorbance spectrum to weigh the effectiveness of the different wavelengths. The action spectra (the relative sensitivity to different UV wavelengths) of adenoviruses and the model organism MS2 phage differed from the action spectrum of DNA as demonstrated by Malley *et al.* (2004) and Mamane-Gravetz *et al.* (2005b). Thus, with different action spectra of model organisms used as biodosimeter different REF values will be calculated for polychromatic UV radiation. Therefore, more information about differences in action spectra between pathogens and potential biodosimeter organisms is required to increase the precision of the fluence determination of MP-systems.

Required fluence table. The accumulated knowledge in this review was used to create the required fluence of LP-lamps for a MIC of 1, 2, 3 or 4 log inactivation for most micro-organisms relevant to microbiological safety of water (Table 6.6). The required fluence is calculated from the k-values presented in Tables 6.1, 6.3 and 6.4 with correction for the increased UV-sensitivity of environmental bacteria and bacterial spores. For five bacteria species this was based on specific literature data and for the other bacteria, this was set at a factor of three, whereas for wastewater disinfection higher correction seems appropriate. Correction for environmental organisms was not necessary for the viruses (see data on FRNA-phages in Figure 6.7) and for the protozoa no data are available. Increased fluence requirement because of DNA-repair did not seem necessary for viruses and protozoa, although for *Giardia* at low fluences of 5–25 mJ/cm^2 results are still conflicting. For bacteria fluence correction for dark-repair is not necessary and further study has to elucidate whether correction for photo-reactivation is required.

Table 6.6 The UV fluence (mJ/cm^2) requirements for a MIC of 1 up to 4 log by monochromatic UV radiation for viruses, bacteria, bacterial spores and protozoan (oo)cysts based on the k-values with or without correction for environmental species; for bacteria in wastewater a higher correction for environmental species is needed and further research has to clarify the need for a higher fluence to account for photoreactivation; for Giardia increased fluence requirement because of dark repair is a factor for further research

	Required fluence (mJ/cm^2)			
MIC required (log):	1	2	3	4
Bacillus subtilis[a]	56	111	167	222
Adenovirus type 40	56	111	167	–[b]
Clostridium perfringens[a]	45	95	145	–[b]
Adenovirus type 2,15,40,41	42	83	125	167
Acanthamoeba[c]	40	71	119	167
Adenovirus[a] (no type 40)	25	50	–[d]	–[b]
Calicivirus canine	10	21	31	41
Rotavirus SA-11	10	20	29	39
Calicivirus feline	9	19	28	38
Coxsackie virus B5	8	17	25	34
Streptococcus faecalis[a]	9	16	23	30
Legionella pneumophila[d]	8	15	23	30
Poliovirus type 1	7	15	22	30
Shigella sonnei[d]	6	13	19	26
Salmonella typhi[a]	6	12	17	51
Hepatitis A	6	11	17	22
Calicivirus bovine	5	11	16	21
E. coli O157[d]	5	9	14	19
E. coli[a]	5	9	14	18
Cryptosporidium USEPA[c]	3	6	12	–[e]
Giardia USEPA[c]	2	5	11	–[e]
Campylobacter jejuni[d]	3	7	10	14
Yersinia enterocolitica[d]	3	7	10	13
Legionella pneumophila[d]	3	6	8	11
Shigella dysenteriae[d]	3	5	8	11
Vibrio cholerae[d]	2	4	7	9

[a]environmental spp.; [b]MIC_{max} < 4 log; [c]no correction for environmental spp. (research needed); [d]corrected for environmental spp.; [e]no value due to tailing

Research items. Based on this review a number of knowledge gaps are identified. More quantitative information is needed to estimate the effect of micro-organism related factors like environmental species, DNA-repair (esp. of *Giardia*) and differences in spectral sensitivity influencing the fluence requirement of

UV disinfection under full-scale conditions. Biodosimetry is a powerful tool to determine germicidal fluence values of CF-systems, but to determine REF for the whole range of relevant micro-organisms with different UV sensitivities, additional model organisms are needed. In the application of medium-pressure lamps with polychromatic UV light, further development of fluence assessment is of importance using biodosimetry with proper weighting for spectral sensitivity in connection with fluence calculation models. Daily UV process control needs further research into accurate description of the distribution of water flows and UV intensity over UV reactors, using CFD, to obtain simple, reliable and cheap *in situ* process control systems. Independent verification with biodosimetry is still essential. Systems to measure the germicidal fluence on-site in water treatment practice, using micro-organisms or compounds 'naturally' present in the water would allow on-site verification of the efficacy of UV systems in practice.

CONCLUSIONS

The accumulated literature data on the inactivation kinetics of disinfection with UV irradiation demonstrate that the process is effective against all pathogenic micro-organisms relevant for the current drinking water practices. The inactivation of micro-organisms by UV could be described with first-order kinetics using fluence-inactivation data from laboratory studies in collimated beam tests. No inactivation at low fluences (shoulder) and no further increase of inactivation at higher fluences (tailing) was observed for some micro-organisms. The former deviation from the log-linear kinetics is included in MIC calculations and the latter was used to determine the maximum (observed) MIC-values. The parameters that were used to describe the inactivation are the inactivation rate constant k (cm^2/mJ), the maximum inactivation demonstrated and (only for bacterial spores and *Acanthamoeba*) the offset parameter. The most persistent organisms known are viruses, specifically Adenoviruses, and bacterial spores. From the protozoa *Acanthamoeba* was highly UV resistant. Bacteria and (oo)cysts of *Cryptosporidium* and *Giardia* are more susceptible with a fluence requirement of $<20\,mJ/cm^2$ for a MIC of 3 log.

Several studies have reported an increased UV resistance of environmental bacteria and bacterial spores, compared to lab-cultured organisms. This means that higher UV fluences are required to obtain inactivation. Hence, for bacteria and spores, a correction factor of two and four was included in the MIC calculation, respectively, and data from the wastewater studies show that a higher correction is required under these conditions. For phages and viruses this phenomenon appears to be of little significance and for protozoan (oo)cysts this aspect needs further attention. For application in drinking water, no correction for repair seems necessary for most pathogens. The results on repair for *Giardia* are conflicting, but no repair occurred at higher fluences ($60\,mJ/cm^2$). For application in wastewater, the occurrence of photo-reactivation of bacteria is a subject for further research. To enable accurate assessment of the effective fluence in continuous flow UV systems in water treatment practice, biodosimetry is of great importance. In the case of MP-lamps more information about differences in spectral sensitivity between pathogens and potential biodosimeter organisms is needed to increase the precision of the fluence determination. For UV systems that are primarily dedicated to inactivate the more sensitive pathogens (*Cryptosporidium*, *Giardia*, pathogenic bacteria), additional model organisms are needed to serve as biodosimeter.

Chapter 7

Elimination of viruses, bacteria and protozoan oocysts by slow sand filtration

Wim A.M. Hijnen[a], J.F. Schijven[b], P. Bonné[c], A. Visser[d] and G.J. Medema[a]

[a]KWR Watercycle Research Institute; [b]National Institute of Public Health and the Environment; [c]Waternet; [d]Dune Water Company South Holland

ABSTRACT

The decimal elimination capacity (DEC) of slow sand filters (SSF) for viruses, bacteria and oocysts of *Cryptosporidium* has been assessed from full-scale data and pilot plant and laboratory experiments. DEC for viruses calculated from experimental data with MS2-bacteriophages in the pilot plant filters was 1.5–2.0 log. *E. coli* and thermotolerant coliforms (Coli44) were removed by full-scale filters and in a pilot plant filter with 2–3 log. At full-scale, *Campylobacter* bacteria removal was 1 log more than removal of Coli44, which indicates that Coli44 is a conservative surrogate for these pathogenic bacteria. Laboratory experiments with sand columns showed 2–3 and >5–6 log removal of spiked spores of sulphite-reducing clostridia (SSRC; *C. perfringens*) and oocysts of *Cryptosporidium*, respectively. Consequently, SSRC is not a good surrogate to quantify oocyst removal by SSF. Removal of indigenous SSRC by full-scale filters is less efficient than observed in the laboratory columns probably due to continuous load of these filter beds with spores, accumulation and retarded transport. It remains to be investigated if this also applies to oocyst removal by SSF. The results additionally showed that the schmutzdecke and accumulation of (in)organic charged compounds in the sand increased the elimination of micro-organisms. Removal of the schmutzdecke reduced DEC for bacteria with ±2 log, but did not affect removal of phages. This clearly indicates that besides biological activity, both straining and adsorption are important removal mechanisms in the filter bed for larger micro-organisms than viruses.

 Quantitative Methods to Assess Capacity of Water Treatment to Eliminate Micro-Organisms. Edited by Wim A.M. Hijnen. ISBN: 9781843393764. Published by IWA Publishing, London, UK.

INTRODUCTION

Recently the Dutch Drinking Water Decree has been revised (Anonymous, 2001). Drinking water companies should demonstrate sufficient elimination of pathogenic viruses, bacteria and (oo)cysts of *Cryptosporidium* and *Giardia* by treatment to comply with an annual infection risk lower than 10^{-4} per person. This requires quantitative knowledge about the elimination capacity of processes. A first quantitative microbial risk assessment at the plants of Amsterdam Water Supply (AWS) and the Dune Water Company South-Holland (DWS) revealed a lack of knowledge about the efficacy of their slow sand filters for the removal of micro-organisms. In collaboration with the National Institute of Public Health and the Environment (RIVM) and Kiwa Water Research they started a project to determine the decimal elimination capacity DEC of slow sand filtration for viruses, bacteria and protozoan oocysts. Additional goals were to investigate the effect of the schmutzdecke and the sand on elimination, and to verify the validity of surrogates used to assess pathogen elimination capacities.

The project was started by evaluation literature for removal of pathogenic micro-organisms. Simultaneously, full-scale data of the elimination of indigenous micro-organisms such as thermolerant coliforms (Coli44), spores of sulphite-reducing clostridia (SSRC) and *Campylobacter*, were collected and reviewed. Both activities did not answer all the questions. Moreover, the low and variable removal of SSRC by full-scale filters seriously doubted the use of these indicators as surrogate for protozoan (oo)cyst removal. Therefore further experimental research on pilot and laboratory scale has been carried out. The results of all these investigations are presented and discussed in this paper.

ASSESSMENT UNDER DIFFERENT CONDITIONS

Full-scale data analysis. The elimination of thermotolerant coliforms Coli44 and spores of sulphite-reducing clostridia SSRC by the SSF at two plants of AWS and one plant of DWS was determined from data of three years of routine sampling ($n = 32$ up to 5184) and of a two-week period in winter and summer with daily large volume sampling ($n \leqslant 20$). Periodically the influent of the filters at the location of DWS contained *Campylobacter* bacteria and simultaneously concentrations of these pathogenic bacteria before and after SSF were measured by large volume sampling. The Decimal Elimination Capacity DEC was calculated from the log transformed average concentration in influent and effluent with $\text{DEC} = \log \overline{C_{in}} - \log \overline{C_{out}}$. The average concentration is the total number of colony forming units divided by the total sampled volume (number of samples $\times$ sample volume) in the selected period.

Pilot plant experiments. Under full-scale situation elimination of viruses could not be determined. Therefore elimination of bacteriophage MS2 by two slow sand filters of the AWS pilot plant Leiduin has been determined with a challenge test. MS2 is an icosahedral phage with a diameter of 27 nm is a conservative surrogate for viruses (Schijven *et al.*, 2003). *E. coli* WR1 was co-injected as a reference to the elimination of indigenous Coli44. The influent of SSF1 was surface water pre-treated by coagulation floc removal, rapid sand filtration 1, dune infiltration followed by an open collection reservoir, rapid sand filtration 2, ozonation, softening and granular activated carbon filtration. SSF2 was supplied with the filtrate of rapid sand filter 2 without additional treatment. DOC, turbidity and pH of the influent of SSF1 was 1.5 mg C/l, 0.1 FTU and 8.0, respectively and of the influent of SSF2 2.1 mg C/l, 0.7 FTU and 8.0, respectively.

The filters (surface of 2.56 m^2, bed depth of 1.5 m and 0.3 mm diameter sand) were operated at a filtration rate of 0.3 m/h. The filter bed porosity determined with a tracer (NaCl) test was 0.27. The schmutzdecke of SSF1 was scraped 12 days before the experiment. SSF2 was tested with a smutzdecke of 81 days. Then the filter was scraped and tested again after 4 days. Temperature ranged from 9.4 up to 11.7°C. Before the start of the spiking experiments MS2 bacteriophages and *E. coli* WR1 were added and well mixed in the water above the filter bed to avoid dilution effects (this was done before each test with different spiking concentrations). Spiking of a low and a high concentration of MS2 and WR1 to the influent of the filters (13 ml of suspension per min) lasted two periods of 24 hours, respectively. The effluent concentrations of both micro-organisms had been monitored during a ten days period.

Column experiments. Due to the relative large scale and possible health risk in the pilot plant, removal of oocysts of *Cryptosporidium* has been determined in columns and compared with elimination of spores of *Clostridium perfringens* D10, a possible surrogate for oocyst removal in treatment (Hijnen *et al.*, 2000b). MS2 and WR1 were co-injected as references with the pilot plant and full-scale results. A full-scale filter of DWS was sampled at four layers (20–40, 40–60, 60–80 and 80–100 cm) and the AWS pilot plant filter at five layers (20–40, 50–70, 80–100, 110–130 and 130–150 cm) for ripened sand. In the DWS and AWS columns (diameter of 9 cm) the 40 cm sand bed was packed with 10 and 8 cm sand of each separate layer of sampled sand, respectively. Each separate layer was introduced in the column with water and mechanically packed by ticking against the Perspex column. Underneath this sand bed a layer of gravel (diameter of 1–2 mm) placed on a rough iron mesh. The influent of the columns was the influent of SSF1 sampled in a large RVS-tank (700 litre) and transported to the experimental location. Temperature of the column influent varied between 8 and 13.5°C. Before the spiking experiment the columns were operated at a filtration rate of 0.08 m/h during 2–3 weeks. 48 hours before the spiking test filtration rate was increased to 0.3 m/h. The water was inoculated with micro-organisms in a separate RVS-vessel located directly above the sand columns (no dilution effect). During two hours this inoculated influent was supplied to the columns and during the following 30 hours the concentration of micro-organisms in the filtrate was monitored. To verify the effect of co-injection two separate columns with AWS sand were tested, one spiked with MS2 and *E. coli* WR1 (AWS1) and the other two (AWS2 and DWS) spiked with a cocktail of MS2, WR1, *Clostridium perfringens* (D10) and *Cryptosporidium parvum*.

Microbiological methods. The analytical methods used for Coli44 and SSRC in the full-scale study were described in detail before (Hijnen *et al.*, 2000). The most probable numbers of thermophilic *Campylobacter* bacteria in the water before and after SSF of DWS was enumerated directly in 1 ml or in 10 ml–100 litres by membrane filtration on 0.22 μm pore size filters. The samples are pre-cultured in Preston-bouillon (Ribeiro and Price, 1984) in 3 × 3 portions of a decimal dilution for 48 h at 42 ± 0.5°C under micro-aerophilic conditions. Each of these pre-cultures are incubated on solid Karmali-medium (Karmali *et al.*, 1986) for 48 ± 2 hrs at 42 ± 0.5°C under micro-aerophilic conditions. Typical colonies are grey and glancing and cell material was microscopically confirmed as mobile and spiral formed cells.

The methods of production and storage of the stock-solutions and enumeration methods of MS2 bacteriophages and *E. coli* WR1 used in the pilot plant study have previously been described (Schijven, 2001). WR1 used for the column tests were pre-cultured in autoclaved tap water with 1 mg/l glucose-C. The production of stock-solution and enumeration method of spores of *C. perfringens* D10 was described previously (Hijnen *et al.*, 2002). *Cryptosporidium parvum* oocysts (Moredun; harvested by sedimentation and differential centrifugation) were spiked from a 1 ml suspension (10^8 oocysts). For oocyst counting, samples of 1–200 ml were analysed directly without concentration using Direct Fluorescence Assay-technique with the chemscan (Chemunex). Samples were filtered and prepared for scanning with the chemscan-kit (Chemunex 200 k0009-01 with IMS) including a mounting medium (85 μl), 25 mm 2.0 μm membrane filters and support Pad. The membrane filters were labelled with 100 μl 1:1 in deionized water diluted monoclonal antibody reagents (Oxoid; diluted 1:1 in deionised water) for 30 min. at 37°C. The filters were scanned and counted spots were microscopically confirmed based on colour, form and size.

ELIMINATION BY FULL-SCALE PROCESSES

The process conditions of the slow sand filters operated at the two locations of AWS and at DWS showed no large differences (Table 7.1). From the concentrations of Coli44 in influent and effluent observed in three separate periods DEC-values of >1.5 up to 3.2 log were calculated (Table 7.1). The average DEC for these indicator bacteria was 2.6 ± 0.5 log. This is higher than the average DEC of 1.9 ± 0.5 log calculated from a number of studies in literature (Poynter and Slade, 1977; Slade, 1978; Cleasby *et al.*, 1984; Bellamy *et al.*, 1985 and Ellis, 1985). Removal of *Campylobacter* bacteria by SSF at the DWS location Scheveningen was 3.4 ± 0.6 log, 1 log more than Coli44 removal by these filters (Table 7.1). This indicates that *E.* coli is a conservative surrogate parameter for *Campylobacter* removal by SSF.

Table 7.1 The decimal elimination capacity (DEC) of full-scale slow sand filters

Location	Weesperkarspel AWS	Leiduin AWS	Scheveningen DWS
Filtration rate (m/h)	0.4	0.25	0.25
Bed length (m)	1.5	1.5	1.0
Surface cleaning freq. (year)	1	2	4–5
Grain size (mm)	0.15–0.6	0.13–0.37	0.3–1.8
DEC ($\log_{10}$) Coli44	2.2[a]; >1.5[b]; >3.7 [c]	3.2[a]; nd[b]; nd[c]	2.9[a]; 2.3[b]; 2.2[c]
DEC ($\log_{10}$) *Campylobacter*	nd[d]	nd	4.1 [a]; 3.0[b]; 3.2[c]
DEC ($\log_{10}$) SSRC	1.6[a]; 1.5[b]; 1.0[c]	(1.8)[a,e]; −0.2[b,e]; −0.1[c,e]	1.8[a]; 0.0[b]; −0.2[c]

[a]three years routine monitoring in small volumes; [b]daily large volume sampling in two weeks in winter; [c]daily large volume sampling in two weeks in summer; [d]nd = not determined; [e]1st DEC high (45/l) and 2nd and 3rd DEC low (0.16/l) SSRC level in influent C_{in}

Elimination of SSRC by SSF varied significantly. The maximum removal was 1.8 log, but at locations Leiduin (AWS) and Scheveningen (DWS) data also showed little to no elimination and even a small increase in the SSRC numbers (negative DEC). Multiplication in anaerobic zones cannot be excluded. However, accumulation and retarded transport seem the most plausible explanations for this phenomenon. Spores not retained irreversibly by straining or attachment to the sand grains, may persist long enough to pass the sand filter eventually in concentrations that may even be higher than in the influent at the time of monitoring. This phenomenon for these spores has also been described by Schijven *et al.* (2003) and indicates that DEC depends on the duration of the contamination. Hence, it explains the positive correlation between elimination and the influent concentration observed at Leiduin. DEC of 1.8 and −0.2 $\log_{10}$ was observed at an average SSRC influent concentration of 45 and 0.16 spores/l, respectively. These results raise serious doubts about the use of these spores as a quantitative surrogate for *Cryptosporidium* oocysts removal by SSF. On the other hand, provided that no irreversible attachment or physical straining occurs, the same phenomenon of breakthrough in the long run may also be the case for protozoan (oo)cysts removal by SSF.

A CHALLENGE TEST UNDER PILOT PLANT CONDITIONS

Figure 7.1 shows the breaktrough curves of MS2 and WR1 in the challenge test. The difference between the average influent and maximum effluent concentration was used to estimate DEC. The tailing after the spiking period of 48 hrs is caused by slow detachment of the attached micro-organisms (Schijven, 2001). The filters SSF1 and SSF2 reduced the concentration of both organisms with 1.7 up to 4.2 $\log_{10}$ (Table 7.2). A concentration increase of about 3 log (Figure 7.1) did not significantly affect DEC. The schmutzdecke of 81 days at SSF2 showed a significant effect on the elimination of WR1. Before cleaning of the filter bed DEC was 3.9–4.2 and after cleaning 2–2.8 log. MS2 elimination however was hardly influenced by the filter bed cleaning and was 2.1 ± 0.6 log. This value is similar to the average removal of indigenous enteroviruses by SSF of 1.9 log observed by others (Slade, 1978 and Ellis, 1985).

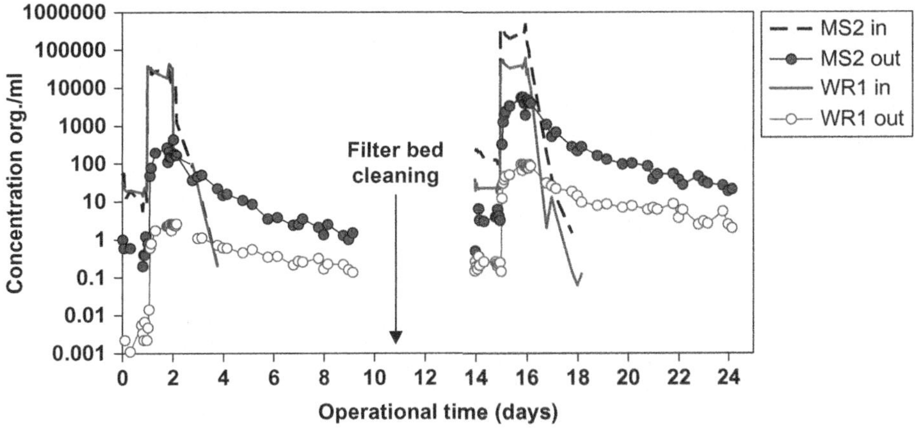

Figure 7.1 The concentrations of bacteriophage MS2 and E. coli WR1 in the influent and effluent of SSF2 during two separate spiking experiments with 24 hours of low and 24 hours of high concentrations; between these experiments, on day 11, the filter bed surface was cleaned (Schmutzdecke was scraped off)

Table 7.2 The Decimal Elimination Capacity ($\log_{10}(C_{in}/C_{max,\ out})$) of the MS2 bacteriophages and E. coli (WR1) of SSF1 and SSF2 of the pilot plant of AWS at 10°C

Filter	Time (days) after schmutzdecke removal	Bacteriophage MS2		E.coli WR1	
		Low C_{in}[a]	High C_{in}	Low C_{in}	High C_{in}
SSF1	12	1,7	1,8	2,1	2,3
SSF2	81	1,8	2,2	3,9	4,2
SSF2	4	1,7	1,9	2,0	2,8

[a]C_{in} = *influent concentration*

A 3 $\log_{10}$ removal of polioviruses by SSF was found by Windle-Taylor (1969). Poynter and Slade (1978) reported a DEC of 3.5 and 2.8 log for MS2 phages and polioviruses. Without a schmutzdecke WR1 removal was 2.3 ± 0.4 log. This value is close to MS2 removal by the same filters and *E. coli* removal observed in full-scale filters and in literature.

COLUMN EXPERIMENTS

Due to the relative large scale and possible health risk in the pilot plant, removal of oocysts of *Cryptosporidium* has been determined in columns packed with ripened sand sampled 20 cm from the top of the filter bed (without schmutzdecke; worst case). Main objective of this part of the study was to compare the elimination of bacteria, viruses and protozoan oocysts under the same conditions; thus MS2 bacteriophages, *E. coli* WR1 and spores of *Clostridium perfringens* D10 were co-injected with *C. parvum*. The influent concentrations of the micro-organisms ranged from 1.0×10^3 to 1.7×10^5 per ml.

The oocyst concentration in the influent of 10^3–10^4 per ml were removed completely (<1 per 200 ml) by both columns with AWS and DWS sand. This resulted in a DEC of at least 5–6 log (Table 7.3). In literature DEC values of SSF for protozoan (oo)cysts assessed with spiking experiments were in the same order of magnitude. For *Giardia* cysts 4–6 log and for *Cryptosporidium* oocysts >4,5 – >6.5 log were described (Bellamy *et al.*, 1985; Schuler *et al.*, 1991 and Timms *et al.*, 1995). Fogel *et al.* (1993) determined concentrations of indigenous *Giardia* and *Cryptosporidium* (oo)cysts in the influent and effluent of a full-scale plant over a period of 1.5 year. From these data a much lower DEC value of 1.2 and 0.3 log for *Giardia* and *Cryptosporidium*, respectively, was calculated.

Table 7.3 The Decimal Elimination Capacity ($\log_{10}(C_{in}/C_{max\ out})$) of the columns compared to DEC assessed for full-scale filters

	E.coli **Coli44**[a]	**MS2**	***C. perfringens*** **SSRC**[a]	***C. parvum***
Column AWS1; 0.4 m	0.2	0.4	ns[b]	ns[b]
Column AWS2; 0.4 m	0.3	0.2	2.3	>5.3
Full-scale AWS; 1.5 m	2.0[c]	1.7	−0.2–1.8[d]	nd[a]
Column DWS; 0.4 m	0.4	0.1	3.2	>6.5
Full-scale DWS; 1.0 m	2.5[e]	nd	−0.2–1.8[d]	nd

[a]indigenous Coli44 and SSRC; [b]ns = no spiking; nd = not determined; [c]average DEC for Coli44 from pilot plant filters in Table 7.2; [d]range of values for SSRC from Table 7.1; [e]average DEC for Coli44 from full-scale filters in Table 7.1

Figure 7.1 shows that MS2 phages and *E. coli* were hardly removed by the AWS2 column (Figure 7.2), a result also observed in column AWS1 and DWS (Table 7.3). MS2 and WR1 results of the AWS columns showed that co-injection with the other organisms did not affect the elimination of MS2 and WR1. The columns removed spores of *C. perfringens* 2–3 log and more efficient than MS2 and WR1 but less efficient than *C. parvum*. This indicates that SSRC is not a suitable quantitative surrogate for *Cryptosporidium* oocyst removal by SSF. Based on comparative spiking studies a similar conclusion was reported by Emelko (2001) for aerobic spores as a surrogate for oocyst removal by rapid granular filtration.

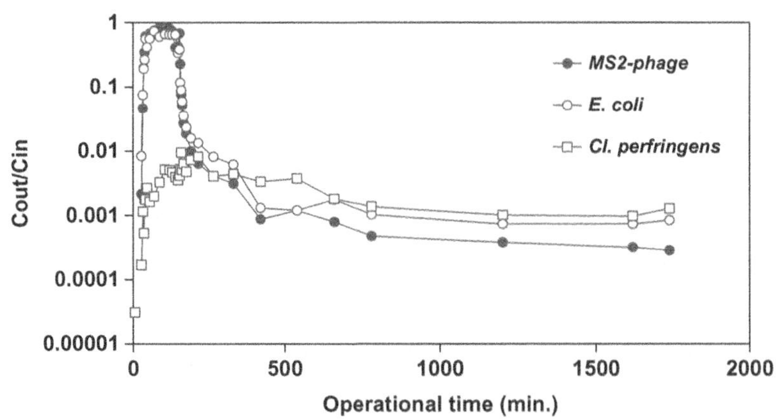

Figure 7.2 Breakthrough curves from column AWS2 (no oocysts of C. parvum detected in filtrate)

The column with DWS sand (grain size of 0.3–1.8 mm) showed more elimination of the bacteria and bacterial spores than the column AWS2 with finer sand (0.13–0.37 mm; Table 7.3). This difference was possibly caused by a higher Carbon and iron oxyhydroxide content in the sand. In the DWS sand concentrations of these components were 0.055% (dry weight) and 2.4 mg/g, respectively, and in the AWS sand <0,002% and 0.027 mg/g, respectively. Thus DWS sand had a larger adsorption capacity, which indicated that attachment plays a significant role in the elimination of bacteria.

In Table 7.3 DEC of the column experiments (0.4 m of ripened sand) were compared with DEC assessed in the pilot and full-scale filters of AWS (1.5 m of filter bed) and DWS (1.0 m of filter bed). This revealed that the removal of MS2-phages and WR1 per unit length in the columns was much less than observed in the slow sand filters. This may be attributed to the fact that ripened sand from the schmutzdecke of the sampled filters have not been included in the columns to investigate elimination under worst case scenario. Moreover, also the higher porosity of the fresh packed columns of about 0.40 compared to the porosity of the ripened filter bed SSF1 of the pilot plant of 0.27 may have contributed to the observed difference in removal capacity.

Although scaling up of the column results to full-scale conditions was not possible, the resemblance of the breakthrough curves of MS2 and WR1 under both conditions substantiates extrapolation of the relative results to full-scale filters. It was concluded that SSF must be very effective eliminating peak concentrations of persistent oocysts of *C. parvum* and spores of *C. perfringens*. However, the low and variable DEC of full-scale filters for SSRC and also for oocysts found by Fogel *et al.* (1993) suggested that overall elimination of both biological particles by SSF assessed over a long period of time is influenced by accumulation and retarded transport. Destructive sampling of the columns for analysis of retained micro-organisms (Schijven *et al.*, 2003) demonstrated that the most significant removal mechanism for the small MS2 bacteriophages was adsorption. The larger *C. perfringens* spores and *E. coli* were removed both by adsorption and physical straining and for the oocysts of *Cryptosporidium* physical straining was the main removal mechanism. From the latter observation one can conclude that elimination in the filter bed is irreversible and retarded transport as proposed for SSRC is not possible. However, unpublished data showed that oocysts might be transported deeper into sandy soil due to changes in pH and conductivity. Also changes in hydraulic conditions may attribute to transport of oocysts through sand beds (Emelko, 2001). The same author emphasised the need for further research to elucidate the significance of seeding experiments with high concentrations as a tool to assess DEC of full-scale filters for *Cryptosporidium*.

CONCLUSIONS

Based on the results of this study with *E. coli* (Coli44) and MS2-bacteriophages DEC of slow sand filters for bacteria and viruses is quantified at 2–3 and 1.5–2 log, respectively. One log higher was the elimination of indigenous *Campylobacter* bacteria by full-scale filters, which indicated that Coli44 is a conservative surrogate for these pathogenic bacteria. MS2-bacteriophages can be regarded as a conservative surrogate for virus removal (low attachment and survival rate). Moreover, viruses are the most critical micro-organisms for the performance of slow sand filters because they are retained the least. The column experiments showed that SSF will have a large efficacy in eliminating peak concentrations of persistent micro-organisms like spores of sulphite-reducing clostridia (>2–3 log) and (oo)cysts of *Cryptosporidium* and *Giardia* (>5 log). This may be more in full-scale filters with a schmutzdecke. On the basis of these results SSRC is not a good quantitative surrogate for protozoan oocyst removal by SSF. Further research is needed to elucidate if and to what extent accumulation and retarded transport, the suggested phenomenon for the low and variable removal of indigenous SSRC by full-scale filters, will affect elimination of oocysts of *Cryptosporidium* in the long run.

Additionally, this study showed that filters with a schmutzdecke have 1–2 log greater capacity to eliminate bacteria, whereas elimination of viruses is not affected by the schmutzdecke. This positive effect of the schmutzdecke on elimination will also count for protozoan oocysts, because straining is an important removal mechanism for these organisms. Furthermore, results also suggested that charged (in)organic components accumulated in the sand might have a significant effect on the removal efficiency of organisms removed by adsorption.

Chapter 8

Removal and fate of *Cryptosporidium parvum*, *Clostridium perfringens* and small-sized centric diatoms (*Stephanodiscus hantzschii*) in slow sand filters

Wim A.M. Hijnen[a], Y.J. Dullemont[b], J.F. Schijven[c], A.J. Hanzens-Brouwer[a], M. Rosielle[d] and G.J. Medema[a]

[a]KWR Watercycle Research Institute; [b]Waternet; [c]National Institute of Public Health and the Environment; [d]Het Waterlaboratorium

ABSTRACT

The decimal elimination capacity (DEC) of slow sand filtration (SSF) for *Cryptosporidium parvum* was assessed to enable quantitative microbial risk analysis of a drinking water production plant. A mature pilot plant filter of 2.56 m^2 was loaded with *C. parvum* oocysts and two other persistent organisms as potential surrogates; spores of *Clostridium perfringens* (SCP) and the small-sized (4–7 μm) centric diatom (SSCD) *Stephanodiscus hantzschii*. Highly persistent micro-organisms that are retained in slow sand filters are expected to accumulate and eventually break through the filter bed. To investigate this phenomenon, a dosing period of 100 days was applied with an extended filtrate monitoring period of 150 days using large volume sampling. Based on the breakthrough curves the DEC of the filter bed for oocysts was high and calculated to be 4.7 log. During the extended filtrate monitoring period the spatial distribution of the retained in the filter bed was determined. These data showed little risk of accumulation of oocysts in mature filters most likely due to predation by zooplankton. The DEC for the two surrogates, SCP and SSCD, was 3.6 and 1.8 log, respectively.

On basis of differences in transport behaviour, but mainly because of the high persistence compared to the persistence of oocysts, it was concluded that both spores of sulphite-reducing clostridia (incl. SCP) and SSCD are unsuited for use as surrogates for oocyst removal by slow sand filters. Further research is necessary to elucidate the role of predation in *Cryptosporidium* removal and the fate of consumed oocysts.

Edited by Wim A.M. Hijnen. ISBN: 9781843393764. Published by IWA Publishing, London, UK.

INTRODUCTION

One of the pathogens of major concern for the drinking water industry is *Cryptosporidium*, a persistent pathogenic protozoan and cause of a number of outbreaks of waterborne diarrhoea documented in the USA and the UK (Richardson *et al.*, 1991 and MacKenzie *et al.*, 1994). Studies have demonstrated the failure of regular water quality monitoring with *Escherichia coli* to indicate the absence of this pathogen in drinking water (Harwood *et al.*, 2005). This shortcoming of current quality control is overcome by the use of Quantitative Microbial Risk Assessment (QMRA) to define the microbiological safety of drinking water (Haas, 1983 and Medema *et al.*, 2003). In the new Dutch Drinking Water Decree 'Waterleidingbesluit' (Anonymous, 2001) a maximum acceptable annual infection risk of 10^{-4} has been introduced for pathogens of fecal origin including *Cryptosporidium*. The required risk level has to be demonstrated with QMRA which is based on dose-response relation in human volunteers and exposure assessment. For exposure assessment quantitative knowledge about the presence of pathogenic micro-organisms in the source water and the capacity of water treatment processes to eliminate these micro-organisms is required, along with data on drinking water consumption.

Slow sand filtration is one of the oldest water treatment processes used to produce microbiologically safe drinking water. Quantitative information of how effective these filters are in removing pathogens, however, turned out to be limited (LeChevallier and Au, 2004). It is only since the last part of the 20th century that some studies were published on the elimination capacity for protozoan (oo)cysts (Bellamy *et al.*, 1985; Schuler *et al.*, 1991; Fogel *et al.*, 1993 and Timms *et al.*, 1995). In most of these studies the actual Decimal Elimination Capacity (DEC; log removal) of slow sand filters for *Giardia* and *Cryptosporidium* (oo)cysts was assessed by dosing peak concentrations, resulting in high DEC-values of 4 to more than 6 log in matured and relatively small sized filter beds. In contrast with these findings, one study (Fogel *et al.*, 1993) determined removal of environmental *Giardia* and *Cryptosporidium* (oo)cysts by full-scale filters by prolonged sampling and showed breakthrough during periods with low temperature ($<5^{\circ}C$) resulting in low average DEC-values of 1.2 and 0.3 log, respectively. In the Netherlands, slow sand filters are operated as the last stage of a multiple barrier treatment. Hence, assessment of the DEC for environmental protozoan (oo)cysts is not feasible. The removal of spores of sulphite-reducing clostridia (SSRC; including *Clostridium perfringens*) as a surrogate for protozoan (oo)cysts (Payment and Franco, 1993 and Hijnen *et al.*, 1997) by full-scale filters was monitored (Hijnen *et al.*, 2000b; 2004d). Results showed low DEC for SSRC, ranging from −0.2 to 1.8 log. Direct comparison of the removal of dosed spores of *C. perfringens* and oocysts of *C. parvum* in a short sand column (0.5 m) showed 2.3–3.2 log removal of the spores and more than 5 log removal of the oocysts, indicating that environmental SSRC is a conservative surrogate for protozoan (oo)cysts (Hijnen *et al.*, 2004a). However, the clear discrepancy between the high removal of dosed spores and the low removal of SSRC by mature full-scale filters sand casts doubts on the validity of translating column study results to the larger full-scale filters. Because spores are very persistent (Medema *et al.*, 1997), it was hypothesized that spores accumulate in the filter bed and they might detach which causes remobilization and delayed breakthrough. This phenomenon was first suggested as cause for the relative low removal of SSRC by granular activated carbon filters (GAC) with high SSRC concentrations in the filter bed and the back wash water (Hijnen *et al.*, 1997). Additional observations supporting this hypothesis have been described in literature for protozoan-sized microspheres in soil (Harvey *et al.*, 1995), 1 μm fluorescent microspheres in GAC and expanded clay filters (Persson *et al.*, 2005), aerobic spores in GAC filters (Mazoua and Chauveheid, 2005) and *C. parvum* oocyst transport in sand columns (Harter *et al.*, 2001; Bradford and Bettahar, 2005 and Hijnen *et al.*, 2005) and *E. coli* and SSRC in the latter study. Furthermore in a column study it was demonstrated that the removal of spores depended on the duration of the seeding (Schijven *et al.*, 2003). As with SSRC, oocysts of *C. parvum* are also persistent, but oocysts are larger by a factor of 3–4 and have potentially different surface properties. As a consequence, it remains unknown whether the phenomenon of reduced elimination by sand filters over time due to accumulation could apply to these pathogens as well.

To investigate this remaining question a study was conducted with the objective of describing and quantifying the removal, accumulation and delayed breakthrough of *C. parvum* oocysts in a slow sand filter of some extent ($2.56\,m^2$) operated under full-scale conditions mimicking as much as possible long term loading conditions. Spores of *C. perfringens* as well as a small-sized algae, the centric diatom *Stephanodiscus hantzschii*, were dosed simultaneously to compare removal of different kind of persistent micro-organisms and to investigate the potential use of both organisms as surrogate for the assessment of protozoan oocysts removal by slow sand filtration (Hijnen *et al.*, 1997 and Akiba *et al.*, 2002).

CONDITIONS OF A LONG TERM DOSING EXPERIMENT

Slow sand filter and the experimental set up. Slow sand filtration in the water treatment Leiduin of Water Company Waternet is the last stage in a treatment train with rapid sand filtration, ozonation and granular activated carbon (GAC) filtration. The source water is pre-treated water from the River Rhine after dune passage and regained by an open collecting system. The filter used for this study was part of a pilot plant, a dummy system of the full-scale treatment. The filter has been in operation for 3 years without surface scraping at a flow rate of $1\,m^3h^{-1}$ and a filtration rate of $0.3\,m\,h^{-1}$ prior to carrying out the present study. The influent, GAC filtrate, had a low DOC ($1.5\,mg\,C\,l^{-1}$) and turbidity (0.1 FTU), a pH of 8.0 and temperature ranging from 8.2 to 18.8°C. The filter bed ($2.56\,m^2$ and depth of 1.5 m) was filled with sand (d50 = 0.28 mm; porosity of 0.41) on top of a support of Nicolon cloth (0.7 mm) and 0.4 m gravel pack. The experimental set up is presented in Figure 8.1 and will be described in detail hereafter.

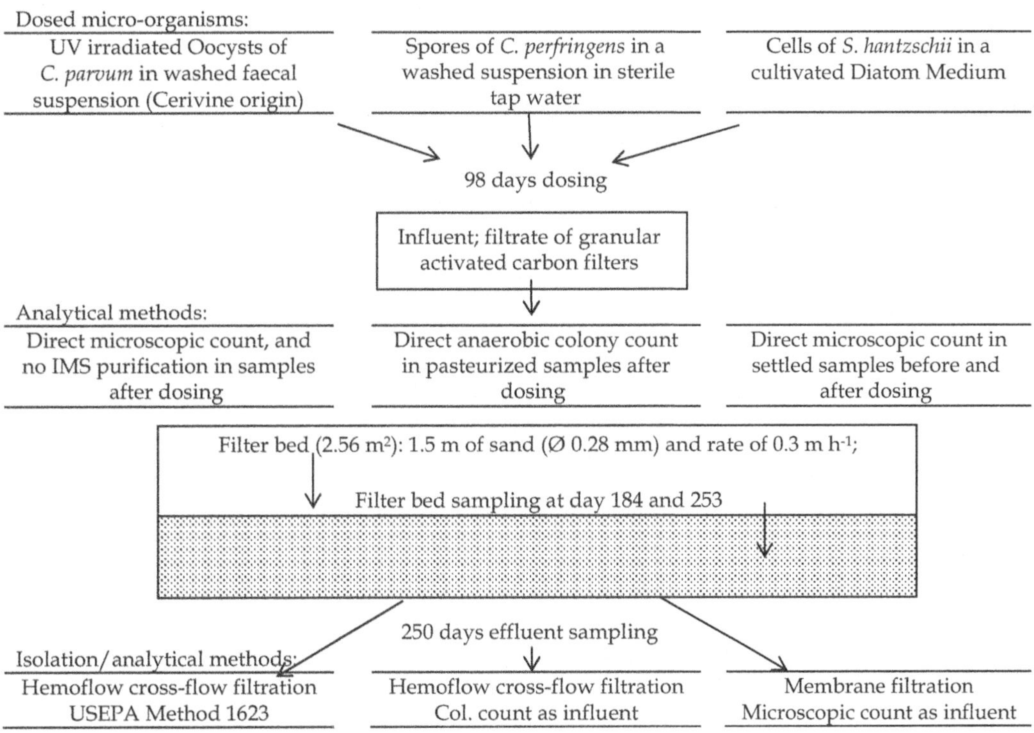

Figure 8.1 Diagram of the experimental set up

Cryptosporidium oocysts. *C. parvum* oocysts suspensions in watery faeces (Cervine origin; Moredun scientific Ltd., Midlothian, UK; 2×10^9) were used. Each tube of suspension was purified by filtering through plankton netting with a mesh width of successively 500, 300 and 100 µm, the netting was flushed with sterile water which was mixed with the filtrate. This suspension was mixed during 30 minutes (CAT S50) and subsequently settled during 30 minutes. The water was decanted and the sediment was flushed twice with sterile water. The accumulated water suspension was centrifuged 10 minutes (4000 RPM) and the pellet was concentrated to a volume of approximately 100 ml with an estimated number of 4.5×10^8. This pre-treated suspension was divided into three portions each dosed during one week to the influent. For safety reasons the oocyst suspensions were inactivated with UV in a collimated beam apparatus (calculated UV fluence of $\geq$10 mJ.cm^{-2}) prior to the dosage.

Clostridium perfringens. A suspension of spores of *C. perfringens* (SCP strain D10; an isolate from a patient suffering from diarrhoea caused by food infection) was prepared with a total number of spores of 2.2×10^9 following a previously described method (Hijnen *et al.*, 2002). The level of sporulation of the suspension was 100% and tested by enumeration on Perfringens-agar-base plates (PAB Oxoid CM587) with and without pasteurization for 15 minutes at 60°C.

Stephanodiscus hantzschii. *S. hantzschii* is a small-sized centric diatom (SSCD) with an average size of 5 µm. SSCD-cells were obtained from a pre-cultured strain of the Culture Collection of Algae and Protozoa (CCAP 1079/4; Dunstaffnage Marine Laboratory, Argyll, UK, Th. Pröschold). The strain was inoculated into a gently stirred 150 ml sterilized Diatom Medium (DM) for fresh water algae (Beakes *et al.*, 1988) and incubated for 2–3 weeks at 20 ± 2°C with a light-dark cycle of 12:12 hours (white light intensity of 50 µE m^{-2} s^{-1}). The growth and conditions of the algae were monitored regularly by light microscope. The concentration of cells in the final suspension of $\pm 1.5 \times 10^8$ ml^{-1} was determined using a haemocytometer.

Dosage procedure. Appropriate volumes of the pre-treated oocyst, the *C. perfringens* and *S. hantzschii* suspensions were added to 150 l of drinking water in a 200-l stainless steel (SS) vessel with cover. This diluted and stirred suspension, which was freshly prepared every week, was dosed continuously into the mains of the influent of the filter at a flow rate of 13 ml min^{-1} during the first two weeks. Because in this vessel biofouling occurred, from week three the oocysts suspension was dosed (0.7 ml min^{-1}) separately from a separate 20 l suspension in drinking water prepared fresh weekly in a 30-l SS vessel with cover and stirrer, cooled in a container with melting ice. Dosing of the SCP suspension was also from this cooled 20 l vessel after seven weeks. The total dosage period was 98 days (14 weeks) in which dosage failed for SCP and SSCD in week 4 and for oocysts in week 5 (tubing not correctly connected; actual dosage period was 91 days).

Large volume sampling. Large volume sampling was considered necessary to determine organism concentration in the filtrate. Water volumes of 1,000 l or more were concentrated on-site using a Hemoflow cross-flow ultrafilter (HF80S; Fresenius, Germany; Simons *et al.*, 2001). The water volumes sampled with the Hemoflow were concentrated to a total volume of 800 ml HF-concentrate. Approximately 5% of the HF-concentrate was used to determine the number of SSRC (including *C. perfringens*); the residual HF-concentrate was used to enumerate the number of oocysts.

The concentration of small-sized centric diatoms (SSCD) in large volumes of effluent was determined using the Membrane Filtration or MF-sampler (Hijnen *et al.*, 2000a). Algal cells were isolated on poly-carbonate membrane filters (2 µm; Sartorius) and re-suspended in tap water by 2 minutes of ultra-sonic treatment (low energetic; Branson 5200, Danbury, USA). The recovery of this method was 53–100% (n = 6) determined with chemically purified centric diatoms in the size range of 6–7 µm (M. Rosielle, personal communications).

Sampling program. The filter bed load of *C. parvum* oocysts and spores of *C. perfringens* as SSRC, was determined weekly or more by sampling the influent of the filter as well as the water volume above the filter bed. The concentration of *S. hantzschii* dosed to the influent was determined as SSCD in the water above the filter bed. Because of the relative high background concentration of SSCD in the influent of the pilot plant, the SSCD concentration was also determined before the dosage point of *S. hantzschii*. The effluent of the slow sand filter was sampled weekly during the dosing period and a period of 22 weeks after the dosage stopped to determine delayed breakthrough.

Analysis. Oocysts of *Cryptosporidium* in the influent and the HF-concentrate were enumerated with the EPA method 1623 (USEPA, 2001). Recovery of the HF-concentrate was tested by addition of 5 ml Colorseed™ (BTF Decisive Microbiology, North Ryde, NSW, Australia) suspension (99 oocysts ±1.3). The HF-concentrate was purified by immuno-magnetic separation (IMS) prior to labeling and microscopic counting. The adsorbed oocysts were eluted from the beads with HCl (0.1 N) and this suspension was neutralized with KOH (1 N). The concentrate was collected on a membrane filter (Millipore, 1,2 μm RTTP) and labeled with FITC (2 hours at 37°C). The membrane was microscopically counted with epi-fluorescence microscopy (Leica DM RXA, magnification 312.5).

The concentration of spores of sulphite-reducing clostridia (SSRC, including *C. perfringens*) was determined either directly or after concentration by membrane filtration or HF-filtration. All samples were pasteurized before enumeration. In the first weeks SSRC in the effluent was measured by membrane filtration (1 or 10 l; (Ø47 mm, 0.45 μm pores; Sartorius 13906-50-ACN) and pasteurization of the membranes (30 min. at 70 ± 1°C; Dutch NEN6567). Given that no SSRC were detected in these samples, SSRC in the effluent was subsequently monitored in the HF-concentrate. Fifty ml of this concentrate was pasteurized (15 min. at 60 ± 1°C; ISO-method 6461/2-1986 for *C. perfringens* spores) prior to cultivation in PAB medium (Oxoid CM 587) under anoxic conditions during 48 ± 4 hour at 37 ± 1°C. A real-time PCR method was used to confirm that the SSRC colonies observed in the PAB medium were spores of *C. perfringens*. Cell material was isolated from the black colonies, suspended in 1 ml sterile water and subsequently tested with the real-time PCR for DNA of *C. perfringens*. The primers which codes with A-toxine gen of *C. perfringens* (Yoo *et al.*, 1997) were used to identify the *C. perfringens*.

For enumeration of the breakthrough of cells of *S. hantzschii* the concentration of SSCD with specific cell morphology and size (4–7 μm) is determined microscopically in concentrates of the influent and the effluent, prepared by sedimentation. Water samples were pretreated with lugol-solution until the sample was light orange to stop biological activity. These samples were pored into Hydro-bios sedimentation chambers (Hydro-bios Apparatenbau, Kiel-Holtenau, GMB) with appropriate size (10–50–100 ml) and settled during 24–72 hours. Water was decanted from these tubes and the sediment was examined microscopically (Olympus IX70, Olympus, Zoeterwoude, NL; magnification – 400×).

Enumeration of retained micro-organisms. After the dosage of micro-organisms was stopped, delayed breakthrough of the dosed micro-organisms was monitored. After 184 days, when oocyst concentration in the filtrate was below the detection limit (DL) of the analysis, filtration was stopped and the water volume above the filter bed was lowered to the surface. At two locations the saturated filter bed was scraped with a scoop to sample the Schmutzdecke (approximately 2–3 mm) and the following 50 cm of the saturated filter bed was sampled with a soil core sampler (veenboor; Eijkelkamp, Giesbeek, The Netherlands). This was repeated after 253 days where sand from the first 5 cm and from layers deeper than 50 cm was collected. At first the presence of *C. parvum* oocysts in the upper part (0–1 cm) was monitored microscopically by fixing sand grains on a Dynal slide. The grains were labeled with FITC (2 hours at 37°C) and completely scanned for oocysts. To determine the concentration of retained micro-organisms in the sand, 3–4 gram of the sampled sand from the two locations in the filter bed was suspended in 100 ml sterile tap water and treated for 2 minutes in a low energetic ultra-sonic cleaner (Bransonic 5510, Branson ultrasonic, Danbury, USA). The number of micro-organisms eluted from the sand was enumerated in this sonicated sand suspension (UTSusp.) with the analytical methods described above. Oocysts and spores in the Schmutzdecke material and the sand of the first cm in the filter bed (0–1) were examined in two separate sub-samples of 3–4 gram. A third sub-sample was treated ultrasonically in 100 ml sterile Laureth-buffer (USEPA, 2001) to verify the elution efficiency of the ultra-sonic treatment for oocyst enumeration. Fifty ml of the UTSusp. was used to enumerate oocysts and spores and the other 50 ml was used to enumerate the SSCD with the methods previously described. Microscopic counting was used to enumerate SSCD in the UTSusp. and because these suspensions had a high content of suspended solids, the collected concentrations must be seen as an indication of the order of magnitude of centric diatoms in the filter bed. Statistical differences between concentrations of retained micro-organisms in the filter bed were analyzed with the Student t-test or the Wilcoxon-test using SPSS (14.0).

ATP and zooplankton analysis in the sand. The concentration of adenosine tri-phosphate (ATP) in the sand was measured (Magic-Knezev and van der Kooij, 2004). The presence of zooplankton in the sand was measured in separate sand samples of approximately 500 gram taken from the filter bed at the end of the operational time. 30 gram of sand was mixed intensively in tap water. The invertebrates were separated from the sediment by separation in a $MgSO_4$ solution in tap water ($49\,g\,l^{-1}$). The sand slurry was tested for the presence of larger invertebrates by filtration over $500\,\mu m$ sieve and subsequently pored in an Anderson glass tube (Anderson, 1981) filled with the $MgSO_4$ solution. After 15–20 minutes of sedimentation the zooplankton sample (± 800 ml) was taken at the upper sampling port and subsequently at the lower sampling port. The suspensions were sieved subsequently through a $30\,\mu m$ sieve and loaded in a counting chamber for microscopic examination (Olympus IX70, Olympus, Zoeterwoude, NL; magnification $100\times$) to count the number and identify the zooplankton in the sand.

Elimination of the micro-organisms and mass balance. The Decimal Elimination Capacity (DEC) of the filter bed for the tested micro-organisms was calculated from the average concentration in the influent $\overline{C}_{in}$ and the effluent $\overline{C}_{out}$, the latter calculated by

$$\overline{C}_{out} = \frac{\sum N_{out}}{\sum V_{out}} \overline{R}_{out} \tag{1}$$

where N_{out} is the number of oocysts counted, V_{out} the sampled volume (l) and R_{out} the recovery in the tested samples.

$$DEC = \log_{10} \frac{\overline{C}_{in}}{\overline{C}_{out}} \tag{2}$$

DEC was also calculated from the mass balance by the following equation

$$DEC_m = \log_{10} \left[\frac{M_d}{M_e} \right] \tag{3}$$

where DEC_m is the elimination capacity on the basis of the mass balance, $M_d = t_d \overline{C}_{in}$ and $M_e = \sum_{i=1}^{n} t_i \frac{C_{out,i} + C_{out,i+1}}{2}$ the total number of micro-organisms dosed to the influent and found in the effluent, respectively, t_d is the time period of dosing (hours) and dt_i is the i-th time interval of n intervals between two successive samples with concentrations $C_{out,i}$ and $C_{out,\,i+1}$.

The mass of micro-organisms (numbers) accumulated in the filter bed M_b after an operational time of 184 and 253 days is derived from

$$M_b = \sum M_{l,n} = \sum (\overline{C}_{l,n}\, 1000)(d_{l,n} A 1000) \tag{4}$$

where $M_{l,n}$ is the total number of micro-organisms and $\overline{C}_{n,l}$ the average concentration of micro-organisms ($N\ ml^{-1}$) in n layers (l) with a thickness of $d_{l,n}$ (m) of the filter bed, respectively with a surface area A of $2.56\,m^2$. $M_{l,n}$ in the un-sampled layers was calculated from the average concentration extrapolated from the concentrations of the next upper and lower layers. Because of the inaccuracy of the concentrations in the filter bed, mass balance calculations was not done for SSCD.

Calculation of Collector and Sticking Efficiencies. Assuming elimination in the filter bed was only due to attachment and detachment, the sticking efficiency of the spores and oocysts was calculated from the colloid filtration model (Yao *et al.*, 1971) described with the equation

$$LN\frac{C}{C_0} = -\frac{3}{2}\frac{(1-\theta)}{d_c}\alpha\eta L \tag{5}$$

where d_c is the diameter of the collector, α the sticking efficiency, η the single collector collision efficiency, and L the length of the column. η was calculated with the optimized equation presented by Tufenkji and Elimelech (2004a). For the calculations the following parameters values were used: bulk water density $999.703\,kg\,m^{-3}$; Hamaker constant for bacterium glass water interface $6.2 \times 10^{-21}\,J$ (Rijnaarts *et al.*, 1995); sizes (m) of spores of *C. perfringens* 1.5×10^{-6} and oocysts of *C. parvum* 4.9×10^{-6} (Medema *et al.*, 1998); and ρ_p *C. perfringens* 1270 (Tisa *et al.*, 1982) and oocysts $1045\,kg.m^{-3}$, respectively (Medema *et al.*, 1998).

REMOVAL OF THE OOCYSTS, SPORES AND DIATOMS

Cryptosporidium removal. Oocysts of *C. parvum* were dosed to the GAC filtrate, the influent of the SSF. Based on the low concentration of environmental oocysts observed in the source water during the experiment (routine monitoring data, not presented) and the preceding removal in the rapid sand filter and GAC-filtration, the background concentration of oocysts in the influent was negligible. The oocyst concentration measured in the influent after dosage and in the water above the filter bed was constant with an average concentration of 314.6 ($\pm$161) $N\,l^{-1}$ (Table 8.1; average recovery of 70.6 ($\pm$13) %; n = 25).

Table 8.1 The concentration of dosed Cryptosporidium oocysts, spores of *C. perfringens* (SCP) and environmental and small-sized centric diatoms SSCD (4–7 μm; incl. *S. hantzschii*) in the influent and effluent of the slow sand filter

Samples	n[a]	n+[a]	Sampled Vol. (l)	Detected organisms	Conc. n/l	STD[c]
Oocysts in[b]	33	33	3.3	733	314.6	161
Oocysts out[b]	22	7	26.613	18	0.0016	ND[c]
SCP in	20	20	2.0	2270	1135.0	1014.8
SCP out	21	16	2.363	336	0.142	0.509
SSCD in BG[d]	21	17	1.51	140	92.9	889.4
SSCD in	17	17	0.98	274	279.6	92.8
SSCD out	29	29	25.94	187	7.22	8.46

[a]n and n+, number of samples and number of positive samples; [b]corrected for recovery; [c]STD = standard deviation; ND = not determined; [d]Background concentration of environmental SSCD (oocysts and SCP below detection limit)

During the dosing period of 98 days and an additional period of 163 days after dosing had stopped, $26\,m^3$ of the filtrate was examined for oocysts. After the first 29 days and during a period of approximately 50 days, 14 oocysts were detected in approximately 50% of the samples of the filtrate (Figure 8.2).

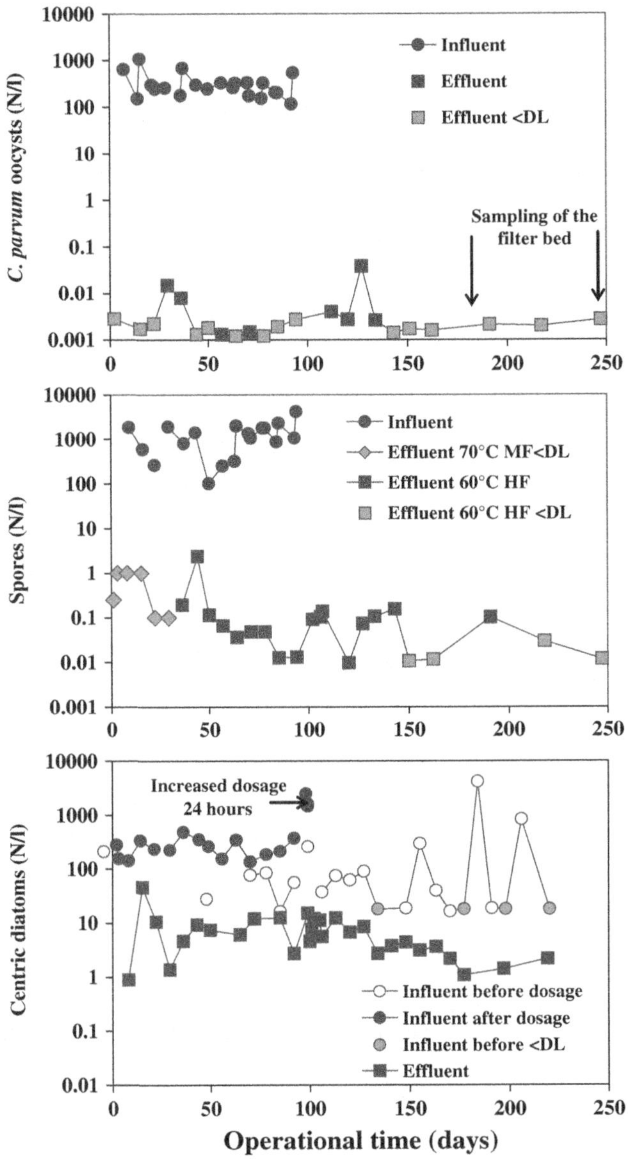

Figure 8.2 Concentration of *C. parvum* oocysts, spores of *C. perfringens* and small sized centric diatoms (SSCD) in the influent and the effluent of the slow sand filter (DL is detection limit)

A second shorter period (25 days) of oocyst breakthrough was observed shortly after dosage was stopped. From the total number of 18 oocysts detected in the volume sampled during the total monitoring period of 247 days an average concentration of 0.0016 oocysts l^{-1} was calculated (Table 8.1; average recovery of 43.6 ($\pm$15.8)%; n = 22) resulting in a DEC-value of 5.3 log.

Removal of spores of sulphite-reducing clostridia. Also for SSRC the background concentration of environmental SSRC in the influent was neglected (data derived from routine sampling and not presented). The average concentration of spores of *C. perfringens* (SCP; confirmed with molecular typing on a few samples) was 1135 (±1015) N l^{-1} (Table 8.1). The temporary decline in influent concentration after 40 days (Figure 8.2) was solved by switching dosage from the large 150-l vessel used for *S. hantzschii* dosage to the smaller and permanently cooled 20-l vessel used for *C. parvum* oocysts. During the first weeks no SCP were observed in effluent sample volumes of 1 and 10 l and breakthrough of these spores was detected starting from day 36 and further (Figure 8.2). The highest breakthrough of SCP was at day 44, 15 days after the first highest oocyst breakthrough (Figure 8.2). Based on the total number of SCP observed in sampled effluent volume of 2.363 l the average concentration was 0.142 (±0.509) SCP l^{-1} (Table 8.1) resulting in a DEC-value of 3.9 log for these organisms

Removal of small-sized centric diatoms. Compared to the *Cryptosporidium* and SCP, SSCD (4–7 μm) were present in higher and more variable concentrations in the source water of this treatment plant. Concentrations ranged from <1 up to $5.4 \times 10^5 l^{-1}$. In early spring or late autumn diatom bloom in the source water reservoir caused peak concentrations. The average SSCD concentration in the source water was a factor of approximately 200 higher than the concentration of SCP and was reduced only slightly (0.4 log) by the rapid sand filters (data not presented). Because of these high levels of SSCD in the source water, SSCD were also determined in the influent before dosage of *S. hantzschii* (Figure 8.2). The concentrations of SSCD in the filter influent were highly variable and only slightly below the level of dosed *S. hantzschii.* Thus, environmental SSCD concentrations contributed 40–50% on average to the total load of the test filter with these micro-organisms (Table 8.1) and could not be separated from cells of *S. hantzschii* in the microscopically enumeration method. The results of SSCD monitoring in the effluent showed a clear breakthrough (Figure 8.2), much larger than observed for *C. parvum* oocyst and *C. perfringens* spores. After an initial high peak in the effluent, breakthrough stabilized. Because of the mixed loading of *S. hantzschii* with environmental SSCD, and the variability herein, the DEC for these micro-organisms was not calculated from the average concentration in the influent and effluent but from the actual concentrations determined at the same day (n = 12). The average DEC was 1.8 ± 0.6 log (range of 0.9–2.6 log).

CONCENTRATION OF MICRO-ORGANISMS IN THE FILTER BED

Concentration of retained micro-organisms. An important objective of the experiment was to monitor delayed breakthrough behaviour of oocysts and its potential surrogates, SCP and SSCD. The filter bed was not disturbed until breakthrough of oocysts was below the detection limit (DL; 0.002 n l^{-1}; Figure 8.2). After 184 days sand samples were taken from the filter bed at two places. This interruption of the filtration process and locally disruption of the filter bed caused no additional oocyst breakthrough (Figure 8.2). The maximum concentration of oocysts per ml sand was 160 (±151, n = 6 and range of 57–426) observed in the top of the filter bed (Figure 8.3a) using the method of elution in sterile water (ultrasonic treated sand suspension; UTSusp.) and declined with increasing depths to a detection limit of 1 oocysts ml^{-1} after 43 cm.

Ultrasonic treatment in 100 ml laureth buffer (LB) applied to the samples of the Schmutzdecke (SM) and the first centimetres did not increase the oocyst recovery from the sand (Figure 8.3b). When the experiment was stopped after 253 days of operation the filter bed was sampled again on two different locations. The concentration of oocysts in the first 5 cm of the filter bed was lower than the concentrations determined at day 184 (Figure 8.3a).

The difference in oocyst concentration observed after 184 and 253 days in the layer 1–5 cm was significant (P = 0.03). Due to the high standard deviations of average concentrations at the top of both samplings the difference was not significant. All oocyst concentrations were corrected individually for the recovery of Colorseed™ seeded in the UTSusp. (24.8 ± 8.1%). The high performance of the elution method in sterile water was also demonstrated by direct microscopic counting of oocysts in the samples. Part of the schmutzdecke samples and sand samples of the first centimetre were examined microscopically for the presence of oocysts. One layer of sand grains was fixed in the well of a Dynal slide with a diameter of 9.26 mm and all fields of

0.39 mm^2 (average grain number of three) were scanned (total of 518 grains on the top and partly the side). No oocysts were detected and on the basis of the average grain size of 0.3 mm which results in a grain number of 70,771 N ml^{-1} of sand and a scan factor of 0.6 (under and part of the sides of the grains could not be scanned) a detection limit of 228 oocysts ml^{-1} of sand was calculated, which was slightly higher than the maximum observed numbers determined by the elution method (Figure 8.3a).

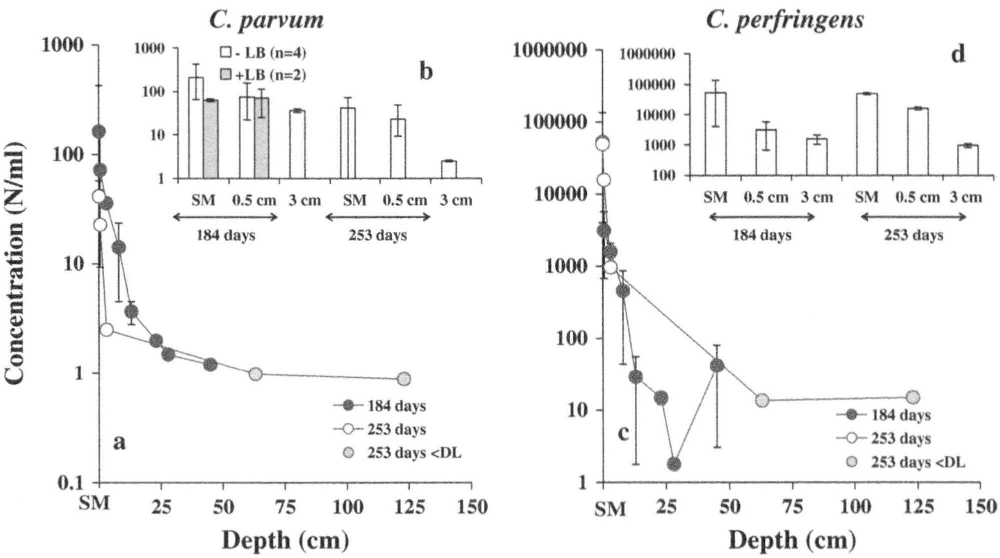

Figure 8.3 The concentration (error bars = range of values; DL is detection limit) of *C. parvum* oocysts and of *C. perfringens* spores over the total filter bed (a and c) and the first 3 cm (b and d) of the filter bed after 184 and 253 days of operation (86 and 157 days after finishing with dosing, respectively) with the additional concentrations of oocysts obtained after laureth buffer elution (+LB; n = 2) during the sampling at day 184 (b)

Appropriate fractions of the UTSusp. were used to determine SCP (1–10 ml) and SSCD (50 ml) concentrations. The SCP concentration in the sand at day 184 was a factor of 100 higher than the oocyst concentration (Figure 8.3a and 8.3c) and declined to 1–10 ml^{-1} in the first 50 cm of the filter bed. The large difference in accumulation in the top of the filter bed between oocysts and spores could not be explained by the difference in influent concentration and observed breakthrough (Figure 8.2). This indicates that there is a difference in fate of the retained oocysts and spores. For SCP the concentrations in the sand sampled at day 253 were in the same order of magnitude as monitored after 184 days (Figure 8.3b). 100% of a total of 90 SCP colonies which were molecularly tested, were identified as *C. perfringens.*

Microscopic counting of SSCD in the UTSusp. was hampered by the high content of suspended solids and thus, the calculated concentrations show the order of magnitude of the presence of these organisms in the filter bed. At day 184, low concentrations of SSCD (<0.3–8.9 N ml^{-1}) were observed in the sand. These concentrations were a factor of 10–1000 below the numbers of oocysts and spores (Figure 8.3) and indicate a relative low adsorption capacity of the sand for SSCD, which agreed well with the high level of breakthrough in this filter (Figure 8.2). At day 253 the estimated concentrations of SSCD in the schmutzdecke material (430–2029 N ml^{-1}) and the sand from the layer of 1–5 centimetre (<0.4–492 N ml^{-1}) were clearly higher than the SSCD concentrations observed at day 184. A relation with the autumn bloom of SSCD in the source water is suspected; day 253 was in the beginning of November. Two of the three peak SSCD concentrations observed in the influent of the filter bed during this period were monitored in between the filter bed sampling at day 184 and day 253 (Figure 8.2).

ATP and zooplankton in the sand. The spatial distribution of ATP in the filter bed showed a higher (micro)biological activity in the schmutzdecke and the first 5 cm of the bed than deeper in the bed (Figure 8.4). ATP concentrations after 184 and 253 days of operation were similar. The determination of zooplankton in the sand of the filter bed revealed high concentrations of *Nematoda, Testacea* (observed species *Centropyxis*, *Cyphoderia* and *Euglypha*), *Oligochaeta*, *Rotatoria* (observed species *Colurella*, *Lecane*, *Lepadella* and other unidentified species), *Nauplii* and *Harpacticoida* and some *Cyclopoida* in the top 5 centimeters of the filter bed (Figure 8.4). Next to these organisms *Ciliata* were detected in lower concentrations at filter depths of 50 and 100 cm.

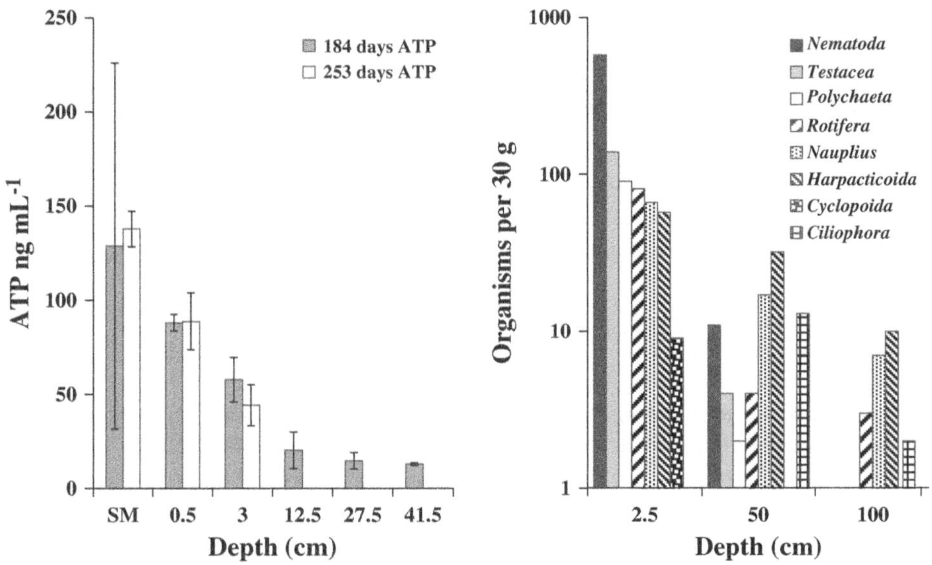

Figure 8.4 The ATP concentration (±SD), after 184 and 253 days of operation (at depth of <50 cm, ATP < 10 ng mL^{-1} of sand) and the zooplankton concentration per 30 gram of sand after 260 days at three depths

MASS BALANCE AND COLLOID FILTRATION THEORY

The mass balance for oocysts and spores (in numbers) was calculated to determine the elimination and accumulation in the filter bed (Table 8.2). From the calculated number of organisms dosed to the influent and passed though the filter bed in the filtrate the DEC_m was calculated. The DEC_m for oocysts and SCP was 4.7 and 3.6 log, respectively, approximately 0.5 log lower than the DEC calculated from the average concentrations. The DEC_m-values are most likely more accurate than DEC calculated from the average filtrate concentration where zero counts are included as zero, thus underestimating the actual average concentration. The accumulated number of *C. parvum* oocysts M_b determined in the filter bed after 184 days was only 1.8% of the total dosed oocyst number and decreased to 0.2% at day 253. For spores of *C. perfringens*, M_b after 184 and 253 days was 32.5 and 45% of the dosed spores, respectively. The difference in M_b between both sampling dates is most likely caused by the variability in SCP concentrations determined on four different locations in the filter bed (Figure 8.3d).

Table 8.2 The mass balance data (dosed organisms (M_d) and organisms in filtrate and filter bed (M_e; M_b) and the colloid filtration parameters (collector efficiency η and collision efficiency α)

	C. parvum	*C. perfringens*
M_d	5.4×10^8	1.9×10^9
M_e	9.8×10^3	5.2×10^5
M_b 184 days[a]	$9.7 \times 10^6 \pm 3.3 \times 10^6$	$7.0 \times 10^8 \pm 4.0 \times 10^8$
M_r 184 days (%)[c]	1.8 ± 0.6	32.5 ± 17.2
M_b 253 days[b]	$1.1 \times 10^6 \pm 2.6 \times 10^5$	$8.9 \times 10^8 \pm 3.5 \times 10^7$
M_r 253 days (%)[c]	0.21 ± 0.05	45 ± 3.3
η	0.0112	0.006
α	0.207	0.310

[a]Number of oocysts and spores in 0.3 m of the filterbed sampled at two locations; [b]Number of oocysts and spores in 0.05 m of the filter bed sampled at two locations; [c]Recovered mass (% of dosed numbers) in filter bed after 184 en 253 days of operation

The collector efficiency η for oocysts in the filter bed of the present study was a factor of two higher than η for spores of *C. perfringens* (Table 8.2). Despite the lower η a higher collision efficiency α was calculated for the removal of the spores, indicating a higher attachment in the filter bed compared to the oocysts.

DISCUSSION

Removal of Cryptosporidium and surrogates. The results of the study clearly demonstrate the high capacity of mature slow sand filters with a well developed schmutzdecke for removal of *Cryptosporidium* from water. During the prolonged loading of the pilot filter very few oocysts passed the filter bed and on the basis of the collected data a Decimal Elimination Capacity DEC_m of 4.7 log was calculated. This removal rate was similar to the removal rates assessed with short term dosing experiments in slow sand filters operated on laboratory and pilot plant scale (Bellamy *et al.*, 1985; Schuler *et al.*, 1991; Timms *et al.*, 1995 and Hijnen *et al.*, 2004a). The present study also showed that slow sand filters have a high capacity to remove spores of *C. perfringens* (SCP; DEC_m of 3.6 log). Consequently, with the used experimental set up low DEC-values for both persistent micro-organisms due to delayed breakthrough could not be demonstrated. The height and duration of the loading with oocysts and spores in the current experiment was probably not enough to achieve sufficient accumulation and consequently increase of concentration in the effluent (Schijven *et al.*, 2003). Moreover, the tested pilot filter was not pre-loaded with environmental oocysts and SSRC prior to the experiment, but a pre-loading with environmental small-sized centric diatoms (SSCD 4–7 μm) is plausible since the influent contained significant amounts of environmental SSCD (Figure 8.2). This might explain the relatively low capacity of the filter bed to remove SSCD (DEC of 1.8 log).

Removal mechanisms and delayed breakthrough. Accumulation and delayed breakthrough of micro-organisms in filters affecting DEC over time is determined by the ruling removal mechanisms. It is not likely that in a fixed filter bed like the slow sand filter, micro-organisms that were removed by straining are remobilized. Delayed breakthrough is restricted to persistent micro-organisms that are removed by reversible attachment and show to some degree detachment. On the basis of the results of this study and observations by others, each of the mentioned processes, straining, attachment and detachment and survival, will be discussed further for the removal of the three tested micro-organisms.

Straining as removal mechanism for biocolloids in filters is expected to depend on the ratio between the colloid and collector size (d_p/d_c) and based on geometric modelling, straining is expected to be of less importance at ratios of <0.05 (Sakthivadivel *et al.*, 1966; 1969) and <0.154 (Herzig *et al.*, 1970). The ratios for oocysts, SCP and SSCD in the pilot filter of the current study are 0.018, 0.005 and 0.014–0.0025, respectively, indicating a minor role of straining. In current literature, however, the significance of straining to colloid removal of the size of bacteria and protozoa is still in debate. One study showed the variability of the role of straining and simple extrapolation of size of grain and microbe is inappropriate (Hijnen *et al.*, 2005). The large difference in removal of SSCD and oocysts, organisms with the same size, in the current study with a slow sand filter containing a well developed schmutzdecke, supports this conclusion. There are studies where straining is presented as a significant removal mechanism in freshly packed columns (Bradford *et al.*, 2003, Bradford and Bettahar, 2005; Tufenkji *et al.*, 2004b and Schijven *et al.*, 2007). The negligible effect of the schmutzdecke on the oocyst removal in slow sand filters observed in other studies (Bellamy *et al.*, 1985; Timms *et al.*, 1995 and Hijnen *et al.*, 2004a) question mark the role of straining in slow sand filtration. In conclusion, straining will have contributed to the overall removal of oocysts, SCP and SSCD in the pilot plant filter with a prolonged pre-filtering period of three years, but is most likely not the dominating removal mechanism.

Attachment and detachment of biocolloids like micro-organisms to sand is described by the colloid filtration theory. The parameter α is the attachment efficiency, calculated from the removal efficiency and the single-collector removal efficiency η, a theoretical parameter based on the physical processes involved in colloid transport and colloid to collector interactions. Assuming complete removal by sorption, the collision efficiency α calculated for the removal of oocysts in the 1.5 m filter bed was 0.207 (Table 8.2). This value is slightly lower than the α values of 0.26–0.37 observed for C. *parvum* oocysts in a column with glass beads of 0.328 mm at approximately the same hydraulic conditions as the present study (Tufenkji and Elimelech, 2005). On the basis of a recent review on *Cryptosporidium*-sand interactions it was concluded that physicochemical filtration plays an important role in oocyst removal in saturated porous media (Tufenkji *et al.*, 2006). The collision efficiency for the removal of spores of *C. perfringens* in the filter bed was 0.310 (Table 8.2) and SCP concentrations at the top were higher compared to the oocyst concentrations (Figure 8.3). The difference in α-value calculated for the removal of spores and oocysts, indicate that attachment of the spores of *C. perfringens* to the sand was more efficient than the attachment of the oocysts of *C. parvum*, an observation described previously (Hijnen *et al.*, 2005). This indicates the different surface properties of spores and oocysts. The zeta-potential of the *C. perfringens* spores in the influent was −32.6 (±2.5) mV (Schijven *et al.*, 2007), whereas for oocysts higher zeta-potentials of −6 mV have been reported by Tufenkji and Elimelech (2005) at pH 8 (similar as pH of the influent of the present study) and also others (Brush *et al.*, 1998 and Dai and Hozalski, 2003). In the top of the filter bed a relatively high Fe-hydroxide (0.55 mg/kg) concentration was determined which will enhance attachment of the mostly negative charged micro-organisms. The low estimated level of SSCD in the filter bed indicates unfavourable surface properties of these micro-organisms for attachment, which is supported by the observed low removal of SSCD in the pilot filter and also in the full-scale rapid sand filters (0.4 log), the first stage of the treatment. Similar low reduction of fluorescent micro algae (0.4–15 μm) of 0.4–1 log in GAC and expanded clay filters was reported in literature (Persson *et al.*, 2005).

Continuation of the breakthrough of oocysts, SCP and SSCD during approximately 50 days after the dosage was stopped (Figure 8.2) and a secondary peak of oocyst concentration in this period, an observation also described by others in column studies (Harter *et al.*, 2001 and Hijnen *et al.*, 2005), demonstrate that the attachment of these organisms in the filter bed is reversible. Davies *et al.* (2005a) demonstrated remobilization of micro-organisms including *C. parvum* oocysts retained in soil in response to changes in ionic strength (from tap water to rain water). The virtually flat tail of the breakthrough curves of all three micro-organisms after dosing was stopped (Figure 8.2) indicates a high detachment rate as was described for SCP (Schijven *et al.*, 2003) and for *C. parvum* oocysts (Bradford *et al.*, 2005) retained in sand column studies.

Finally, for delayed breakthrough the removed micro-organisms must survive (negligible inactivation rate and no mass reduction) in the filter bed. The oocysts, spores and SSCD were primarily retained in the Schmutzdecke and the first 5–10 cm of the filter bed of the present study (Figure 8.3), the part with the increased microbiological activity (Figure 8.4) and the highest accumulation of (in)organic components (data not presented). Timms *et al.* (1995) showed similar spatial distribution of oocysts in a slow sand filter bed monitored immediately after a dosing experiment and 80% of the oocysts being present in the top layer of the filter bed. Our study revealed, however, a low recovery of oocysts in the filter bed determined 86 days after dosing was stopped (Table 8.2). In the next 69 days of filtration a further decline in mass of retained oocysts was observed. In contrast with these results, the recovery of *C. perfringens* spores in the filter bed was significantly higher and decline of the retained numbers of spores was observed (Table 8.2). The mass accumulation and the log-linear inactivation or mass reduction rate of spores and oocysts, respectively in the filter bed (Figure 8.5) was estimated from the mass load, observed removal rate and numbers retained in the filter bed. The estimated inactivation rate of SCP was 0.005 d^{-1} (Figure 8.5), similar to the inactivation rate assessed for environmental SCP in river water (Medema *et al.*, 1997). The estimated mass reduction rate of retained oocysts was 0.014–0.02 d^{-1} (Figure 8.5), values lower than estimated from a study where the mass reduction of retained oocysts in soil exposed to faecal wastes inoculated with *C. parvum* oocysts was determined (Hutchison *et al.*, 2005). From the published oocyst mass reduction data in the soil of the latter study log-linear mass reduction rates of 0.136 and 0.153 log d^{-1} were estimated. King *et al.*, (2005) showed lower mass reduction rates of *C. parvum* oocysts in untreated surface water from a reservoir; estimated rates were 0.0027, 0.0069 and 0.0076 d^{-1} at 15, 20 and 25°C, respectively. The same authors noticed no reduction occurred in autoclaved samples of the same water and from their results King *et al.*, (2005) hypothesized an influential role of predation in oocysts mass reduction.

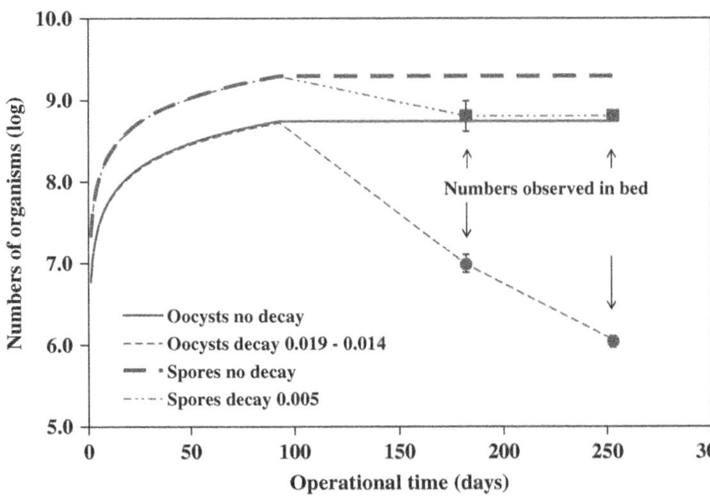

Figure 8.5 Simulated accumulation of mass of spores and oocysts (numbers) in the filter bed with and without decay rates estimated from the mass determined after 184 and 253 days of operation

Disintegration, predation and inactivation of oocysts. The observed reduction of oocysts in the sand over time can be caused by disintegration or predation. Microscopic evidence for disintegration of the retained oocysts in this study, though, was not obtained. The observed oocysts in the filter bed were intact and only to some degree, differential staining of the oocysts in the schmutzdecke with the monoclonal antibody was noticed. From the results of the zooplankton analysis (Figure 8.4) it was deduced that predation of oocysts was a possible cause of the large oocyst reduction in the filter bed. Eight different species were observed in the filter bed in relative high numbers. Two of these species, *Testacae* (Testate amoebae) and *Rotatoria*, are family of zooplankton species for which predation of oocysts have been documented (Fayer *et al.*, 2000 and Stott *et al.*, 2001, 2003). The results of the mass recovery of spores in the filter bed show that spores of sulphite-reducing clostridia are less susceptible to predation.

The presence of zooplankton like protozoa, Rotifera, Nematoda and Oligochaeta in slow sand filters and the significant role of predation in the removal of bacteria was shown by several studies (i.e. Lloyd, 1996 and Weber-Shirk and Dick, 1997). Predation as an oocyst reduction mechanism depends on the presence of suitable predators which is not always the case as demonstrated by Davies *et al.* (2005b). In air dried and sieved surface soil samples from drinking water supply catchment areas inoculated with *C. parvum* oocysts they observed hardly any reduction in total numbers during 180 days of incubation at 4, 20 and 35°C. In the absence of predation, inactivation of the retained oocysts is the only process which will diminish the risk of delayed breakthrough of reversible attached infectious oocysts. Inactivation rates of oocysts in the soil samples of the study of Davies *et al.* (2005b) estimated by loss of viability (fluorescence in situ hybridisation, FISH) ranged from 0.015–0.022 d^{-1}. Using cell culture-Taqman PCR assay, King *et al.* (2005) demonstrated higher rates in loss of infectivity of *C. parvum* oocysts inoculated in surface water of 0.01, 0.045 and 0.049 d^{-1} at 15, 20 and 25°C, respectively. In addition this inactivation rate adds to the observed mass reduction rate in reducing the risk of remobilization and delayed breakthrough of infectious oocysts.

SSRC and SSCD as surrogates. The results of this study demonstrate clearly differences between the behaviour of *Cryptosporidium* oocysts, SSRC and SSCD in slow sand filters. Based on geometric considerations removal of oocysts and SSCD by straining is similar and higher than removal of SSRC by straining, but the importance of this removal mechanism is uncertain. The present study indicates that attachment is of more importance for the removal of SSRC and oocysts than for the removal of SSCD, but for all three micro-organisms a high degree of detachment is observed. Finally, oocysts are less persistent than SSRC and SSCD because of a higher inactivation rate and the susceptibility to predation. Therefore the risk of accumulation of infectious oocysts in biological active slow sand filters is expected to be low. This implicates that a lowering of DEC of slow sand filters for infectious oocysts as observed for SSRC and SSCD, is less likely to occur. Furthermore, it is assumed that the environmental oocysts observed by Fogel *et al.*, 1993 in the filtrate of full-scale slow filters after delayed breakthrough and causing low DEC were most likely not infectious. From these considerations it is concluded that SSRC and to a larger extent SSCD, are too conservative surrogates for the assessment of capacity of slow sand filters to eliminate *Cryptosporidium* oocysts due to differences in size, surface properties and persistence. The results of this study suggest that the higher persistency of both surrogates due to low inactivation rates and susceptibility to predation is the major basis for this conclusion.

CONCLUSIONS

The results of this study demonstrate the high capacity of mature slow sand filters to remove *Cryptosporidium* oocysts and spores of *C. perfringens*. The observed decimal elimination capacity DEC_m was 4.7 and 3.6 log, respectively. The relative low DEC of 1.8 log observed for the small sized centric diatoms (SSCD) with a similar size as the oocysts was attributed to low attachment and accumulation followed by delayed breakthrough, a phenomenon also observed for SSRC. On the basis of this study, the risk of delayed breakthrough of infectious oocysts is low because of a rapid decline in oocyst concentration in the filter bed most likely caused by predation. The results of this study indicate that environmental SSCD and SSRC persist longer in the filter bed. From this finding in combination with the difference in observed DEC of the filter bed for the three organisms, it was concluded that SSCD and SSRC are too conservative parameters to be useful as surrogates for the assessment of the elimination capacity of slow sand filters for *Cryptosporidium* oocysts. Further investigations are necessary to elucidate the role of predation and the ultimate fate of the ingested oocysts for *Cryptosporidium* removal and the effect of temperature and filter bed scraping on this removal mechanism.

Chapter 9

Transport of MS2 Phage, *Escherichia coli*, *Clostridium perfringens*, *Cryptosporidium parvum* and *Giardia intestinalis* in a gravel and a sandy soil

Wim A.M. Hijnen[a], Anke J. Brouwer-Hanzens[a], Katrina J. Charles[b] and Gertjan Medema[a]

[a] *KWR Watercycle Research Institute;* [b] *University of New South Wales, UNSW-Sydney, Australia*

ABSTRACT

To define protection zones around groundwater abstraction wells and safe setback distances for artificial recharge systems in water treatment, quantitative information is needed about the removal of micro-organisms during soil passage. Column experiments were conducted using natural soil and water from an infiltration site with fine sandy soil and a river bank infiltration site with gravel soil. The removal of phages, bacteria, bacterial spores, and protozoan (oo)cysts was determined at two velocities and compared with field data from the same sites. The microbial elimination rate (MER) in both soils was generally >2 log, but MER in the gravel soil was higher than that in the fine sandy soil. This was attributed to enhanced attachment, related to higher metal-hydroxides content. From the high sticking efficiencies and the low influence of flow rate on MER it was deduced that straining played a significant role in the removal of *Escherichia coli* and *Cryptosporidium parvum* oocysts in the gravel soil. Lower removal of oocysts than the 4–5 times smaller *E. coli* and spores in the fine sand indicates that the contribution of straining is variable and needs further attention in transport models. Thus, simple extrapolation of grain size and particle size to the extent of microbial transport underground is inappropriate. Finally, the low MER of indigenous *E. coli* and *Clostridium perfringens* observed in the soil columns as well as under field conditions and the second breakthrough peak found for *Cryptosporidium* and spores in the fine sandy soil upon a change in the feedwater pH indicate a significant role of detachment and retardation to microbial transport and the difficulty of extrapolation of quantitative column test results to field conditions.

 Quantitative Methods to Assess Capacity of Water Treatment to Eliminate Micro-Organisms. Edited by Wim A.M. Hijnen. ISBN: 9781843393764. Published by IWA Publishing, London, UK.

INTRODUCTION

Soil passage is frequently used as pretreatment in production of drinking water in The Netherlands in river bank filtration (RBF; 5%) and artificial recharge (AR) in open basins (13%) or deep wells (1%) by several water suppliers. It is an intensive filtration process with long contact times and an effective barrier for pathogenic micro-organisms such as viruses, bacteria, and protozoa. How effective it is, however, is not known and is a question of growing interest since the introduction of quantitative microbial risk assessment for drinking water safety (Haas *et al.*, 1999). In 1980 a minimum water travel time of 60 days as a protection zone around groundwater abstraction wells was formalized in The Netherlands (Anonymous, 1980b). This travel time was assumed to cause sufficient die off of pathogenic bacteria from contamination sources (Knorr, 1937). In the past decades, however, viruses, and more recently protozoa like *Cryptosporidium* and *Giardia*, have been recognized as pathogens of major concern in the water industry (Craun *et al.*, 1997; MacKenzie *et al.*, 1994 and Gerba *et al.*, 1990). These organisms have been related to waterborne diseases because of their persistence in the environment, resistance to water treatment, and high infectivity. These organisms are different from bacteria in survival, surface properties, and size. Moreover, it has become clear that die off in groundwater is not the only process that governs the transport of micro-organisms. For viruses it was demonstrated that attachment to soil particles was more important than survival in the groundwater (Schijven, 2001). Therefore, viruses and maybe protozoa could be transported over longer distances in soil and thus be more significant to the microbial safety of groundwater. To verify the safety of the 60-days set back distance guideline, but also to assess the microbial safety of vulnerable groundwater systems and of RBF and AR systems with shorter water travel times, more quantitative information is needed on the elimination capacity of soil passage systems.

A number of field studies have been carried out that established either removal of indigenous micro-organisms or lab-cultured seeded micro-organisms (Schijven *et al.*, 1999, 2000, 2001; Van Olphen *et al.*, 1993 and Medema and Stuyfzand, 2002). These studies showed that soil passage poses a very effective barrier to micro-organisms, but critical situations may arise (Medema and Stuyfzand, 2002). Such situations are intrusion of contaminations to unconfined aquifers above groundwater wells, water abstraction during RBF from a gravel aquifer, with increased risk during high flow events, or short circuiting during recollection in AR systems. Field studies are valuable but hampered by some drawbacks. The concentration of pathogens in the field is generally too low to assess removal, and only non hazardous model micro-organisms (*Escherichia coli*, bacteriophages, and spores of clostridia) can be used in spiking studies (Schijven *et al.*, 2000). Moreover, the removal process is complex and influenced by a range of factors which vary considerably between sites. Hence, the effect of specific conditions such as soil characteristics, water velocity, and water quality variations are difficult to assess under field conditions. Some of these disadvantages can be overcome by column-based studies with spiked micro-organisms. Information can be collected on microbial transport in different soil types under well defined and standardized conditions as described in the literature. The importance of attachment and the surface properties of bacteriophages, bacteria, and soil and of water quality parameters has been elucidated by column experiments (Burge and Enkiri, 1978; Sobsey *et al.*, 1980; Bales *et al.*, 1991; Jin *et al.*, 1997; Goldschmid *et al.*, 1972; Fletcher and Marshall, 1982; Scholl *et al.*, 1990 and McCaulou *et al.*, 1994). More recently, transport of the oocysts of *Cryptosporidium* in soil columns was studied (Harter *et al.*, 2001; Logan *et al.*, 2001; Bradford and Bettahar, 2005 and Tufenkji *et al.*, 2004a), and results indicate the importance of straining on the removal of these larger organisms. The significance of column studies increases when results are related to field conditions of the selected soils and validated by field studies, as described for phage MS2 in dune sand by Schijven (2001). The number of such studies is still limited. Furthermore, there is a dearth of comparative studies describing elimination of the spiked model organisms relative to pathogens and of lab-cultivated versus indigenous strains of these spiked model organisms. The objective of the present study was to compare the elimination rate of FRNA phage MS2 as a model for viruses (Havelaar, 1993), indicator bacteria *E. coli* and spores of *Clostridium perfringens* as models for bacterial and protozoan pathogens (Payment and Franco, 1997 and Hijnen *et al.*, 2000a), and *Cryptosporidium* and *Giardia* (oo)cysts in two different soil systems: an AR site with fine dune sand and a RBF site with a gravel aquifer. For the AR site elimination of MS2 phages in soil columns was compared with elimination under field conditions as published by Schijven *et al.* (1999). For the RBF site elimination of the lab-cultivated phages and indicator bacteria was compared with the elimination of indigenous FRNA phages and indicator bacteria under column as well as field conditions, the latter published by Medema and Stuyfzand (2002).

COLUMN EXPERIMENT: COMPARISON OF SOIL TYPES

Soil and Water from the Infiltration Sites. The AR infiltration site is located in Castricum, The Netherlands, which is situated in the coastal dunes, where pretreated river Lek water is recharged in open basins without unsaturated zones. Saturated soil was collected at a depth of 3 m from a flow path 3 m from the edge of the open infiltration basin. The other infiltration site studied is the RBF site located in Roosteren, The Netherlands, where river Meuse water is abstracted through a fluvial gravel aquifer. During low river flow soil from the river bank was collected from a drill at a depth of 0.05–0.65 m. The soil samples were stored for 1.5 months in the dark at 5ºC before the experiment. The water used for the experiment was the natural water of both infiltration sites: for Castricum this is pretreated river Lek water (coagulation, sedimentation, and filtration) and Roosteren untreated river Meuse water. A 600 L water sample was taken in two separate stainless steel (SS) vessels and stored for 1 week at approximately 10ºC. River Meuse water was stirred continuously to avoid sedimentation.

Soil and Water Analysis. After treatment of the soil with HCl and H_2O_2 to remove cementing materials, the grain size distribution of the <2 mm fraction was measured using a FRITSCH Laser Particle Sizer A22 (Laval Lab inc., Canada). The clay and silt fractions were converted to the traditional grain size analysis with pipet and sieve analysis (Konert and Vandenberghe, 1997). Total organic material and carbonate were determined using the methods of Stuyfzand and van der Jagt (1997). Soil pH and the electrical conductivity (EC) were determined after shaking 20 g for 2 h with 50 mL of ultrapure water in the decanted fluid at 20ºC. Free (amorphous and exchangeable) Fe, Mn, and Al were determined in 2.5 g of soil after shaking for 2 h in the dark with 100 mL of an ammonium-oxalate/oxalic acid mixture (16.1 and 10.9 $g.L^{-1}$, respectively) at pH 3. Mn and Al were quantified by ICP-MS and Fe by AAS-flame photospectroscopy. The effective cation exchange capacity (CEC) of the soil was assessed by the AgTU method (Stuyfzand and van der Jagt, 1997). Water quality of the infiltration water in the storage tanks was determined using standard methods.

Column Setup. The soil column setup consisted of four Perspex columns (Ø9 cm; height 50 cm), located in a temperature-controlled dark room (16 ± 1ºC). Two columns were filled with Castricum soil and two with Roosteren soil. The columns were packed with humid soil and placed on a shaker for 0.5 and 1 min. at amplitudes of 6 and 10, respectively. The bed settled 3–5 cm; consequently, the surface was replenished with soil to eliminate the headspace. Subsequently, the columns were closed and supplied with water from 30 L SS vessels connected to the columns with 5 mm PVC piping. These vessels were regularly refilled with Lek water and Meuse water supplied to the Castricum and Roosteren columns, respectively. For the experiment with spiked river water 30 L SS vessels were filled with river water, inoculated with the micro-organisms, and continuously mixed with a stirrer. The columns were operated at a constant flow rate or fluid approach velocity U (i.e., volume of water per unit time per unit cross sectional area of media) with a peristaltic pump (Gilson, minipuls 2; Gilson inc., Middleton) located at the outlet of the filters. One column of each site was operated at $U = 0.9\,m.day^{-1}$. U under field conditions at the Castricum site can be lower, and at the Roosteren site higher rates have been observed during high-flow events. Therefore, the second column of the Castricum and Roosteren sites was operated at 0.5 and 2.5 $m.day^{-1}$, respectively.

Tracer Test and Breakthrough Modelling. The medium dispersivity α_L, porosity θ, and pore velocity v were determined with a tracer test using sodium chloride (275 and 310 mg L^{-1} NaCl for Castricum and Roosteren columns, respectively). The conductivity was monitored continuously with two sensors connected to data log units (Campbell 21X, Campbell Scientific Inc., Utah). A one-dimensional convection-dispersion model (CDE) was used to describe the breakthrough curves with the following equation (CXTFIT; Toride *et al.*, 1997)

$$\frac{\delta C}{\delta t} = D.\frac{\delta^2 C}{\delta x^2} - v\frac{\delta C}{\delta x} \tag{1}$$

where C is the electrical conductivity, D the coefficient of hydrodynamic dispersion ($m^2.h^{-1}$), x the distance in the columns (m), v the average pore velocity ($m.h^{-1}$), and t the time (hours).

Removal of Micro-organisms. Water flow was started 7 days prior to the experiment to equilibrate the columns. In the days before the challenge experiment, concentrations of indigenous F-specific RNA bacteriophages (FRNA phages), thermotolerant coliforms (Coli44), and spores of *C. perfringens* were determined in the inlet water and outlet water of the columns ($C_{out,indigenous}$). For the experiment 10 L of Lek water and 26 L of Meuse water in the 30 L SS vessels were inoculated with lab-cultured micro-organisms and the average inlet concentration during the spike C_{in} was measured in a 1 L sample of inoculated water stored next to the setup. The following organisms were spiked: FRNA phage MS2 ($C_{in} = 1.6 \times 10^9 \pm 6.3 \times 10^8$ plaque-forming particles PFP.L^{-1}), *E.coli* (WR1) ($C_{in} = 1.2 \times 10^6 \pm 6.6 \times 10^5$ colony-forming particles CFP.L^{-1}), *C. perfringens* (D10) ($C_{in} = 7.9 \times 10^5 \pm 1.1 \times 10^5$ CFP.L^{-1}), ocysts of *C. parvum* ($C_{in} = 1.4 \times 10^6 \pm 7.5 \times 10^5 L^{-1}$) and cysts of *G. intestinalis* ($C_{in} = 1.6 \times 10^6 \pm 5.0 \times 10^5 L^{-1}$). The inoculated water in the vessel was stirred continuously with a magnetic stirrer. For 24 h the columns were challenged with these suspensions, and thereafter the supply of unspiked water was restored. Breakthrough of the spiked organisms was determined by sampling complete column filtrate in separate containers of different volumes over 2 weeks, which were analyzed for each organism. Breakthrough curves were constructed by plotting the ratio between C_{out} and C_{in} ($n.L^{-1}$) equal to C/C_o against the number of pore volumes, corrected for the time interval of the sampling. The micro-organisms elimination rate (MER; log) was calculated based on the mass balance calculated by

$$MER = \log_{10}\left(\frac{t_{ch} * Q * \bar{C}_{in}}{\sum(t_{s,i} * Q * (C_{out} - C_{out,indigenous}))}\right) \tag{2}$$

where t_{ch} and $t_{s,i}$ are periods (h) of challenge test and sampling for sample i and Q is the volume load ($L.h^{-1}$).

Micro-organisms Suspensions and Enumeration Methods. Methods for production and storage of stock solutions MS2 phages and *E. coli* WR1 and for enumeration of FRNA phages and Coli44 have been described previously (Schijven, 2001). Stock solution procedures for spores of *C. perfringens* D10 and the enumeration method were described by Hijnen *et al.* (2002). *C. parvum* oocysts (Moredun; harvested by sedimentation and differential centrifugation) were spiked from a 2 mL suspension (10^8 oocysts; viability 70%, PI staining) and *G. intestinalis* cysts (H3 isolate, Waterborne, harvested by sedimentation and differential centrifugation) from a 50 mL suspension (10^8 H3 isolate; viability 58%, PI staining). (Oo)cyst were counted in samples of 1–3.000 mL using a direct fluorescence assay technique with the Chemscan-RDI (Chemunex SA, Ivry-sur-Seine Cedex, France). Samples were filtered and prepared for scanning with the Chemscan kit (Chemunex 200 k0009-01 with IMS) including a mounting medium (85 μL), membrane filters (25 mm 2.0 μm), and support Pad. The membrane filters were labeled with 100 μL diluted monoclonal antibody reagents (Oxoid Ltd.,

Hampshire, U.K.) 1:1 in deionized water for 30 min. at 37°C. The filters were scanned, and counted spots were microscopically confirmed based on fluorescence, morphology, and size.

Statistical Methods. The data were analyzed with the nonparametric Sign test using SPSS (SPSS Inc., Chicago, IL) to test for median differences, assuming a non-symmetric distribution.

Modelling Removal Data. The conceptual model used in modelling micro-organisms removal has been described by several authors (Schijven *et al.*, 1999; Bales *et al.*, 1991; McCaulou *et al.*, 1994; Toride *et al.*, 1997 and Šimůnek *et al.*, 1998). On the basis of the results, this was only possible for MS2 phage breakthrough. The Hydrus-1D model (Šimůnek *et al.*, 1998) was used to model the breakthrough curves. It consists of a first-order dispersion model for water transport supplemented with equations for attachment and detachment of MS2 to soil and a first-order inactivation kinetic with different rate coefficients for the free and attached phages. The concentration in the water (C; $PFP.L^{-1}$) was described by

$$\frac{\delta C}{\delta t} = D\frac{\delta^2 C}{\delta x^2} - v\frac{\delta C}{\delta x} - k_{att}C - \mu_1 C + k_{\det}\frac{\rho_B}{\theta}s \tag{3}$$

The concentration of attached MS2 (S; $PFP.g^{-1}$) is described by

$$\frac{\rho_\beta}{\theta}\frac{\delta S}{\delta t} = k_{att}C - k_{\det}\frac{\rho_B}{\theta}s - \mu_s\frac{\rho_B}{\theta}s \tag{4}$$

Subject to boundary conditions $C = C_o$ at $x = 0$ and $\delta C/\delta x = 0$ at $x = \infty$. Here D is the hydrodynamic dispersion coefficient ($m^2.h^{-1}$), x is the distance (m); k_{att} and k_{det}, are the attachment and detachment rate coefficients (h^{-1}), respectively, for a one kinetic site, μ_l and μ_s (h^{-1}) are the inactivation rate of the free and attached phages, respectively, ρ_B is the dry bulk density ($kg.m^{-3}$), and θ is the porosity.

Calculation of Collector and Sticking Efficiencies. Yao *et al.* (1971) presented the first model for the transport of colloidal particles from the pore fluid to the vicinity of the porous medium described with the following equation

$$LN\frac{C}{C_0} = -\frac{3}{2}\frac{(1-\theta)}{d_c}\alpha\eta L \tag{5}$$

where d_c is the diameter of the collector, α the sticking efficiency, η the single collector collision efficiency, and L the length of the column. Rajagopalan and Tien (1976) developed a semiemperical equation to solve η, an equation recommended by Logan *et al.* (1995) who compared different filtration models. Recently, Tufenkji and Elimelech (2004a) optimized the RT model by including the influence of hydrodynamic and van der Waals interactions, resulting in the following equation 6

$$\eta_0 = 2.4A_s^{1/3}N_{Pe}^{-0.715}N_R^{-0.081}N_{vdW}^{0.052} + 0.55A_S N_R^{1.675}N_A^{0.125} + 0.22N_R^{-0.24}N_G^{1.11}N_{vdW}^{0.053}$$

where the Happel porosity dependent parameter $A_s = 2(1-\gamma^5)/(2-3\gamma+3\gamma^5-2\gamma^6)$ and $\gamma = (1-\theta)^{1/3}$; the peclet number $N_{Pe} = Ud_c/D$ with the Diffusion coefficient $D = k_B(T+273)/(3\pi d_p\mu)$, Boltzmann's constant $k_B = 1.38 \times 10^{-23}\,J.K^{-1}$, temperature T (°C), the colloid diameter d_p (m), and the dynamic viscosity μ ($M.L^{-1}.T^{-1}$); interception number $N_R = d_p/d_c$; van der Waals number $N_{vdW} = A/k_B(T+273)$ with A being the Hamaker constant (J); attraction number $N_A = A/(12\pi\, d_p^2 U)$; gravity number $N_g = 2/9 d_p^2(\rho_p - \rho_B)g/\mu U$, where ρ_p is the particle density in $kg.m^{-3}$ and gravitational constant $g = 9.8806\,m.s^{-2}$. For the calculations the following parameters values were used: bulk water density $999.703\,kg.m^{-3}$; Hamaker constant for bacterium glass water interface (Rijnaarts *et al.*, 1995) 6.2×10^{-21} J; sizes (m) of the micro-organisms MS2 2.1×10^{-8} (Havelaar, 1993), *E. coli* 1.5×10^{-6}, *C. perfringens* 1.5×10^{-6}, protozoan (oo)cysts *C. parvum* 4.9×10^{-6} and *G. intestinalis* 10.8×10^{-6} (Medema *et al.*, 1998); and ρ_p of MS2 and *E. coli* $1085\,kg.m^{-3}$ (Bouwer and Rittman, 1992), *C. perfringens* 1270 (Tisa *et al.*, 1982) and (oo)cysts 1045 and $1036\,kg.m^{-3}$, respectively (Medema *et al.*, 1998).

MICRO-ORGANISM REMOVAL BY TWO SOIL TYPES

Soil Characteristics and Water Quality. Geochemical analysis of the soils revealed that Castricum soil was a calcareous sand with high uniformity ($U_c = 1.6$) and a fine grain size. It had a low content of organic matter and metal hydroxides (Table 9.1). In contrast, Roosteren soil was coarser with a high estimated coarse fraction (>2 mm) of 25% and thus a lower uniformity. This soil was richer in organic matter and metal hydroxides (Fe- and Al-ox) which resulted in a higher CEC. The Ca content and pH of both soils were high. Levels of most water quality parameters in the Meuse water were higher than those in the pretreated Lek water (Table 9.1), which corresponds to the higher content of organic and inorganic matter in Roosteren soil.

Table 9.1 Water quality of the infiltration waters and geochemical characteristics of the soils

Parameter	Water Quality		Soil characteristics		
	Lek water[a,b]	Meuse water[a,b]	Parameter	Castricum[a]	Roosteren[a]
EC ($\mu S.cm^{-1}$)	759–768	569–574	EC ($\mu S.cm^{-1}$)	95	98.3
pH	7.7–8.4	8.0–8.4	pH	8.73	8
HCO_3^- ($mg.L^{-1}$)	162–170	163–180	CEC (meq/kg dw)[c]	19.3 (14)[d]	67
NH_4 ($mg\ N.L^{-1}$)	0.04–0.57	0.29–0.37	Gravel (%; >2 mm)	0	25
Turbidity (Ftu)	0.1–0.37	1.18–1.0	Grain size (d50; mm)	0.18 (0.21)[d]	0.5
Mg ($mg.L^{-1}$)	11.4–11.7	8.2–8.3	Clay (%; ≤2 μm)	2.43 (0.85)[d]	4.2
Ca ($mg.L^{-1}$)	76.6–76.7	72–72.6	U_c (d_{60}/d_{10})	1.6 (0.12/0.19)	5.3 (0.11/0.56)
DOC ($mg.L^{-1}$)	1.8–1.9	2.5–2.5	Ca (%dw)	2.5	2.1
Fe ($mg.L^{-1}$)	<0.1–<0.1	0.14–0.13	Org. matter (%dw)	0.6 (0.1)[d]	2.7
Coli44 ($CFP.L^{-1}$)	2–10	2,400–3,200	Fe-ox. (g/kg dw)	0.58 (1.2)[d]	5.11
C. perfr. ($CFP.L^{-1}$)	20–48	3,000–6,200	Al-ox. (g/kg dw)	0.22 (0.32)[d]	1.11
FRNA ($PFP.L^{-1}$)	<100	700–1,000	Mn-ox. (g/kg dw)	0.02 (0.02)[d]	0.19

[a]Lek water = Castricum; Meuse water = Roosteren; [b]two samples, before and after challenge period; [c]dw = dry weight; [d]between brackets values of Castricum soil sample studied by Schijven (1999, 2001)

The hydrodynamic parameters of the four columns were calculated from the tracer tests (Table 9.2). The porosity was 36% and 32% for Castricum and Roosteren soil, respectively. The low uniformity of the Roosteren soil caused a 10 times higher dispersivity compared to the Castricum soil column at a velocity of $0.9\,m.day^{-1}$.

Transport of FRNA Phages. Prior to the spiking experiment, the concentration of FRNA-phages in the feedwater and the filtrate of the columns was determined. These phages were observed in the Meuse water (700–1000 $PFP.L^{-1}$), but not in the filtrate of the Roosteren columns supplied with this water. The MS2 spike concentration was 1000 times higher than that of the other micro-organisms, which resulted in higher concentrations in the filtrate and clear breakthrough curves (Figure 9.1). The micro-organisms elimination rate, MER, observed in the columns is presented in Table 9.3. Modelling the breakthrough curves of MS2 with the Hydrus-1D model resulted in a good fit of the observed data for most columns (Figure 9.1; Table 9.2). The evident scatter in Roosteren water at 0.9 m d^{-1} resulted in a poor fit (Figure 9.1; $R^2 = 0.63$). The rate of inactivation in the feed water μ_l was assumed to be equal to the rate of decrease of the maximum breakthrough concentration during the challenge test and calculated using linear regression of the natural log-transformed data.

Table 9.2 Hydrodynamic parameters of the columns and Hydrus-1D parameters (±Standard Deviation) for MS2 transport in the Castricum and Roosteren soil

	Castricum 0.5	Castricum 0.9	Roosteren 0.9	Roosteren 2.5
Velocity U ($m.d^{-1}$)	0.46	0.92	0.90	2.49
Velocity v ($m.d^{-1}$)	1.27	2.54	2.81	7.77
Porosity θ (%)	36	36	32	32
Dispersivity α_L (cm)	0.238 (±0.002)	0.167 (±0.399)	2.287 (±0.013)	18.44 (±0.005)
R^2 tracer	0.89	0.97	0.99	0.99
μ_1 (h^{-1})	−0.120 (±0.030; $R^2 = 0.93$)		−0.106 (±0.020; $R^2 = 0.92$)	
μ_s (h^{-1})	−0.056 (±0.017)	0.098 (±0.016)	−0.069 (±0.021)	0.076 (±0.010)
k_{att} (h^{-1})	0.724 (±0.018)	0.622 (±0.018)	2.385 (±0.048)	4.509 (±0.424)
k_{dett} (h^{-1})	0.004 (±0.003)	0.009 (±0.006)	0.004 (±0.003)	0.004 (±0.001)
R^2	0.91	0.95	0.63	0.92

For Castricum water μ_l was 0.120 h^{-1} (Table 9.2; column 0.9 $m.day^{-1}$) and for Roosteren water $\mu\gamma_1$ was 0.106 h^{-1} (column 2.5 $m.day^{-1}$). In addition, the estimated inactivation rate of the attached phages μ_s (Table 9.2) was lower than μ_l.

Blanc and Nasser (1996) reported lower μ_s in loamy soil and sandy soil saturated with groundwater at 10 and 23°C of 0.007–0.013 and 0.018–0.019 (h^{-1}), indicating that the values in Table 9.2 are relatively high. Calculated from the empty pore volume contact time and μ_l, it was estimated that inactivation accounted for 13.6% and 11.4% of the overall removal of MS2 in the Castricum columns 0.5 and 0.9 $m.day^{-1}$, respectively.

In the Roosteren columns 0.9 and 2.5 $m.day^{-1}$ contribution of inactivation to the overall elimination was less, 5.7% and 4.6%, respectively. From this consideration it was deduced that attachment was the most important process in overall MS2 removal in both columns, while detachment rates were a factor of 100–1000 lower (Table 9.2). In Roosteren soil operated at 0.9 $m.day^{-1}$, the attachment rate coefficient k_{att} was calculated at $2.385 \pm 0.048\,h^{-1}$, 4 times higher than the k_{att} calculated for the Castricum column operated at the same velocity. Despite the lower collision efficiency η of Roosteren soil, the sticking efficiency α in this soil was a factor of 10 higher than that calculated for Castricum soil (Table 9.3).

Adsorption of micro-organisms to solid surfaces is governed by electrostatic forces when surfaces are charged, and by intermolecular forces governed by the hydrophobic or hydrophilic composition of both surfaces (Ryan and Elimelech, 1996). For micro-organisms the isoelectric point (IEP), defined as the pH where net surface charge of suspended colloids is zero, is generally below 7.0 (Fletcher and Marshall, 1982; Harden and Harris, 1953). Thus, at neutral pH, micro-organisms are negatively charged and they preferably attach to the positively charged sites on the negatively charged mineral grains of the soil. pH values of the Lek and Meuse water were comparable. However, the relatively high pH of the Castricum soil compared to the pH of Roosteren soil (Table 9.1) is one possible explanation for the differences in k_{att} values between both soils.

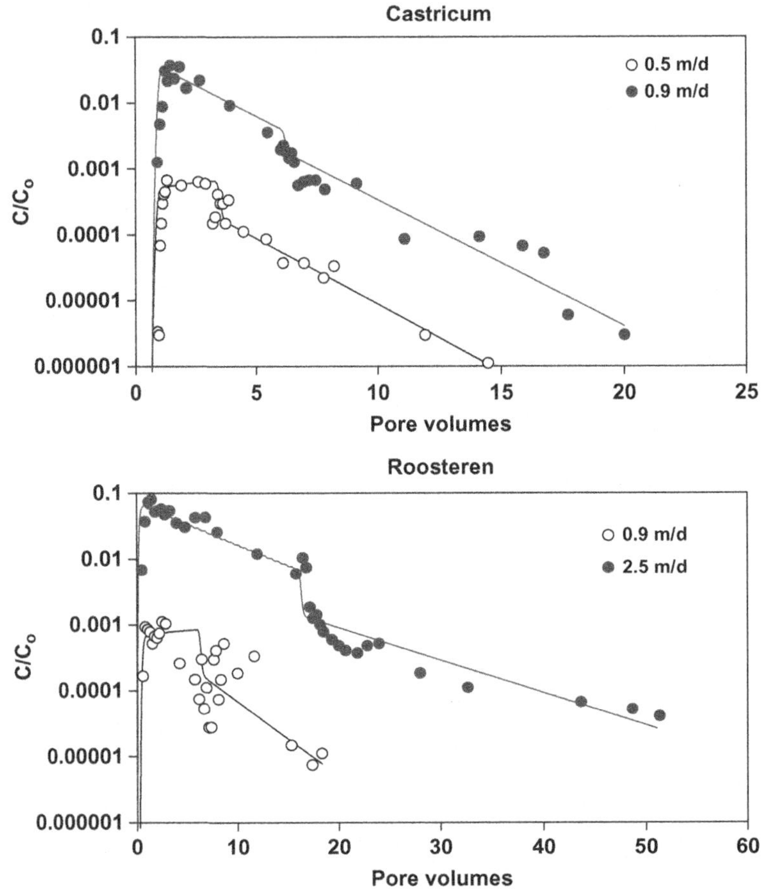

Figure 9.1 The breakthrough curves of MS2-phages in the Castricum and the Roosteren columns (open and closed circles are the observed values and lines indicate the Hydrus-1D model fit)

Table 9.3a Micro-organisms Elimination Rate (MER; log) of the Castricum Soil Columns for Spiked Micro-organisms, and the Calculated Collector Efficiency η and Sticking Efficiency α (Tufenkji and Elimelech, 2004a)

	column 0.5 m.d^{-1}		column 0.9 m.d^{-1}	
	MER	η;α	MER	η;α
MS2[a]	3.3	1.035;0.003	2.2	0.680;0.003
E. coli	4.7	0.048;0.085	4.2	0.032;0.113
C. perfringens	≥5.0	0.067;0.064	≥4.5	0.042;0.092
C. parvum	3.9	0.084;0.040	3.3	0.059;0.048
G. intestinalis	6.2	0.250;0.021	>6.2	0.172;0.031

[a]*corrected for elimination caused by inactivation (10%)*

Table 9.3b Micro-organisms Elimination Rate (MER; log) of the Roosteren Soil Columns for Spiked (Tufenkji and Elimelech, 2004a)

	column 0.9 m.d^{-1}		column 2.5 m.d^{-1}	
	MER	$\eta;\alpha$	**MER**	$\eta;\alpha$
MS2[a]	3.4	0.390;0.019	1.8	0.188;0.021
E. coli	4.8	0.019;0.562	4.1	0.009;1.023
C. perfringens	≥2.4	0.032;0.169	≥3.0	0.013;0.516
C. parvum	>6.7	0.038;0.402	>7.2	0.017;0.959
G. intestinalis	>7.4	0.120;0.140	>6.8	0.051;0.299

[a]*corrected for elimination caused by inactivation (5%)*

Hydroxides of iron, aluminum, and manganese are the most common sources for positively charged sites at neutral pH. While the IEP of quartz is 2.0, the IEP of Fe-hydroxide-coated quartz is 8.5 (Stumm and Morgan, 1981). The positive effect of metal hydroxide coating on bacterial and viral attachment has been demonstrated experimentally (Scholl *et al.*, 1990 and Ryan *et al.*, 1999). Castricum and Roosteren markedly differed in metal hydroxide content (Table 9.1). The higher metal content in Roosteren soil yields more attachment sites and thus a higher k_{att}-value. Roosteren soil was also higher in organic matter, but the role of organic matter in attachment of micro-organisms to soil is undecided and strongly depends on the nature of the accumulated organic matter, which influences the surface charge as well as the hydrophobic/hydrophilic character. Consequently, the higher MER observed in the Roosteren soil for MS2 can probably be attributed to the presence of a more positively charged surface (metal oxides) and possibly to the presence of more hydrophobic organic matter.

The results clearly demonstrate that the MER of MS2 was lower at a higher water velocity (Table 9.3). Mass transport to the soil surface and micro-organism-surface interactions depends on water flow velocity, diffusion characteristics and surface properties of the micro-organisms and soil. It has been proposed that based on the colloid filtration theory, colloid removal in filters is proportional to the water velocity $v^{-2/3}$ (Yao *et al.*, 1971), neglecting dispersion, inactivation and detachment. This ratio has been observed in experiments for poliovirus and bacteriophage ΦX174 (Schijven, 2001; Jin *et al.*, 1997 and Wang *et al.*, 1981). On the basis of this ratio and the MER observed at 0.9 m.day^{-1}, the MER for the lower velocity in the Castricum is estimated at 3.3 log and for the Roosteren soil columns with higher velocity 1.7 log. These values were almost similar to the observed MER-values for MS2 in these columns (Table 9.3).

Schijven *et al.* (2003) also studied MS2 removal in Castricum soil columns and determined less than 1 log removal in a bed of 1.4 m operated at a higher velocity of 1.5 m.day^{-1} and lower temperature of 5 ± 3°C than in the current study. Model parameters μ_1, μ_s, and k_{att} calculated from the experiments of Schijven *et al.* (2003) were 0.003, 0.004, and 0.071 h^{-1}, respectively, clearly lower than the values presented in Table 9.3. Because decay explains only 10% of the overall removal, we attribute the higher elimination of MS2 in the current study with Castricum soil to increased attachment. Interception and sedimentation are most likely processes of minor importance to contact of small phages with soil particles. More relevant is diffusion. The difference in diffusion constant D for MS2 calculated for the temperatures of both studies (16 ± 1 and 5 ± 3°C) was a factor of 1.4, explaining only a small part of the large difference in k_{att} observed between both studies. Geochemical characteristics of the soil samples tested in both studies are presented in Table 9.1. This showed that, besides more inactivation and attachment at higher temperature in the current study, the Castricum soil studied by Schijven was coarser with a lower organic content and CEC-value, which may have contributed to the higher attachment of MS2 in our columns.

Transport of Escherichia coli. Indigenous thermotolerant coliforms (Coli44) in the Lek water (Table 9.1) were removed completely by the Castricum soil columns, however, after the Roosteren soil columns low numbers of Coli44 (3–97 CFP.L^{-1}) were observed. On the basis of the concentration of Coli44 in Meuse water (Table 9.1), the MER of unseeded Coli44 was 2.8 and 1.5 log in the columns operated at 0.9 and 2.5 m.day^{-1}, respectively (C/C_0 of 0.0012–0.0021 and 0.03–0.04, respectively; Figure 9.2).

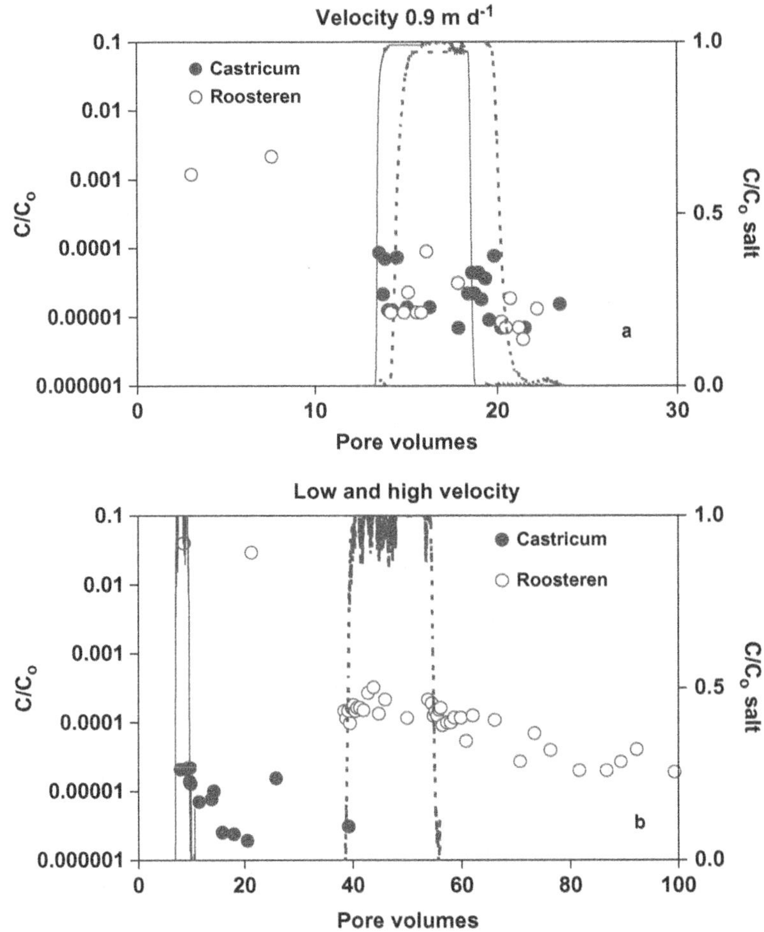

Figure 9.2 Breakthrough curves of *E. coli* and the tracer (C/Co; EC, solid line Castricum and dashed line Roosteren); (a) columns with similar velocity and (b) columns with low (Castricum 0.5 m day^{-1}) and high velocity (Roosteren 2.5 m day^{-1}).

Spiked *E. coli* WR1 was eliminated to a larger extent by both soils than MS2. The relatively low spike concentrations of these bacteria (10^6 CFP.L^{-1}) combined with high removal caused little breakthrough; the outlet concentrations of the Roosteren columns increased only slightly ($p = 0.007$ and 0.018). As observed for MS2, Roosteren soil columns removed WR1 to a higher extent than Castricum soil columns ($p = 0.015$), despite the lower η values (Table 9.3) due the higher grain size. The calculated sticking efficiencies α of 1.023 for Roosteren soil at a flow rate of 2.5 m.day^{-1} was slightly above unity. Assuming all parameters used to calculate η were valid, the high MER must be caused by an additional removal mechanism not accounted for in the colloid filtration model, i.e., straining. A parameter related to straining is the interception number NR, reflecting the ratio between the collector and colloid sizes. It is described in the literature that straining becomes important when this ratio is >5% (Herzig *et al.*, 1970), which was true only for the largest of the tested micro-organisms, *G. intestinalis*, in the finer Castricum soil (6%).

However, more recently Bradford *et al.* (2002, 2003) and also Tufenkji *et al.* (2004b) showed the importance of straining when this ratio is <5%, even as low as 0.2% (Bradford *et al.* 2003). NR is calculated from the average diameter of the grains and there is no correction for the grain size distribution. Straining occurs in the small sized pores. Pore size distribution is influenced by the uniformity and shape of the grains as well as the content of sediments. Roosteren soil has a lower uniformity and porosity with higher sediment content of organic and inorganic nature compared to Castricum soil (Table 9.1), resulting in a higher presence of small sized pores. This will result in lower permeability or hydraulic conductivity of the soil and consequently inhibition of microbial transport.

Horizontal and vertical hydraulic conductivity of a representative Roosteren soil sample with U_c of 2.4 was 5 and 1 m.day^{-1}, respectively, and lower than the horizontal and vertical hydraulic conductivity of about 12 and 6 m.day^{-1} observed for the Dutch dune sand areas (Stuyfzand; personal communication).

The relatively low influence of flow rate on the removal of *E. coli* in both soils also point to an additional removal by straining. Though significantly higher removal was observed in Castricum and Roosteren soil with decreased flow rate ($p = 0.018$ and <0.001, respectively), the observed increase in MER was low compared to the effect observed for MS2 and not proportional to the water velocity $v^{-2/3}$. Straining is expected to be independent of the flow rate.

Transport of Spores of C. perfringens. Concentrations of indigenous *C. perfringens* spores in both river waters were of the same order of magnitude as those observed for Coli44 (Table 9.1). However, these indigenous micro-organisms were observed after both soils in higher concentrations than Coli44: 9–18 CFP and 76–293 CFP.L^{-1} after Castricum and Roosteren columns, respectively. Castricum columns at 0.9 and 0.5 m.day^{-1} showed 0.4 and 0.6 log removal of the indigenous spores, respectively. The MER calculated for the indigenous spores in the Roosteren columns at 0.9 and 2.5 m.day^{-1} were 1.2 and 1.6 log, respectively. During the challenge experiment *C. perfringens* D10 was spiked in a concentration of about 100–10,000 times the concentration of indigenous spores in the river water, but this had no significant effect on the *C. perfringens* concentrations in the outlet, showing a high removal of seeded spores. The minimum MER-values are presented for spiked *C. perfringens* spores in Table 9.3.

Transport of Protozoan (Oo)cysts. The high MER values obtained for the protozoan (oo)cysts (Table 9.3) indicate low mobility of these organisms in soil. Neither oocysts of *C. parvum* nor cysts of *G. intestinalis* were observed in the column filtrate of Roosteren soil, demonstrating removal of >6.7–>7.4 log. In the filtrate of the Castricum soil columns, however, *C. parvum* oocysts were observed frequently (Figure 9.3) and a *G. intestinalis* cyst was detected once (1 per 20 mL). The difference in (oo)cyst breakthrough between both soils is significant. In 34 and 25 samples with a total sampled volume of 9.75 and 10.4 L for Castricum and Roosteren soil in both columns 410 and 0 oocysts were counted, respectively. η values calculated for the columns with a similar fluid approach velocity of 0.9 m.day^{-1} showed a slightly higher η value in the Castricum soil than in the Roosteren soil. For the α values this was the other way around: a 10 times higher α value for the Roosteren soil than for the Castricum soil (Table 9.3).

As discussed for *E. coli*, the higher MER for *C. parvum* in Roosteren soil compared to Castricum soil again indicates that the average grain size is not a proper generic predictor for microbial transport in soil. Previous studies (Harter *et al.*, 2001 and Logan *et al.*, 2001) demonstrated a positive correlation of the level oocysts breakthrough with the grain size in a higher grain size range of 0.3 mm up to 1.4–2.4 mm (U_c of 1.7 to 2.1). Oocysts breakthrough was approximately 0.2% (MER of 2.7 log) in a 0.1 m column with 0.42 to 0.5 mm sieved and acid-treated sand (Harter *et al.*, 2001). Compared to this breakthrough, 0.01% oocyst breakthrough observed in a 0.5 m column with fine and highly uniform Castricum soil (Table 9.1), is relatively high. The contribution of attachment to the high removal of oocysts in Roosteren soil is illustrated by the high α values (Table 9.3), but, as discussed for *E. coli*, an α value close to unity for *G. intestinatlis* indicates a supplementary effect of straining to the removal. Breakthrough of the oocysts (C/C_0) in the Castricum soil column was higher with elevated water velocity (Figure 9.3), but the effect on MER was small (Table 9.3). Enhancement of *C. parvum* transport by increased flow rates has been described before (Harter *et al.*, 2001), but this study used a medium coarse and coarse sands (0.42 to 2.4 mm). Since NR for this Castricum soil is 0.033 (3.3%) the contribution of straining to the removal process in this soil (Bradford *et al.*, 2003) is the most likely explanantion for the low influence of flow rate on the MER.

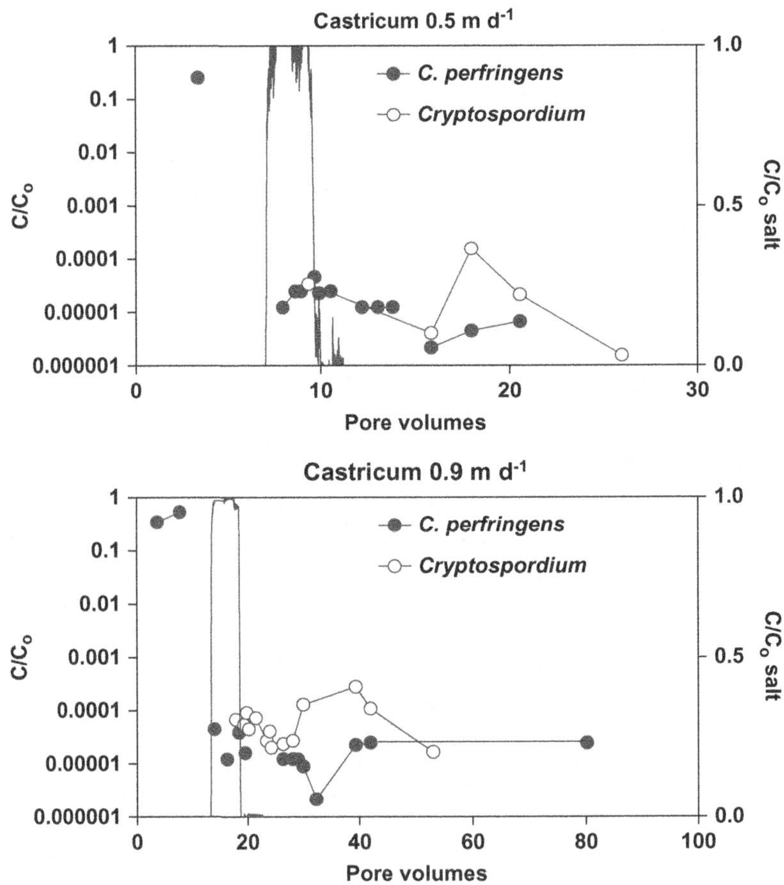

Figure 9.3 Breakthrough of tracer (C/C_0; EC solid lines) and of *C. parvum* oocysts and *C. perfringens* observed in the Castricum sandy soil

A second breakthrough with a considerably higher maximum number of oocysts was observed in the period after the challenge test in both columns. The difference in the observed number of oocysts in the subsamples of the Castricum columns was a factor of 100, the concentrations showed a subsequent increasing and decreasing trend and this trend was observed in both columns. Simultaneously, an increase in *C. perfringens* concentration in the filtrate was observed (Figure 9.3). This second breakthrough of oocysts coincided with an increase of 0.7 pH unit in the Lek water used to feed the columns (Table 9.1). As described before, water quality influences the geochemical characteristics of the soil (Table 9.1). Besides this indirect effect of water quality on microbial transport, pH and ionic-strength influences the attachment/detachment of colloids in granular media, as demonstrated by several authors (Burge and Enkiri, 1978; Sobsey *et al.*, 1980; Goldschmid *et al.*, 1972 and Ryan *et al.*, 1999). An increase in pH increases the negative charge on the surfaces of oocysts (Drozd and Schwartzbrod, 1996) and soil, and enhances the electrostatic repulsion between attached oocysts and soil. This may have caused an increased detachment rate of reversibly attached oocysts and most likely the same accounts for the increase in spores in the column effluent. A secondary breakthrough was also observed by Harter *et al.* (2001) and they concluded that long-term, low-level elution is a potential source of oocyst transport. Similar observations of high retardation and a multi-peaked breakthrough of labeled flagellates and protozoan-sized microspheres were described by Harvey *et al.* (1995). They attributed this to grain-surface interactions in the subsurface transport behavior of these kinds of particles. More recently, the role of electrosteric repulsion in oocysts adhesion to quartz surfaces has been proposed. This repulsive force cannot be captured by the classic DLVO theory and is attributed to the presence of proteins on the surface of the oocysts extending in the solution (Kuznar and Elimelech, 2004).

COMPARISON OF MICRO-ORGANISMS

Elimination of bacteriophages, bacteria, spores and protozoan oocysts in the soil of the two infiltration sites was determined simultaneously in this comparative study. The Microbial Elimination Rate (MER) in both soils was generally >2 log but the sequence of the MER for the different sized organisms differed. In the gravel soil of the RBF site of Roosteren the sequence was *Giardia* ≥ *Cryptosporidium* ≥ *C. perfringens* ≥ *E. coli* > MS2 (Table 9.3). In the sandy soil of the AR site, however, the sequence was *Giardia* > *C. perfringens* > *E. coli* > *Cryptosporidium* > MS2 phages (Table 9.3), indicating that particle and grain sizes do not universally govern removal of micro-organisms through soils. From two observations it was deduced that this difference in sequence of organism removal between both soils was caused by increased straining in the Roosteren soil. First, the sequence in Roosteren soil corresponds to the sequence in size of these organisms, and second, the difference in MER between Castricum and Roosteren soil increased with size of the particles (Table 9.3). The ratio between values of the different sized organisms was as previously described (Logan *et al.*, 1995). The sticking efficiencies indicated lower attachment of MS2 compared to the other organisms, although an additional effect of straining in this soil too cannot be excluded. This emphasizes the complexity of the involved processes and the relative contribution of straining, inactivation and attachment/detachment to the overall removal in different soils. Furthermore, it demonstrates that the semiempirical filtration models must be extended with parameters that are related to the permeability of the soil such as pore size distribution and hydraulic conductivity.

Indigenous Micro-organisms and Retarded Transport. Remarkably, the MER observed for the indigenous Coli44 in the Roosteren soil columns (2.8 and 1.5 log) determined prior to the challenge test, was a factor of 1.7 to 2.7 lower than the MER determined for spiked *E. coli* WR1 in these columns. For *C. perfringens* the difference in removal of indigenous and spiked spores was even larger (>1.9 up to

11-fold) and these MER values were lower than those observed for (seeded) MS2. These observations indicate that the elimination rate in soil assessed with peak loading of spiked micro-organisms most likely over-estimates the elimination of indigenous micro-organisms under field conditions. It is hypothesized that this observation is caused by the presence of high concentrations of indigenous Coli44 and *C. perfringens* accumulated in the collected soils from both infiltration sites used for this experiment. Detachment of a fraction of these accumulated micro-organisms probably caused the low MER for unseeded Coli44 and *C. perfringens*. Similar observations of accumulation of *C. perfringens* spores, delayed breakthrough (retardation), and as a consequence a low elimination rate, have been described for slow sand filters (Hijnen *et al.*, 2004a). Clostridial spores are persistent and survive longer than *E. coli* (Medema *et al.*, 1997), which might explain why indigenous *C. perfringens* showed the lowest MER values in three of the four columns. The results of a column study (1.4 m of length; 1.5 $m.day^{-1}$) by Schijven *et al.* (2003), using the same soil and *C. perfringens* spores (D10), support the observations in the current study. They determined a MER (log C/C_o) of approximately 4 log during the peak dose (24 h), which decreased and leveled off to ±2.3 log in the following 23 days of filtration. Modelling of these data revealed a high detachment rate (Schijven *et al.*, 2003). A high detachment rate will cause a high level of remobilization and further transport of the spores. The second breakthrough peak for *Cryptosporidium* oocysts in this study and the study of Harter *et al.* (2001) also indicates that detachment may play an important role in retarded transport for these microbes in the underground.

TRANSLATION OF COLUMN DATA TO FIELD CONDITIONS

The quantitative data of micro-organisms removal during soil passage obtained in this study came from column experiments. It is important to evaluate how these results relate to micro-organisms removal under field conditions. Therefore, column data for MS2, *E. coli*, and *C. perfringens* (indigenous and spiked) were compared to field observations at the same sites, as described by others (Schijven *et al.*, 1999, 2001, 2003 and Medema and Stuyfzand, 2002), plotting elimination rates against the travel time (Figure 9.4).

A field study in Castricum AR site at low temperature (5 ± 3°C; Schijven *et al.*, 1999), demonstrated that removal of MS2 declined as a logarithmic function of travel time ($R^2 = 0.96$; $n = 7$) (Figure 9.4a). MS2 removal in Castricum soil columns determined under the same temperature (Schijven *et al.*, 2003) fitted well in this relationship. The high MER of MS2 observed in the current column study with Castricum soil at 16°C did not fit in the presented relationship (Figure 9.4a). As discussed before, a higher inactivation rate, μ_1, as well as a higher attachment rate, k_{att}, in our columns probably caused this difference.

The MER of the indigenous *E. coli* and *C. perfringens* observed in the Roosteren soil columns and observed under field conditions (Medema and Stuyfzand, 2002) also fitted in a logarithmic function with travel time (Figure 9.4b,c; $R^2 = 0.93$ and 0.94, respectively). This was not the case for spiked MS2 (Figure 9.4a), *E. coli* and *C. perfringens* (Figure 9.4b,c) in the Roosteren soil columns, however. The following differences in the experimental conditions of the column and field study may have contributed to this difference. Soil for the column study was collected from the first meter of the travel distance in the fluvial gravel aquifer. This part of the river bank is heavily loaded with untreated river Meuse water and most likely contains larger amounts of organic and inorganic matter (Table 9.1), compared to the deeper layers in the gravel aquifer. This result in lower permeability of this part of the infiltrated soil compared to the deeper parts. Moreover, field observations were done during extreme flood peaks in winter (Medema and Stuyfzand, 2002). Under these circumstances high flow rates decrease the travel time to the production well from 45–65 days to 10–14 days.

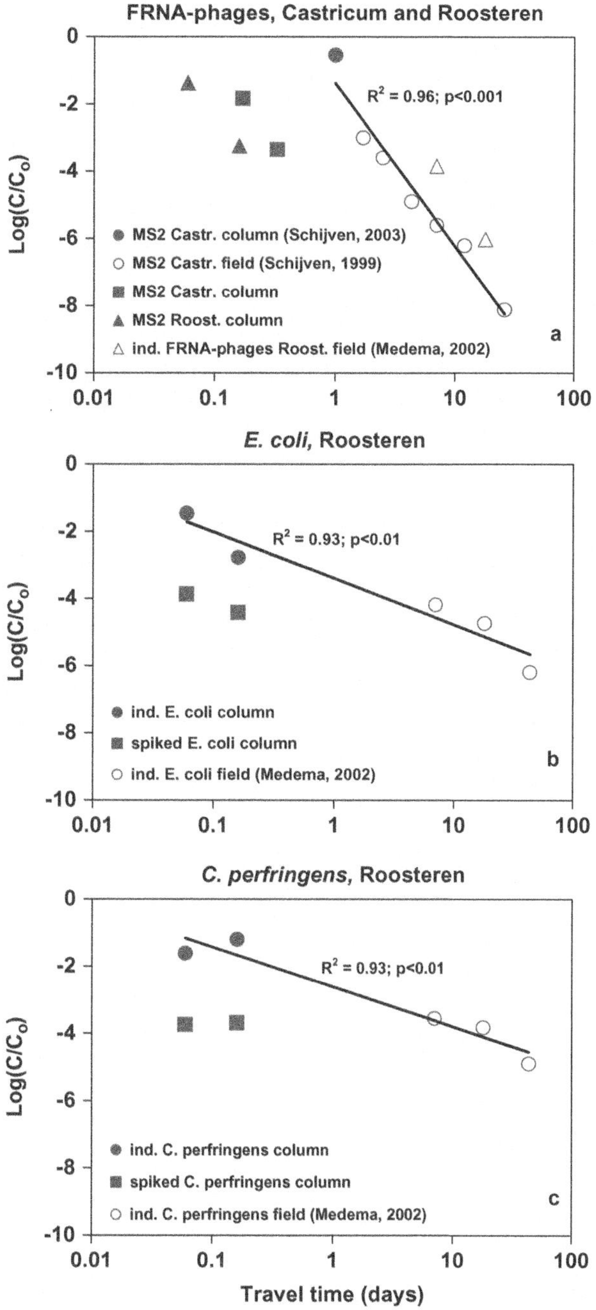

Figure 9.4 Breakthrough of MS2 and FRNA-phages, *E. coli* and spores of *C. perfringens* as function of the contact time (days; notice log-scale) in columns (closed symbols) and field studies (open symbols) combined in a logarithmic relation

Despite these differences, the relatively low elimination rates of indigenous *E. coli* and *C. perfringens* in both columns and in the Roosteren aquifer can be seen as additional support for the occurrence of accumulation, survival and remobilization during microbial transport in the underground. Furthermore, the sequence of MER observed under field conditions was FRNA-phages > *E. coli* > *C. perfringens*, indicating the role of the persistence of these micro-organisms as discussed before. Additionally, it shows that quantitative data about the removal of micro-organims over distance or travel time obtained with columns experiments using spiked micro-organisms cannot simply be extrapolated to the field.

By studying transport of multiple micro-organisms in two different soils with column experiments and comparing theses data with field behavior from the same infiltration sites, the variability of microbial transport in the underground is demonstrated. The largest travel distances can be predicted for viruses from the MS2 results. Despite a relatively high inactivation rate of MS2 in the water, modelling the breakthrough curves showed a high contribution of attachment to the overall removal of these phages. The variability of the contribution of attachment and straining and the importance of the hydraulic conductivity to the transport of micro-organisms in soil was demonstrated by the larger MER-values observed in the coarse gravel soil compared to the fine sandy soil. Furthermore, higher breakthrough of *Cryptosporidium* oocysts compared to *E. coli* and spores of *C. perfringens* in the sandy soil, demonstrate the importance of surface properties of the micro-organisms to the transport behavior. Finally, the significance of detachment and retardation to microbial transport in the underground was deduced from the low MER of indigenous micro-organisms, observed in the columns as well as under field conditions and of the second breakthrough peak of *Cryptosporidium* found after the sandy soil. These results show that the sequence in transport of viruses, bacteria and protozoan (oo)cysts in the underground is governed by the size of the organisms (viruses > bacteria > protozoa) and the hydraulic conductivity of the soil (straining), as well as by the surface properties of both micro-organism and soil, related to their attachment/detachment behavior, and the inactivation rate. Thus, modelling microbial transport in the underground to assess travel times and to set protection zones requires knowledge of the hydrological and geochemical field conditions and quantification of process parameters related to straining, attachment and detachment and inactivation.

Chapter 10

GAC adsorption filters as barriers for viruses, bacteria and protozoan (oo)cysts in water treatment

Wim. A.M. Hijnen[a], Trudy M.H. Suylen[b], Jan A. Bahlman[b], Anke Brouwer-Hanzens[a] and Gertjan Medema[a]

[a]KWR Watercycle Research Institute; [b]Evides Water Company

ABSTRACT

Granular Activated Carbon (GAC) adsorption filtration is commonly used in drinking water treatment to remove NOM and micro-pollutants and on base of the process conditions a certain capacity to eliminate pathogenic micro-organisms was expected. The experiences with the mandatory quantitative microbial risk assessment of Dutch drinking water revealed a lack of knowledge on the elimination capacity of this process for pathogens. The objective of the current study was to determine the capacity of GAC filtration to remove MS2, *E. coli* and spores of *C. bifermentans* as process indicators for virus, *Campylobacter* and protozoan (oo)cyst and more directly of UV treated (oo)cysts of *C. parvum* and *G. lamblia*. Challenge tests with fresh and loaded GAC were performed in pilot plant GAC filters supplied with pre-treated surface water at a contact time which was half of the contact time of the full-scale GAC filters. MS2 phages were not removed and the removal of *E. coli* and the anaerobic spores was limited ranging from ≤0.1–1.1 log. The (oo)cysts of *C. parvum* and *G. lamblia*, however, were removed significantly (1.3–2.7 log). On base of the results of the experiments and the filtration conditions the removal of the indicator bacteria and (oo)cysts was largely attributed to attachment. The model of the Colloid Filtration Theory was used to describe the removal of the dosed biocolloids in the GAC filters, but the results demonstrated that there is a lack of quantitative knowledge about the influence of collector characteristics on the two major CFT parameters, the single collector and the sticking efficiency.

 Quantitative Methods to Assess Capacity of Water Treatment to Eliminate Micro-Organisms. Edited by Wim A.M. Hijnen. ISBN: 9781843393764. Published by IWA Publishing, London, UK.

INTRODUCTION

Microbiological monitoring for faecal indicator bacteria (coliforms, *E. coli*) in drinking water has been the major tool in managing the microbiological safety of drinking water over the last century. In the last decade of the 20th century, however, the limitations of this concept were demonstrated in literature (Dutka, 1973 and Craun *et al.*, 1997). Waterborne outbreaks occurred under conditions where water quality complies with the standards *E. coli* and coliforms and water treatment failed to eliminate peak concentrations of a persistent pathogen such as *Cryptosporidium* (Dykes *et al.*, 1980 and Richardson *et al.*, 1991). To anticipate on these findings a maximum acceptable annual infection risk of 10-4 for pathogenic micro-organisms of faecal origin in drinking water was introduced in the revised Dutch Drinking Water Decree (Anonymous, 2001). Compliance with this requirement has to be demonstrated by means of Quantitative Microbial Risk Assessment (QMRA), including source water monitoring for index pathogens (enteroviruses, *Campylobacter*, *Cryptosporidium* and *Giardia*) and assessment of the elimination capacity of the water treatment (Wetsteyn, 2005). In the last decade much effort was put into the assessment of the elimination capacity of the major water treatment processes applied for water safety i.e. conventional water treatment, disinfection processes and slow sand filtration (Hijnen, 2009). These data have been used in the first QMRA projects of the Dutch Water Companies. These projects showed a lack of quantitative knowledge on the elimination capacity of Granular Activated Carbon filtration (GACF; Schijven *et al.*, 2009). These filters are primarily used as adsorption filters to remove micro-pollutants and natural organic matter (NOM). The assumption in the first QMRA projects was that the elimination capacity of GACF was negligible. Filtration rates and the granular material of GACF, however, are comparable to the rates and grain sizes in rapid sand filtration (RSF). Contact times in GACF are usually longer than in RSF with lower back wash intensity (lower frequency and washing rate). Surface characteristics of GAC also differ from those of sand with a higher adsorption capacity for organic compounds and higher biofilm concentrations (Magic-Knezev and Van der Kooij, 2004). These different conditions may implicate that these filters do eliminate micro-organisms to some extent. Water Company Evides produces drinking water from surface water using conventional treatment, ozonation or UV disinfection, GACF and post-disinfection. In order to complete their QMRA and to enable completion of QMRA at other locations with GACF, knowledge on this process as an additional barrier for pathogenic micro-organisms in treatment facilities had to be qualified.

Objective of this study was to determine the elimination capacity of GACF with fresh and pre-loaded GAC for multiple organisms at pilot plant scale and to determine the removal mechanisms under these conditions mimicking full-scale conditions as much as possible. Elimination of MS2 phages, *E. coli*, spores of Clostridium bifermentans and (oo)cysts of *C. parvum* and *G. lamblia* was studied in a challenge test with a GACF pilot plant. MS2, *E. coli* and the spores were studied as process indicators for enteric viruses, *Campylobacter* and protozoan (oo)cyst removal, respectively (Wetsteyn, 2005 and Hijnen, 2009). The influence of the use of pre-cultured organisms in the study on the assessed elimination capacity was studied by determining the elimination of indigenous *E. coli* and spores of sulphite-reducing clostridia (SSRC) prior to the challenge test.

PILOT PLANT EXPERIMENT UNDER FIELD CONDITIONS

Pilot plant GAC filters. The pilot plant in this study consisted of two duplicate columns (Ø = 0.15 m; height 1.35 m; Figure 10.1) loaded with one meter of fresh GAC (Chemviron F400) for the first test and with one meter of loaded GAC (40.000 bed volumes of filtering) for the second test. The sieve analysis of the materials showed that the median grain size of both GAC samples was 1 mm (0.8–1.1). The loaded GAC had a d10 of 0.65 mm and was slightly more uniform (Uc = d60/d10 = 1.7) than the fresh GAC (d10 = 0.55; Uc = 1.9) with a lower percentage of lower sized (<0.425 mm) material (1.5 versus 5.4% in the fresh GAC). The freshly loaded columns were backwashed prior to the challenge test to create a GACF with the proper hydraulic conditions (in accordance with the full-scale filter beds). In seven steps the backwash velocity was increased from 0 to $35\,m.h^{-1}$ for a bed expansion of 30%. The washing of the fresh GACF was kept at this rate until the wash water was visually clear (free of fines) and the washing of the loaded GACF was kept at this rate for 5 minutes; thereafter the rate was gradually reduced to $0\,m.h^{-1}$. The columns were operated with a constant filtration rate of $5\,m.h^{-1}$ (pore water velocity of $15.6\,m.h^{-1}$; porosity 32%) at a pressure drop of 35.3 ± 6.2 and 21.8 ± 2.3 mBar in the fresh and loaded GACF, respectively. Under these conditions a contact time of 12 minutes was achieved which was half the contact time applied in the full-scale filters of 4 m operated at a rate of $9–10\,m.h^{-1}$.

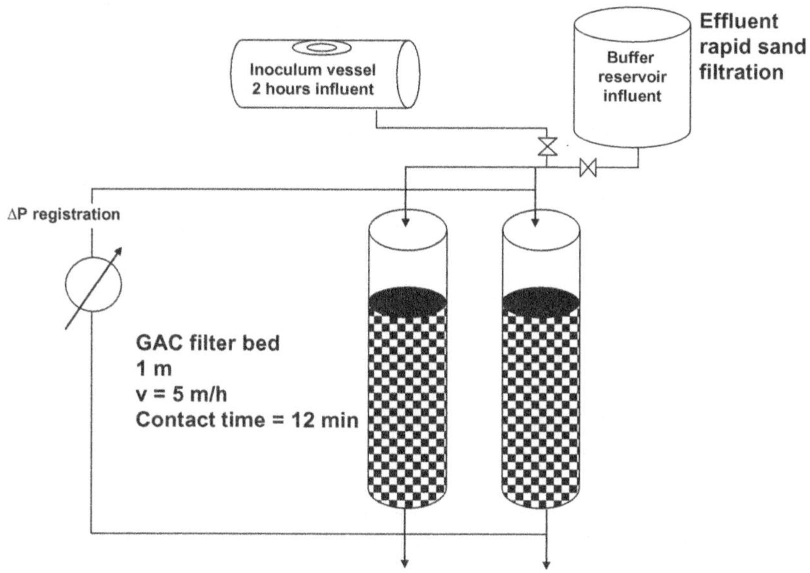

Figure 10.1 Schematic overview of the GACF pilot plant

Water quality. The influent of the GACF was river Meuse water after impoundment reservoirs, coagulation and rapid sand filtration. The physical and chemical water quality was measured (n = 16) during the tests. Turbidity of the influent was 0.09 ± 0.02, pH and conductivity were 8.1 ± 0.0 and $44.8 \pm 0.5\,mS.m^{-1}$, respectively, non-purgable dissolved organic matter (NPOC) was $2.4 \pm 0.1\,mg\ C.l^{-1}$. Temperature (n = 5) during the test with fresh GAC was 14.5 ± 0.8°C and with loaded GAC 10.0 ± 0.8°C. The fresh GAC reduced the NPOC of the water to 0.5 ± 0.1 and the loaded GAC to $2.1 \pm 0.0\,mg\ C.l^{-1}$ in the effluent, showing the difference in adsorption capacity between both GACF.

Challenge tests and sampling procedure. With the pilot plant filters challenge tests were performed. The influent volume for two hours of operation was inoculated with high concentrations of MS2 phages, *E. coli*, spores of *C. bifermentans*, (oo)cysts of *C. parvum* and *G. lamblia* in a separate vessel (Figure 10.1). Prior to the tests the elimination of indigenous indicator bacteria *E. coli* and spores of sulphite-reducing clostridia (SSRC) was determined using large volume sampling (two samples each column). Concentration of indigenous indicator bacteria and the low concentrations of dosed organisms in the tail of the breakthrough curves were determined with large volume sampling using a Hemoflow cross-flow ultrafilter (HF80S; Fresenius, Germany; Hijnen *et al.*, 2007; Veenendaal and Brouwer-Hanzens, 2007). Influent Hemoflow (HF) sample size ranged from 12–40 l and effluent HF sample size ranged from 214–614 l. During the challenge test the influent and effluent was sampled intensively and the effluent sampling was continued for four additional days to assess the tail of the breakthrough curves.

Suspensions of micro-organisms and concentrations in the influent. A suspension of MS2 (ATCC#15597-B1) phages of 1012 PFU.ml^{-1} (GAP EnviroMicrobial Services) was used. This 1 ml MS2 suspension was directly dosed to the 360 l influent vessel. During the test influent samples were taken after 10, 30, 60, 12 minutes from both columns and analyzed for all dosed micro-organisms (n = 4 in every column). The MS2 measured influent concentrations were 1.2 × 109 ± 1.9 × 108 (s.d.) and 7.2 × 106 ± 1.0 × 106 PFU.l^{-1} in the influent of the fresh and loaded GAC columns, respectively. Since MS2 was not removed in the fresh GAC and the removal of phages in filter beds is independent from the influent concentration (Hijnen, 2009), the MS2 influent concentration in the experiment with loaded GAC was lower to obtain a practical range of dilutions. An *E. coli* (WR1) suspension (1.1 × 106) stored at 5 ± 3°C was prepared in a 1 litre of Mineral medium with 1 mg Glucose-C.l^{-1} after growth at 22°C. 500 ml of this suspension was added to the influent vessel resulting in measured *E. coli* concentrations in the influent of the fresh and loaded GAC of 1.1 × 106 ± 3.1 × 105 and 6.0 × 105 ± 2.6 × 104 CFU.l^{-1}, respectively. For the spores of sulphite-reducing clostridia (SSRC) a pure culture of *C. bifermentans* (CP1) was produced as described before (Hijnen *et al.*, 2002). 25 ml of the spore suspension (1.3 × 107.ml^{-1}) was dosed in the influent vessel to obtain measured concentrations of 1.7 × 106 ± 3.9 × 105 (fresh GAC) and 2.0 × 106 ± 5.1 × 105 (loaded GAC) CFU.l^{-1} in the influent of the columns. For the protozoan (oo)cysts *Cryptosporidium parvum* (Holstein calves origin) and *Giardia lamblia* (rodent origin) suspensions were obtained from Waterborne Inc. (108. 8 ml^{-1}; ±10%). These suspensions were UV irradiated before dosing to the influent vessel at measured influent concentrations of 1.1 × 105 ± 3.1 × 104 and 1.6 × 105 ±4.5 × 104 *C. parvum* oocysts and 9.8 × 104 ± 4.2 × 104 and 4.8 × 104 ± 2.4 × 104 *G. lamblia* cysts per litre in the influent of the fresh and loaded GAC columns, respectively.

Microbial analysis. MS2 phages were enumerated with the method NEN-ISO 10705-1 using host strain WG49 and the double layer agar method (coefficient of variation (VC) of inter-laboratory quality control of 25%). For *E. coli* enumeration samples were directly or after membrane filtration incubated on Sodium Lauryl-sulphate Agar (LSA, Oxoid MM0615) for 5 ± 1 hours at 25 ± 1°C and for 14 ± 2 hours at 44 ± 2°C (VC of 19%). For the indigenous *E. coli* yellow colonies were confirmed for indol formation at 44 ± 2°C. Spores of sulphite-reducing clostridia (SSRC; including *C. bifermentans*) were enumerated in pasteurized samples (15 min. 70–75°C) after isolation on 0.2 μm sterile membrane filters which were incubated anaerobically in SA agar (SFP-agar base; BD-Difco 281110) as described before (Hijnen *et al.*, 2002) for 48 hours at 37 ± 2°C (VC of 23%). The concentration of protozoan (oo)cysts including analytical recovery assessment with Colorseed (BTF Decisive Microbiology, North Ryde NSM, Australia) for every individual sample, was determined as described before (Hijnen *et al.*, 2007).

Spatial distribution of the biomass and retained micro-organisms. The biomass concentration in the GAC material (ng.ml^{-1}; volumetric weight of 0.66 g.ml^{-1}) was determined by measuring the concentration adenosine tri-phosphate (ATP) in the GAC by an optimized method (Magic-Knezev and Van der Kooij, 2004) using high energetic sonication (HES; 2 minutes at 45%) to extract ATP from 5 gram of GAC in 100 ml of sterile tap water. The same extraction method was used to determine the concentration of retained micro-organisms in the fresh GAC (CFU or (oo)cysts.ml^{-1}). To determine the extent of attachment to the fresh GAC, selected GAC sample were treated with successively 2 minutes of hand shaking, 2 minutes of vortex mixing, 2 minutes of low energetic sonication (LES) and 2 minutes of HES in separate 20 ml suspensions. This increased the mass recovery of (oo)cysts considerably and therefore this optimized extraction method (separate suspensions combined to one overall suspension) was used to determine the concentration of retained micro-organisms in the loaded GACF. Due to the high (oo)cyst concentrations in the extraction suspensions analytical recovery assessment with Colorseed was not feasible in these samples; to correct for the recovery of the analytical method, the average analytical recovery determined with Colorseed in the analysis of the water samples was used.

Calculations and statistics. The Decimal Elimination Capacity of the GACF for the indigenous indicator bacteria (*DECi*; logremoval) was calculated from the influent and effluent concentrations (Cin and Cout) of the two columns by

$$DEC_i = Log_{10}\left(\overline{C}_{In}/\overline{C}_{out}\right) \tag{1}$$

The DEC for the dosed organisms (*DECd*) was calculated for the separate columns when the influent and effluent concentrations were found to be statistically different (Mann-Whitney test results in one-tailed $p < 0.05$). Basis for the calculation is the mass of organisms in the influent and the filtrate from the Mass balance calculations

$$\sum M_i = \sum M_f + \sum M_c \tag{2}$$

$$DEC_d = \log_{10}\left[\frac{\sum M_i}{\sum M_f}\right] \tag{3}$$

where M_i, M_f and M_c are the total number of organisms in respectively influent and filtrate calculated with $M_i = t_d\overline{C}_iQ$, $M_f = \sum_{i=1}^{n} t_iQ\frac{C_{f,i} + C_{f,i+1}}{2}$ and in the GAC columns $M_c = \sum M_{l,n} = \sum d_{l,n}A\frac{(C_{l,n} + C_{l,n+1}) \times 10^6}{2}$, where t_d = dosing time (hours), Q = volume rate ($l.h^{-1}$), C_i = average concentration in the influent ($n.l^{-1}$), $C_{f,i}$ = concentration in the filtrate of the ithe interval of n intervals in time t_i (hours), $d_{l,n}$ = depth (m) of the n^{the} layer in the column with length l, A = surface of the column (m^2) and $C_{l,n}$ = concentration micro-organisms ($n.ml^{-1}$) in the n^{the} layer. These data were calculated for the two duplicate columns. In the figures the arithmetic average concentrations of both columns are presented with the error bars as the range between the values of the two columns to demonstrate the variability.

Assuming that the elimination in the filter bed was only due to attachment and detachment, the sticking efficiencies of the micro-organisms were calculated from the colloid filtration model (Yao *et al.*, 1971) described with the equation

$$LN\frac{C}{C_0} = -\frac{3}{2}\frac{(1-\theta)}{d_c}\alpha\eta L \tag{4}$$

where d_c as the diameter of the collector, α the collision efficiency, η the single collector collision efficiency, and L the length of the column. η is calculated with the optimized equation presented by Tufenkji and Elimelech (2004a). For the calculations the following parameters values were used: bulk water density $999.703\,kg.m^{-3}$; Hamaker constant for bacterium glass water interface 6.2×10^{-21} J (Rijnaarts *et al.*, 1995); sizes (m) of the MS2 phages 2.1×10^{-8} (Havelaar, 1993), *E. coli* and spores of *C. bifermentans* 1.5×10^{-6}, oocysts of *C. parvum* 4.9×10^{-6} and of cysts of *G. lamblia* 10.8×10^{-6} (Medema *et al.*, 1998); and density ($kg.m^{-3}$) of MS2 and *E. coli* 1085 (Bouwer and Rittman, 1992), spores of *C. bifermentans* 1270 (Tisa *et al.*, 1982) and (oo)cysts of *C. parvum* and *G. lamblia* respectively, 1045 and 1036 (Medema *et al.*, 1998).

REMOVAL OF MICRO-ORGANISMS

Removal of indigenous indicator bacteria. In the two HF samples of the influent and the effluent of the GACF the counts of indigenous indicator bacteria *E. coli* and SSRC were high (>10). Due to these high counts the uncertainty in the assessed concentrations from these two samples is low (Emelko *et al.*, 2008). The concentrations in the influent and effluent calculated from these data were 1–10 and 0.1–2 CFU.l^{-1}, respectively (Figure 10.2). From these data the *DECi* was calculated. Fresh GACF reduced indigenous *E. coli* with 0.4–0.4 log, whereas the GACF with loaded GAC reduced these bacteria with 1.0–1.1 log (Table 10.1). For indigenous SSRC the removal in the fresh GACF was 1.0–1.1 log and in the loaded GACF 0.9–0.9 log.

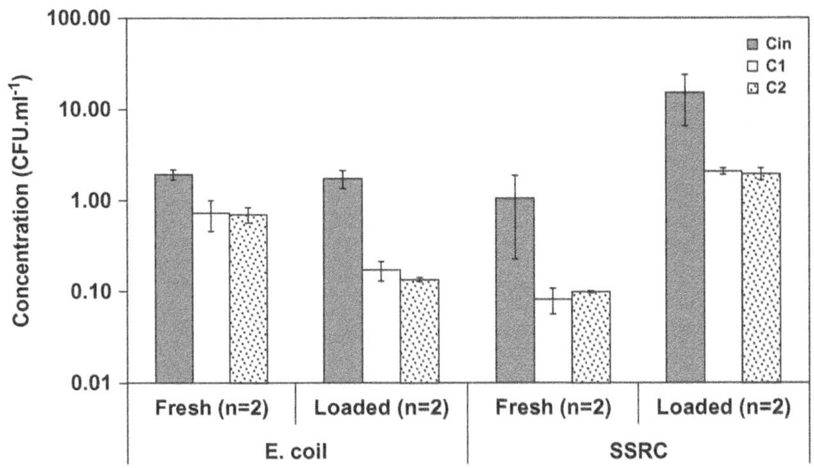

Figure 10.2 Concentration (n.l^{-1}) of indigenous indicator bacteria in the influent and effluent of the columns (error bar denote the range of concentration)

Table 10.1 The Decimal Elimination Capacity (*DEC*; log) of GACF with fresh and loaded GAC for indigenous indicator bacteria and the dosed micro-organisms

Organisms[a]	Fresh GAC (C1; C2)[b]		Loaded GAC (C1; C2)[b]	
	DEC_i[c]	DEC_d	DEC_i[c]	DEC_d
MS2	nd[d]	0.0; 0.0	nd	0.0; 0.0
E. coli	0.4; 0.4	0.0; 0.1	1.0; 1.1	0.3; 0.3
C. bifermentans/*SSRC*	1.1; 1.0	0.4; 0.6	0.9; 0.9	0.4; 0.4
Cryptosporidium	nd	2.7; 2.7	nd	1.3; 1.1
Giardia	nd	2.1; 2.0	nd	2.1; 2.2

[a]no removal ($p > 0.05$); other organisms $C_{in} > C_{out}$ ($p < 0.05$); [b]values of column 1 and column 2; [c]DEC_i elimination of indigenous indicator bacteria; [d]nd = not determined

***Removal of dosed MS2 phages,* E. coli *and spores of* C. bifermentans.** The degree of dispersion of the dosed phages, bacteria and spores in the influent of both columns during the two tests equals the normal analytical variations evaluated with the CV-values. In the fresh and loaded GACF no removal of MS2 phages was observed (Figure 10.3); the influent and effluent concentrations were not significantly different (Table 10.1). The concentration of dosed *E. coli* bacteria and spores of *C. bifermentans*, however, was reduced in the GAC columns (Figure. 10.4a,b; C/C_o<1.0). The removal of *E. coli* in the fresh GAC was limited (≤0.1 log;

Table 10.1). The calculated *DEC* for the dosed indicator bacteria in both tests ranged from <0.1 to 0.6 log (Table 10.1) which was lower than the indigenous indicator bacteria. Despite this difference in DEC values, the trends observed for the removal of dosed and indigenous indicator bacteria were similar: higher *E. coli* removal in the loaded GACF and no significant difference in SSRC removal in the fresh and loaded GACF.

Removal of the protozoan (oo)cysts. A considerable removal of *C. parvum* and *G. lamblia* (oo)cysts was observed in the GACF with both fresh and loaded GAC (Figure 10.5a,b; $C/C_o << 1.0$). The concentration used for the calculations was corrected for the recovery of the analytical method determined for the individual samples; the average analytical recovery for *C. parvum* was about 13% and for *G. lamblia* about 7% (Table 10.2).

Based on these data the DEC values for *C. parvum* were 2.7 and 1.2 log for the fresh and loaded GACF, respectively (Table 10.1). The average removal rate of *G. lamblia* cysts by both fresh and loaded GACF was 2.1 log.

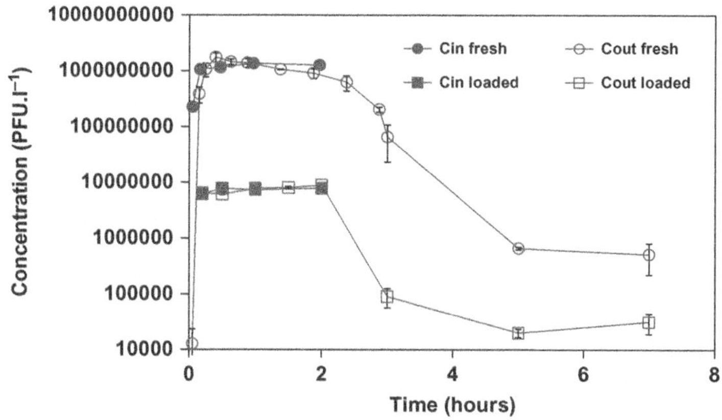

Figure 10.3 Concentration MS2 phages in the influent and effluent of the GACF (error bar denote range)

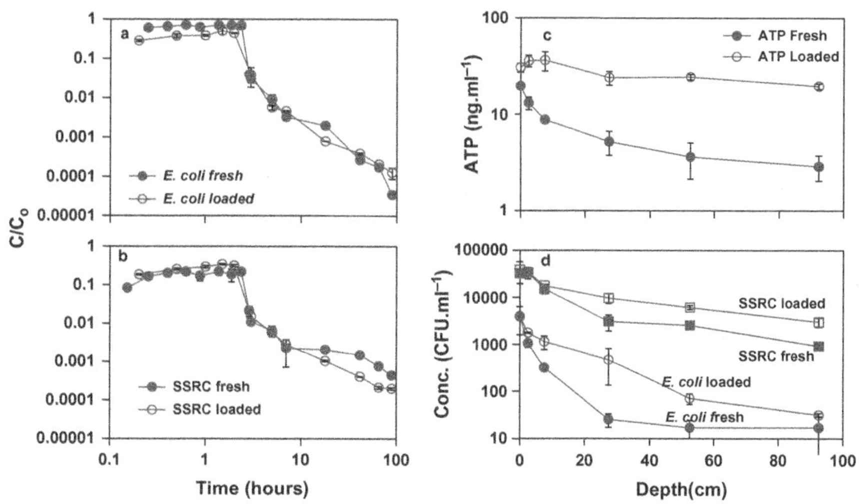

Figure 10.4 Breakthrough curves (BTC) of E. coli and spores of *C. bifermentans* from the challenge tests with fresh and loaded GACF and the corresponding spatial distribution of ATP and retained micro-organisms in the filter beds (error bar denote range)

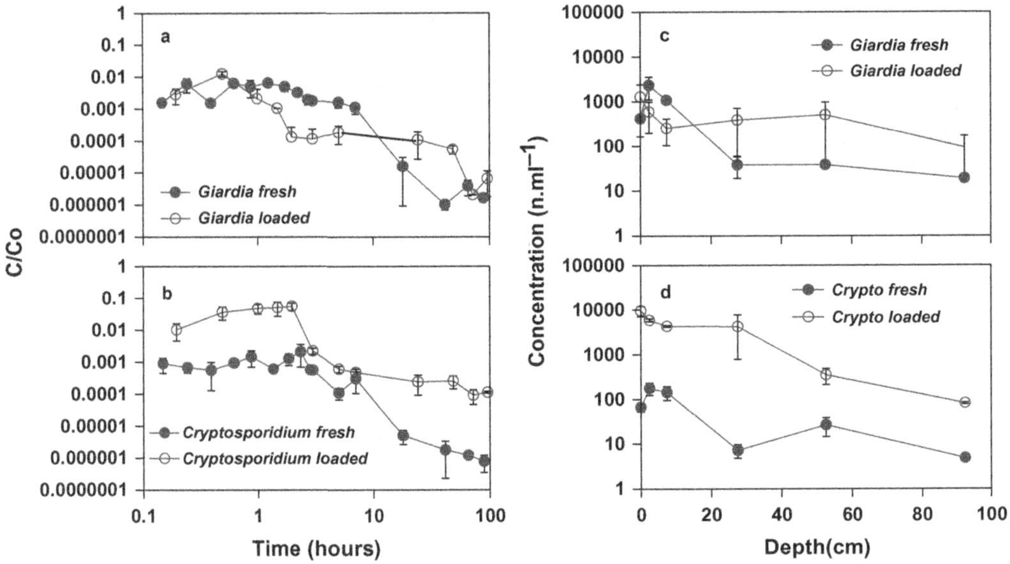

Figure 10.5 Breakthrough curves (BTC) of (oo)cysts of C. parvum and G. lamblia from the challenge tests with fresh and loaded GAC and the corresponding spatial distribution of retained micro-organisms in the filter beds (error bars denote range)

Table 10.2 Analytical recovery percentages of (oo)cysts isolated directly from influent and effluent water samples and from concentrates of the large volume sampling with the hemoflow (HV)

Organisms	Isolation	Fresh GAC[a]	Loaded GAC[a]
C. parvum	Direct	13.9 ± 6.0 (n = 26)	13.1 ± 9.2 (n = 16)
C. parvum	HV	13.0 ± 13.7 (n = 7)	12.3 ± 5.8 (n = 7)
G. lamblia	Direct	6.6 ± 7.9 (n = 26)	7.4 ± 7.8 (n = 16)
G. lamblia	HV	7.9 ± 7.1 (n = 7)	4.1 ± 2.3 (n = 7)

[a]*±s.d.; n = number of samples*

MICRO-ORGANISMS IN THE GAC COLUMNS

Spatial distribution of biomass and retained micro-organisms. One week after the challenge tests the spatial distribution of the concentrations of active biomass and retained micro-organisms in the filter bed was determined. The active biomass concentration in the fresh GAC determined with ATP was high (19.6 ± 0.8 ng.ml^{-1}) in the first 10 cm of the filter bed and decreased rapidly as function of the depth (Figure 10.4c). The ATP concentration in the loaded GAC was significantly higher than in the fresh GAC and showed only a limited decrease over the filter bed depth. The spatial distribution of retained *E. coli* and spores of *C. bifermentans* depicted in Figure 10.4d shows a hyper-exponential decline of concentration as function of the bed depth. For both micro-organisms a higher concentration was observed in the loaded GAC compared to the concentration in the fresh GAC. For *E. coli* this agrees with the higher removal in the loaded GAC (Figure 10.4a), however, for SSRC this was not the case. The DEC of the fresh and loaded GACF were similar. A same result was observed for the protozoan oocysts. The spatial distribution of these organisms in the fresh and loaded GAC is presented in Figure 10.5c,d and showed higher concentrations of retained (oo)cysts in the loaded GACF than in the fresh GACF. This is remarkable especially for *C. parvum*, considering the higher retention in the fresh GACF compared to the retention in the loaded GACF (BTC; Figure 10.5b). Higher concentrations of spores and (oo)cysts in the loaded GAC than in fresh GAC was caused by the use of an optimized extraction method used for the loaded GAC samples as explained next.

Attachment of the (oo)cysts to the GAC. On base of the high removal rate of the fresh GACF for (oo)cysts and the low mass recovery of attached (oo)cyst from the GAC samples of these columns, a strong attachment of these micro-organisms to the GAC was suspected. To investigate the attachment of (oo)cysts to the GAC, selected samples of the fresh GAC were tested with a number of separate extraction methods with increasing intensity. Two mixed GAC sample of fresh carbon derived from the first 30 cm of the filter bed of column 1 and 2 were analyzed by a number of successive conducted extraction methods: hand shaking, vortex mixing and low energetic ultrasone treatment (LES). The (oo)cyst concentration assessed in the separate suspension after hand shaking was in the same order of magnitude or higher than the (oo)cyst concentration in the separate suspension after the single HES extraction method initially applied (Table 10.3; average concentration calculated from the separate samples of the first 30 cm). A further increase in extracted (oo)cysts from the same sample was observed after vortex mixing, but especially after LES extraction of the same sample the number of released (oo)cysts increased considerably (Table 10.3). Based on the cumulative numbers of (oo)cysts in the separate suspensions it was calculated that LES yielded 91.2 – 63.7% of the *Cryptosporidium* oocysts in the GAC of column 1 and column 2, respectively and 66.7 – 57.2% of the *Giardia* cysts in the GAC of both columns. These results of the fresh GACF demonstrated that a single step HES treatment was not appropriate to determine the concentration of retained (oo)cysts in the GAC. Consequently, the GAC samples from the loaded GACF were analyzed by means of the combined extraction method (hand shaking, vortex mixing, LES) followed by a final HES treatment. This yielded higher (oo)cysts concentrations in the loaded GACF as presented in Figure 10.5c,d.

Table 10.3 Concentration of *C. parvum* and *G. lamblia* in fresh GAC (n.ml^{-1}) sampled from the first 30 cm filter bed of column 1 and column 2 determined in the separate suspension of a single HES extraction and in the separate suspensions of three successive extraction methods

		C. parvum		*G. lamblia*	
		C1	C2	C1	C2
Average 0–30 cm single	HES extraction	93.9	104.8	712.6	1224.3
Successive extraction	Hand shaking (H)	77.1	541.5	556.4	4402.7
	Vortex mixing (V)	98.3	576.2	1393.1	7371.7
	LES (L)	1811.4	1959.8	3899.1	15738.5
	Cum. (H+V+L)	1986.7	3077.4	5848.6	27512.9

MASS BALANCE AND COLLOID FILTRATION THEORY

Mass balance and CFT calculation. Mass balance calculations were conducted using the concentrations of detected micro-organisms in the filtrate and the GAC. The results revealed a high mass recovery of MS2 (114.9–129.4%) and of *E. coli* in the fresh GAC experiment (94.1%; Table 10.4). The high recovery (>100%) for MS2 cannot be caused by growth or introduction of indigenous phages. The contact time was too short and under the test conditions growth of phages has not been reported in literature. Furthermore, the observed concentrations of indigenous F-RNA phages in the water used as the influent (data not presented) are far below the concentrations of dosed phages in the test. A higher dispersion of the dosed phages due to the filtration process or analytical variations (CV of 25%) resulting in an inaccurate mass estimate are more likely explanations. The mass recovery of *E. coli* in the loaded GAC experiment was 63.1% and lower most likely because of decay during the time delay between the dosing experiment and the sampling of the filter bed. The mass recoveries of spores of *C. bifermentans* in the fresh and loaded GAC were 74.3 and 83.9%, respectively. The Because of the low concentrations of retained (oo)cysts in the fresh GAC, the recoveries for these organisms of the test with these columns were low; 3.6 and 30.5% for *C. parvum* and *G. lamblia*, respectively (Table 10.4). The mass recovery percentages for these organisms in the tests with loaded GAC assessed with the optimized extraction method were considerably higher (81.3 and 64.6%, respectively, Table 10.4).

Assuming that all micro-organisms were removed by attachment, the single collector efficiencies η and collision efficiencies α were calculated using the removal data (Table 10.1) and the specified characteristics of the GACF and different micro-organisms (size and density). Except for *E. coli* removal in the fresh GAC columns, all α values were above unity (Table 10.4). In the fresh GAC the highest α was calculated for *C. parvum* and for the loaded GAC the stickiest micro-organism was *C. bifermentans*.

DISCUSSION

GAC filters as microbial barriers in water treatment. This study showed that GAC absorption filters are no barrier in water treatment for the elimination of viruses. MS2, an icosahedrische F-specific bacteriophage (Ø 21 nm and iso-electric point (pI) of 3.5 (Penrod *et al.*, 1996)) regarded as a conservative process indicator for enteric virus removal (Schijven *et al*, 2003), was not removed in the fresh and loaded GACF. Other authors have reported a low capacity of GACF to eliminate viruses and bacteriophages (0.2–0.7 log; Guy *et al.*, 1977; Scott *et al.*, 2002 and Persson *et al.*, 2005).

The presented data showed that GACF have a limited capacity to eliminate pathogenic bacteria; *E. coli* can be regarded as a process indicator for the removal of pathogenic bacteria such as *Campylobacter* in filtration processes (Hijnen, 2009). DEC of the GACF process for indigenous *E. coli* assessed with large volume sampling (Hijnen *et al.*, 2000a) was higher than the DEC for the pre-cultured *E. coli* bacteria. This observation was consistent for all experiments and also for the removal of SSRC. This can be caused by more favourable surface properties and/or particle association of the indigenous indicator bacteria. The DEC values for indigenous *E. coli* of 0.5 and 1.1 log for fresh and loaded GACF were similar to the removal rates reported by others (Gray *et al.*, 1997 and Hijnen and Medema, 2007). Spores of sulphite-reducing clostridia SSRC used in the Dutch regulations as a process indicator for the removal of protozoan (oo)cysts (Wetsteyn, 2005) were removed to the same extent as *E. coli* bacteria in the loaded GACF (Table 10.1). Low average elimination capacity of 0.1 log with sometimes negative values was observed for indigenous SSRC in a pilot plant filter bed operated for about one year under full-scale conditions with backwashing (Hijnen *et al.*, 1997). This low and variable removal was attributed to accumulation and delayed breakthrough. Higher removal rates for SSRC were observed at full-scale GAC filters ranging from 0.4–1.1 log (Hijnen and Medema, 2007). Rice *et al.* (1996) and Gray *et al.* (1997) presented data on the removal of spiked aerobic spore forming bacteria (ASFB) with DEC values in the same range (0.3–1.4 log). The use of indigenous ASFB as process indicator, however, has limitations because there are serious indications of regrowth in GACF (Galofre *et al.*, 2004 and Mazoua and Chauveheid, 2005).

The current study yielded high removal rates of (oo)cysts of *C. parvum* (1.1–2.7 log) and *G. lamblia* (2.0 log) in GACF. For removal of *Giardia* cysts by GACF operated at a rate of 6.8 m.h^{-1}, a higher DEC of 3.3 ± 0.3 log was reported by Logsdon *et al.* (1995). Patania *et al.* (1995), however, described lower DEC values of 0.4–1.3 log for *Cryptosporidium* and *Giardia* (oo)cysts removal by GACF operated at high filtration rates of 12–18 m.h^{-1}. Comparison of these removal rates with the removal rate determined for the anaerobic bacterial spores showed that these spores are too conservative process indicators for protozoan (oo)cyst removal by GACF.

Table 10.4 The mass recovery (%) of the different dosed micro-organisms calculated from the mass balance data from the BTC and the spatial distribution after one week and the single collector efficiency η and collision efficiency α calculated from the CFT model equation using two different grain size characteristics

	Recovery (%)[b]	Collector size d50 (1.0 mm)		Collector size d10 (0.55; 0.65 mm)[a]	
		Collector eff. η^b	Collision eff. α^b	Collector eff. η^b	Collision eff. α^b
MS2	114.9 ± 32.9; 129.4 ± 4.0	0.0074; 0.0067	nd	0.0108; 0.0112	nd
E. coli	94.0 ± 28.6; 63.1 ± 2.9	0.0005; 0.0004	0.227; (1.679)	0.0010; 0.0008	0.062; 0.570
C. bifermentans	74.3 ± 15.7; 83.9 ± 16.9	0.0006; 0.0004	(2.011); (2.042)	0.0011; 0.0008	0.674; 0.714
C. parvum	3.6 ± 0.0; 81.3 ± 18.6	0.0014; 0.0013	(4.212); (2.077)	0.0036; 0.0027	0.935; 0.653
G. lamblia	30.5 ± 17.6; 64.6.4 ± 72.7	0.0043; 0.0042	(1.076); (1.245)	0.0108; 0.0081	0.242; 0.381

[a]*fresh and loaded GAC;* [b]*fresh and loaded GAC; average of two columns ±s.d.*

Removal mechanisms: straining. The sequence in DEC values observed in the current study for the different groups of micro-organisms was more or less associated with the size of the micro-organisms: bacteriophages < bacteria ~ bacterial spores < protozoan (oo)cysts. Based on geometric modelling Sakthivadivel (1966; 1969) showed that straining as a removal mechanism in granular material becomes of significance at a size ratio between colloid and collector (dp/dc) of 0.05. From experimental observations Bradford *et al.* (2004) and Xu *et al.* (2006) suggested lower straining threshold ratios of 0.005 and 0.008, respectively. The ratios calculated for the removal conditions of MS2 phages and both indicator bacteria (*E. coli* and SSRC) in the GACF of the current study are 0.00002 and 0.00150, respectively. Compared to the previously mentioned threshold ratios, this indicates no or a minor contribution of straining in the removal of these organisms. The ratios for *C. parvum* and *G. lamblia* in the current study are 0.0049 and 0.0108, respectively, which is around or higher than the lowest suggested threshold ratios for straining. Higher removal of the smaller *C. parvum* in the fresh GAC compared to *G. lamblia* (Table 10.1) and the low recovery of both protozoan (oo)cysts after hand shaking of 3.9–17.6 and 9.0–16.6%, respectively (Table 10.3), are indications for a minor contribution of straining in the removal of these micro-organisms by the GACF.

Removal by attachment. Driving forces in the attachment of micro-organisms or colloids to surfaces are electrostatic forces at longer distances and van der Waals attraction at shorter distances. MS2 removal in granular materials is under most conditions solely related to attachment and these bacteriophages were not removed in the GACF. This indicates that the attachment conditions for these micro-organisms in these filter beds are not favourable. This may be explained by the relatively high pH of the influent of these filters of 8.0 which will enhance electrostatic repulsion due to increased negative charged surfaces of the phage. These conditions in the GAC filters most likely seriously hamper hydrophobic surface interaction which is of importance for MS2 retention in soil (Bales and Li, 1993) and on membrane filters (Farrah, 1982). Adsorption experiments have shown that MS2 can attach to GAC F-400, but these observations were obtained in laboratory batch tests and small scale column tests performed at lower pH 7.4 and filtration rate of $0.9\,m.h^{-1}$ (Powell *et al.*, 2000).

The significant role of electrostatic forces and positively charged carbon surfaces on bacterial adhesion to GAC was demonstrated recently (Rivera-Utrilla *et al.*, 2001 and Busscher *et al.*, 2006). After an acid treatment of GAC grains to remove positively charged metals from the surface, the adsorption capacity of this GAC for *E. coli* was reduced (Rivera-Utrilla *et al.*, 2001). The higher attachment rate of *E. coli* bacteria and Clostridium spores compared to MS2 indicates either lower electrostatic repulsion or a higher hydrophobic interaction, despite the lower η value (Table 10.4). The increased removal of indigenous *E. coli* and SSRC compared to the dosed pure cultures of these bacterium species is most likely caused by a more neutral surface charge of the indigenous bacteria due to the application of coagulation with Fe2(SO4)3 of the influent prior to the GACF.

The contribution of attachment in the removal of *C. parvum* and *G. lamblia* was deduced from the data of the extraction experiment (Table 10.3) where after a preceding hand shaking and vortex mixing of a GAC sample, most of the (oo)cysts (57.2–91.2%) were released from the GAC by a subsequent more intensive ultrasone LES treatment. The relatively low mass recovery of the more intense single HES extraction applied to the fresh GAC samples was attributed to the release of carbon fines which negatively affected the IMS recovery of the (oo)cyst analysis.

The CFT calculations using the average characteristics of the collector and the biocolloids and assuming both particles are spherical, yielded collision efficiencies α that were high and mostly above unity. The CFT calculations are based on ideal conditions with spherical collector and colloids. The used GAC is Chemviron F400, which is a re-agglomerated carbon grain, non spherical, angular and thus irregular of shape. Furthermore, the grain size is not homogeneously distributed over the GACF depth after back washing, with the smallest grain sizes on top of the filter bed. The shape of the collectors has a large impact on the single collector efficiencies η as recently presented (Saiers and Ryan, 2005) with η values that differ more than a factor of two. No data of the shape of the GAC grains are available and therefore we used the effective size of this material which approximates the d10 value for further calculations; this yielded more realistic α-values (Table 10.4). Saiers and Ryan (2005) suggested on base of their theoretical considerations that the effect of non-spherical shaped particles on the calculated CFT parameters for colloids of 4 µm or larger is more significant than for the smaller viruses and bacteria. They also discussed the impact of surface roughness on the colloid deposition and suggested that this characteristic has a larger influence on α than on η and is of more importance for biocolloids of 1 µm or larger. These findings demonstrate that the CFT model is designed for ideal conditions and application of the model to environmental conditions is hard without considering the variability of the studied system as presented before in literature for MS2 and microspheres transport in sand (Knappett *et al.*, 2008). This statement is also supported by the observed spatial distribution profiles in the columns (Figure 10.4 and 10.5). Under conditions of uniform particle characteristics (colloid and collector) and stable filtration conditions the deposition rate in the filter bed is constant yielding an exponential decrease of retained micro-organisms in the filter bed (Tufenkji and Elimelech, 2004b and Tufenkji, 2007). This results in a linear decline of the concentration in the filter bed in a log-linear plot as used in Figure 10.4 and 10.5. The spatial distribution of bacteria, bacterial spores and (oo)cysts in the fresh and loaded GACF, however, was non-linear and hyperexponential. For this deviation from the CFT model the heterogeneity of the colloids and collector conditions was suggested in literature (Tufenkji and Elimelech, 2004a, 2005 and Tong and Johnson, 2007). As a result of this heterogeneity the filtration conditions in the filter bed are not constant which is clearly demonstrated by a higher microbial density measured with ATP (Magic-Knezev and Van der Kooij, 2004) in the upper part of the filter bed (Figure 10.4). Comparison of the spatial distribution of ATP in the fresh and loaded GAC with the spatial distribution of *E. coli* and SSRC suggests an enhanced attachment of bacteria and bacteria spores at higher biofilm concentrations. Unfortunately for protozoan (oo)cysts this comparison is not conclusive since the spatial distributions assessed in the fresh GACF was not reliable as discussed before. The data of the BTC curves, however, showed a lower removal of *Cryptosporidium* oocysts in the loaded GACF with higher biofilm concentration, whereas the removal of *Giardia* cysts was not affected by the differences between fresh and loaded GAC. An enhancement of colloid removal due to increased microbial activity in the filter bed was also presented by Morales *et al.* (2007). They calculated an increasing α value for the attachment of Laponite RD colloids in a filter bed fed with gluconate, an easily assimilable organic compound. Further research is necessary to elucidate the effect of biofilms in GAC on the attachment and removal of pathogenic micro-organisms.

Fate of the retained micro-organisms. The fate of retained micro-organisms in the GACF was an aspect of concern in the early days of the use of GAC in water treatment with the focus on association with GAC fines leaving the filter bed (Camper *et al.*, 1985) causing increased resistance against post-disinfection (LeChevallier *et al.*, 1984). A role of predation in a slow sand filter on the fate of retained persistent (oo)cysts was suggested more recently (Hijnen *et al.*, 2007). The role of predation in water treatment as an aspect in pathogen transmission through drinking water has been reviewed recently (Bichai *et al.*, 2008) and is also explored further in connection with the data of the current study by Bichai *et al.* (2009, this issue). In the present study the GAC filter beds were not washed, whereas under full-scale operations at the treatment plant Berenplaat back washing is required at a variable frequency. Back washing of GAC filter beds enhances removal and/or further remobilization of retained micro-organisms. On the base of the high observed DEC values for protozoan (oo)cysts it is of importance to study the removal of these pathogens by GAC filters operated under normal operations which includes back washing.

CONCLUSIONS

The results of this pilot plant study demonstrate that GAC adsorption filters are no barrier for viruses in water treatment. For pathogenic bacteria these filters contribute to the overall elimination in water treatment but the process can not be regarded as a major barrier. The elimination of pre-cultured indicator bacteria (*E. coli* and

SSRC) was lower than the elimination of indigenous indicator bacteria assessed with large volume sampling (0.5–1 log). For pathogenic protozoan (oo)cysts the results clearly showed that GAC adsorption filters can be serious barriers in water treatment. The observed DEC values ranged from 1.3 to 2.7 log under the applied experimental conditions. On base the differences in removal rates Clostridium spores is a too conservative process indicator for protozoan (oo)cyst removal by these filters. Attachment of bacteria, bacterial spores and protozoan (oo)cysts appeared to be the dominant removal mechanism in the GACF. Increased biomass concentrations in the GACF enhance bacterial removal, but for the oocysts of *C. parvum* the attachment conditions were less favourable at higher biomass concentrations. Further studies are recommended to verify the influence of GAC type, biofilms and back washing on the elimination capacity of full-scale GAC adsorption filters for protozoan (oo)cysts.

Chapter 11

Survival of *Clostridium* spores in river water and in sand from a slow sand filter

Wim A.M. Hijnen, Leo Heijnen and Gertjan Medema

KWR Watercycle Research Institute

ABSTRACT

Spores of sulphite-reducing clostridia (SSRC; incl. C. perfringens) have been introduced in water quality monitoring for microbial safety of drinking water because of their higher resistance to water disinfection processes than E. coli. Interpretation of this parameter, however, is hampered by lack of quantitative data on the spore decay rates. We determined decay rates of environmental Clostridium spores in river water and in filter sand in a twelve year microcosm study at different temperatures. Decay was not significant at 3°C and limited at 15°C. First order decay rates in water and sand were in the same order of magnitude and ranged from of 0.001 up to 0.0023 d^{-1}. Molecular typing of isolates of surviving SSRC showed diversity in the species at different temperatures. From the results in combination with literature data it was concluded that there is heterogeneity in the longevity of environmental SSRC. There is a need for more information on the survival of C. perfringens spores which is essential for interpretation of this parameter in water microbiological quality control.

INTRODUCTION

The use of faecal indicator bacteria in the control of waterborne diseases was introduced at the end of the 19th century. *Escherichia coli* (Escherich, 1885) was the first bacterium of faecal origin introduced in water quality monitoring. Later the use of *Bacterium enteritidis sporogens*, renamed *Clostridium perfringens*, was proposed as faecal indicator bacterium (Klein and Houston, 1898–1899). In 1984 this indicator bacterium was introduced in the Dutch Drinking Water Decree in 1984 (Anonymous, 1984) because of their higher persistence

 Quantitative Methods to Assess Capacity of Water Treatment to Eliminate Micro-Organisms. Edited by Wim A.M. Hijnen. ISBN: 9781843393764. Published by IWA Publishing, London, UK.

against disinfection processes and environmental stress than *E. coli*. Persistent pathogenic micro-organisms of interest for microbiological safety of drinking water are *Giardia* and *Cryptosporidium*. The analytical method for *C. perfringens* was based on the measurement of the non-specific group of spores of sulphite-reducing clostridia (SSRC; Anonymous, 1985a). Initiated by the introduction of more preventive strategies in water safety the SSRC parameter was additionally proposed (Hijnen *et al.*, 1997; 2009) and implemented (Wetsteijn, 2005) as an process indicator for the removal of protozoan (oo)cysts. Although there is literature showing that *Clostridium* species other than *C. perfringens* are not necessarily of faecal origin, there are also studies which demonstrate that these anaerobic spore forming bacteria can often been found in the intestinal tract of mammals (Collins *et al.*, 1994; Duncan *et al.*, 2007 and Song *et al.*, 2002; 2004). For interpretation of data on the prevalence of *Clostridium* spores in the environment with respect to faecal pollution and water safety, quantitative information on the decay rate of these spores is needed. The longevity of bacterial spores in the environment has been deduced from several studies showing survival in culture media or environmental soil samples after periods of 3–30 years (Evans and Curran, 1959 and Hofer and Davis, 1972) but decay rates were not reported. A number of recent studies showed hardly any decay of *C. perfringens* spores in river water, groundwater and sediments from the sea during relatively short periods of 42 up to 120 days (Medema *et al.*, 1997; Schijven *et al.*, 2000 and Davies *et al.*, 1995). The objective of the current study is to determine the decay rate of indigenous SSRC in river water and sand from a slow sand filter over a longer period (>10 years) at different temperatures. The most persistent *Clostridium* species were isolated form these samples after twelve years of incubation and identified by molecular typing.

SURVIVAL STUDY UNDER LABORATORY CONDITIONS

River water and sand samples. The decay of environmental SSRC and a pure culture of *Clostridium bifermentans* was determined in river Rhine water sampled from the Lek canal at Nieuwegein. The turbidity of the water was 18 NTU, the pH was 7.8 and the TOC-content 2.6 $mg.L^{-1}$. Decay of environmental SSRC was measured in sand sampled from a slow sand filter which was in operation for 34 years. Two sand samples were used, a sample form the top of the filter bed (TL sand) and a sample from a layer 40–70 cm below the top (40–70 sand).

Microcosm experiments. River Rhine water (600 ml) was incubated in 1 Liter Pyrex glass flasks with sterile air exchange (0.22 μm membrane filter) stored in a refrigerator at 3 ± 2°C and in incubators at 15 ± 1°C and 25 ± 1°C in the dark. Sand samples of 50 ml were stored in sterilized 100 ml glass flasks either saturated with sterile tap water or as unsaturated moist sand and incubated under the same conditions as the water samples. During the incubation, samples of water and sand were removed aseptically from the flasks for microbiological analysis. Variable water volumes (1–500 ml) were taken from the river water for SSRC analysis (volume depended on the number in the water). The sand sample size ranged from 3–10 gram of sand (volumetric weight 1.5 $g.ml^{-1}$). The sand samples were suspended in 100 ml sterile tap water and treated for 3 minutes in a low energetic ultra-sonic cleaner (Bransonic 5510, Branson ultrasonic, Danbury, USA) to separate the SSRC from the sand. The microcosms were sampled more frequently during the first four years than in the subsequent period. After eight years one sample was analyzed and during the eleventh and twelfth year of incubation the microcosms were sampled until exhaustion.

Microbiological methods. Production of the spore suspension of *C. bifermentans* inoculated in the river water was described before (Hijnen *et al.*, 2002). SSRC numbers in water samples and in the sand/water suspensions were determined after pasteurization (30 min. at 70 ± 1°C) directly (samples of ≤1 ml) or after membrane filtration (>1 ml) by culturing in Perfringens-Agar-Base medium (PAB; Oxoid CM 587; Anonymous, 1985a). The samples were mixed with the liquefied medium in the cover of a petri dish and covered with the bottom to create anoxic conditions; this method is prescribed in the Dutch Drinking water Decree (Anonymous, 1985a and Havelaar *et al.*, 1983). Black colonies were enumerated after 48 ± 4 hour of incubation at 37 ± 1°C. The variation coefficient (SD/avg) of the SSRC analysis in water of 27.6% was derived from repetitive analysis of an internal laboratory standard (pure culture). Variation coefficient of the SSRC analysis in sand samples was 25.6%, determined for indigenous SSRC population in a separate sand sample (n = 10).

Molecular typing of SSRC. At the end of the study SSRC isolates were obtained from the PAB medium. Colonies were isolated by means of a small sterile pipette from the medium and suspended in 0.5 ml sterile DNA-free tap water. This isolate suspension was mixed intensively and inoculated and incubated in PAB medium under anoxic conditions. From these isolates, only the black colonies were recovered and cell material from the colonies was suspended in 0.5 ml DNA-free sterilized water. PCR (Polymerase Chain Reaction) was used on 10 μl of this suspension to amplify a ca. 1380 bp fragment of the 16S rRNA gene using primers 8F (5′-AGAGTTTGATC(A/C)TGGCTCAG-3′) and 1392R (5′-ACGGGCGGTGTGTACA-3′) each in number of 200 nM in a reaction mixture containing 25 μl 2X PCR master mix (Promega, Madison, USA) in a reaction volume of 50 μl. The temperature cycles for PCR were: 3 min. 95°C, followed by 30 cycles of 45 sec. 95°C, 1 min. 57 °C and 2 min. 72 °C. Sequence analyses of part of the fragments was performed by Baseclear (Leiden, The Netherlands) using the 8F primer and verified using primer 907R (5′-CCGTCAATTCCTTTRAGTTT-3′) primers. The Blast tool (Altschul *et al.*, 1990) at the National Center for Biotechnology Information (NCBI) was used to find matching sequences in the DNA-sequence databases.

Decay rate calculation. The decay is described by a first order process with the following equation:

$$\ln \frac{C_o}{C_t} = -kt$$

where C_o and C_t are the numbers microorganisms at time t_0 and t (days) and k is the first order inactivation rate constant (day^{-1}). The data of the survival curves were analyzed with the non-parametric Mann-Whitney test (SPSS 16.0). The linear regression analysis of the decay curves were calculated with the Microsoft Excel Analysis ToolPak.

SURVIVAL UNDER DIFFERENT CONDITIONS

Decay in river water. The SSRC number during the first 4 years of incubation in the river water with and without inoculation of *C. bifermentans* spores documented as a species of faecal origin (Duncan *et al.*, 2007) showed no decline in the samples at 3°C (Figure 11.1). In the samples stored at 15 and 25°C a slight decline of the SSRC number was observed during these 4 years, but this decline was not significant deduced from the variability presented in Figure 11.2. In Figure 11.2 the average numbers (±SD) in the samples calculated from four consecutive incubation periods are presented. The observed variation coefficients during these separate periods for indigenous SSRC and the pure culture *C. bifermentans* in the river water were 52% (33–82) and 55% (16–84), larger than derived from a positive control in the internal quality control (27.6%). A significant decrease in SSRC numbers was observed when incubated at 25°C (Figure 11.2a); average number after 11–12 years of incubation was significantly lower than in the first year ($P < 0.05$). Statistical tests showed that the average SSRC number in the river water after twelve years of incubation at 3 and 15°C was not significantly different from the initial numbers. The *C. bifermentans* spore number in the river water after twelve years of incubation was significantly ($P < 0.05$) lower than the number in the first year of incubation at all temperatures (Figure 11.2b).

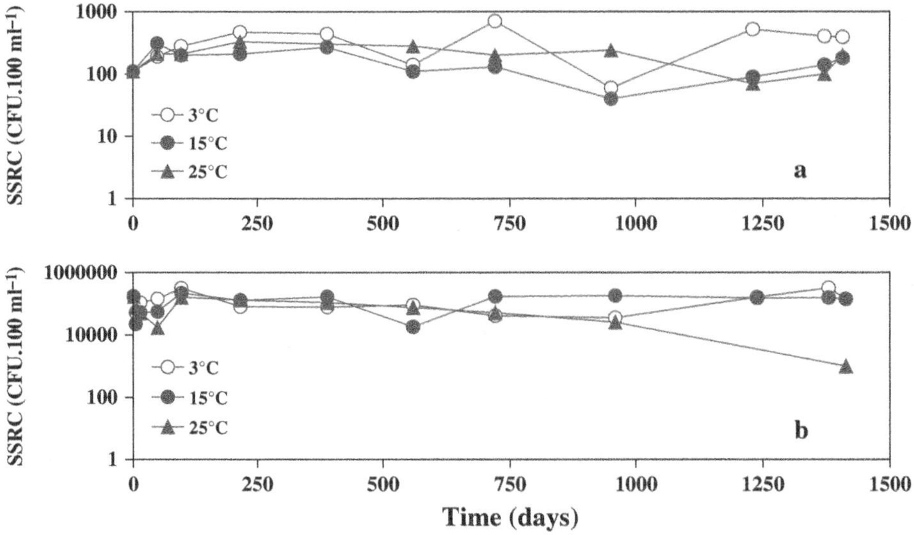

Figure 11.1 SSRC number in the river water without (a) and with (b) inoculation of *C. bifermentans* spores during the first 4 years of incubation at different temperatures

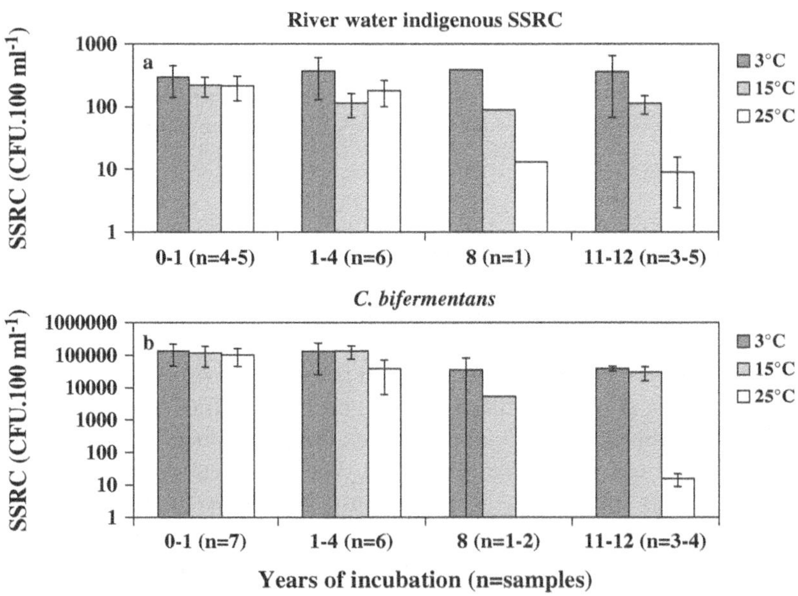

Figure 11.2 The average SSRC (a) and *C. bifermentans* (b) spore number (error bar = SD) in the river water measured in four consecutive periods of incubation at three different temperatures

Survival in sand. The number of SSRC in the sand from the different microcosms during the incubation period showed a significant variation (Figure 11.3). The average variation coefficient observed in the consecutive incubation periods was 75% (39–121). This was higher than observed in the water and than assessed for indigenous SSRC in a separate sample (25.6%). Except in the 40–70 sand incubated at 25°C, no significant decay was observed in the saturated sand during the first four years of incubation (Figure 11.3a,b). For the total period of 11–12 years of incubation of the saturated sand a significant decrease in spore numbers was observed at 25°C in both samples (TL and 40–70 sand) and in the 40–70 sand at 15°C (Figure 11.3b); the average numbers after this period was significantly lower than the average numbers in these samples in the first year ($P < 0.05$). After 11–12 years the SSRC numbers in the saturated sand incubated at 3 and 15°C were not significantly lower than the average number in the first year. The SSRC number in the unsaturated sand samples showed more decay than in the saturated sand samples (Figure 11.3c,d). A significant drop in SSRC number was observed after 8 years at all temperatures, probably due to complete dehydration of the sand (visual observation). In all unsaturated sand samples the average SSRC number after 11–12 years was significantly lower ($P < 0.05$) than the average number during the first year of incubation.

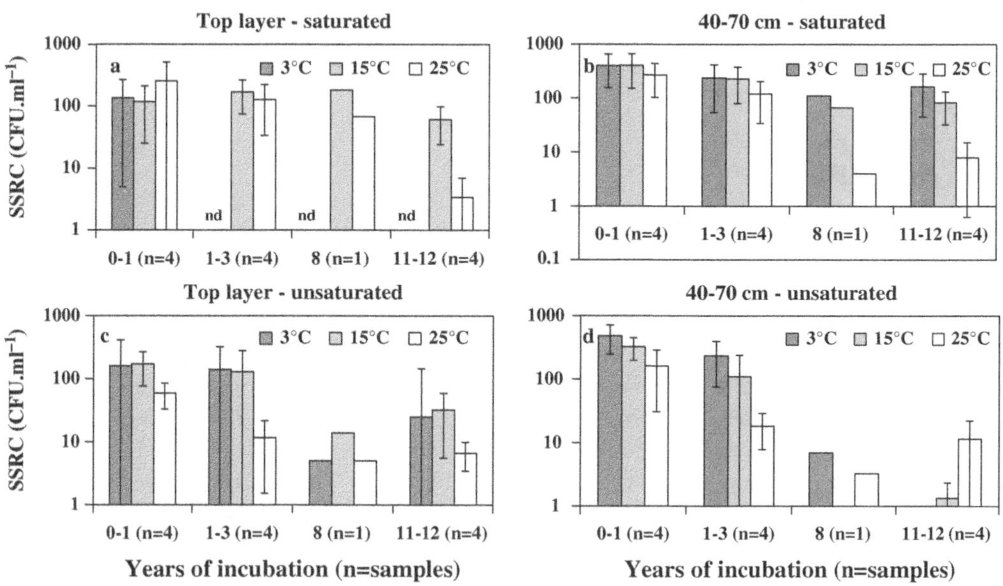

Figure 11.3 The average SSRC number (error bar = SD) in the different sand samples incubated water saturated (a,b) and unsaturated (c,d) and measured in four consecutive periods of incubation at three different temperatures (nd = not determined)

DECAY RATES

Decay rates (d^{-1}) of SSRC in the water and sand microcosms (Table 11.1) were calculated from those survival curves when (i) a statistical significant decrease in average number was observed during the total period, (ii) the level of significance was high enough ($P < 0.01$) and (iii) the decay could be described by the first order decay rate. The decay rates of indigenous SSRC and spores of *C. bifermentans* in the water samples incubated at 25°C were respectively 0.0010 and 0.0023 d^{-1}. The decay rates of SSRC in the two saturated sand microcosms (TL and 40–70 sand) incubated at 25°C were in the same order of magnitude as observed for SSRC in the river water (Table 11.1). The decay rates of *C. bifermentans* in river water and SSRC in 40–70 saturated sand at 25°C during the first four years of incubation (Figure 11.1) were 0.0023 and 0.0019 d^{-1} ($P < 0.01$), respectively. Thus, supporting the decay rates calculated from the somewhat clustered data from total period (Table 11.1). From the SSRC numbers in the unsaturated sand samples no decay rates could be calculated. The decay kinetics in these samples showed a two phase decay, most likely due to the dehydration (Figure 11.3c,d) and did not fit a first order process.

Table 11.1 The first order decay rate constant (ln.d^{-1}) for SSRC and spores of *C. bifermentans* in river water (RW) and for SSRC in water saturated sand samples

	Decay rate (ln.d^{-1}; 95% CI; r^2; $P < 0.001$)	
	15°C	**25°C**
RW SSRC	–*	0.0010 (0.0007–0.0012; 0.83)
RW *C. bif.*	–*	0.0023 (0.0020–0.0027; 0.94)
TL Sand	–*	0.0009 (0.0013–0.0006; 0.79)
40–70 Sand	0.0004†(0.0007–0.0001; 0.54)	0.0010 (0.0014–0.0006; 0.79)

**no significant decay $P > 0.01$; †$P < 0.01$*

MOLECULAR TYPING

The results of the 16S rRNA sequencing of the 24 river water and 34 sand isolates, respectively (Table 11.2), collected in the last year of incubation showed more different *Clostridium* species in the unspiked river water than in the sand. In the river water *C. bifermentans* (31% of all isolates), *C. butyricum* (23%), *C. bartletti, C. paraputrificum, C. tertium* and *C. sporogens/botulinum* (all 8%) were identified. In the river water inoculated with spores of *C. bifermentans* incubated at 3 and 15°C no other spp. than *C. bifermentans* were observed. At 25°C, however, two of the three surviving isolates in the water were identified as *C. barati* and *C. sporogenes/botulinum* (both 9% of the isolates). *C. frigidicarnis* was dominant in the sand samples (71% of all isolates), especially in the sand from the top layer (94%). *C. thiosulfatireducens* was observed in 27% of the sand samples with the highest percentage (56%) in the samples of the 40–70 cm layer. In the sand from this layer *C. frigidicarnis* was also observed (44%). One isolate from the top layer incubated at 15°C was identified as *C. subterminal* (6%).

DISCUSSION

SSRC decay in water and sand. The results of this study showed the longevity of spores of sulphite-reducing *Clostridium* species in water and sand. Spores survived for more than ten years. Long survival periods are described previously by several authors for these anaerobic spores in media and lake sediments on basis of incidental observations and indirect deduction from anaerobic heterotrophic spore counts (Hofer and Davis 1972 and Rothfuss *et al.*, 1997). Accumulation and delayed breakthrough of SSRC in slow sand filtration was attributed to the longevity of these spores in the filter bed (Hijnen *et al.*, 2007), which can be in operation for more than 30 years. A number of short term experiments already demonstrated the low decay rates of these spores (Davies *et al.*, 1995; Medema *et al.*, 1997 and Schijven *et al.*, 2000), but to the knowledge of the authors the presented study is the first to describe survival of anaerobic sulphite-reducing *Clostridium* spores in river water and sand during a long term microcosm experiment.

Table 11.2 Molecular typing of *Clostridium* isolates from water and sand, based on partial 16S rRNA gene sequencing. Matching of almost all isolates was based on >98% sequence identity

°C	River water n = 13	River water inoculated n = 11	TL sand n = 18	40–70 sand n = 16
3	bifermentans 3 butyricum 1 bartletti 1 unclass. species 2	bifermentans 3	frigidicarnis 5	thiosulfatireducens 4
15	bifermentans 1 butyricum 2 paraputrificum*1	bifermentans 5	frigidicarnis 8 subterminale 1	thiosulfatireducens 3 frigidicarnis 4
25	tertium 1 sporog./botulin. 1	bifermentans 1 barati 1 sporog./botulin. 1	frigidicarnis 4	thiosulfatireducens 2 frigidicarnis 3

**C. paraputrificum had 94% sequence identity*

Molecular identification of the Clostridium species. Molecular identification of *Clostridium* isolates was applied in the last phase of the study to characterize the surviving *Clostridium* population in the samples. These results revealed a higher number of species in the unspiked river water than in the sand samples after the long term incubation. The *Clostridium* population in the river water after twelve years of incubation consisted of *Clostridium* species also isolated from the human intestinal tract (Song *et al.*, 2002 and Duncan *et al.*, 2007). Spores of *C. perfringens* known as the more typical indicator for faecal pollution were not observed in the samples at the end of the experiment. The shift in species found in both the unspiked and spiked river water samples which were incubated at different temperatures (Table 11.2), indicate that the genus *Clostridium* consists of species with different persistence at different temperatures. The absence of *C. bifermentans* and the presence of two other *Clostridium* species in the spiked river water at 25°C after this long incubation period is a clear indication for this hypothesis. The dominant isolate in the sand was *C. frigidicarnis,* described as a psychrotolerant, proteolytic and saccharolytic bacterium associated with meat spoilage (Broda *et al.*, 1999). The other dominant species in the sand was *C. thiosulfatireducens* recently described as a more mesophylic and solely proteolytic *Clostridium* bacterium found anaerobic sludge blanket reactor in water treatment (Hernández-Eugenio *et al.*, 2002). These results indicate that the *Clostridium* spp. in the sand originated not specifically from a faecal pollution but can be identified as environmental species.

Decay rate of Clostridium spores. The high observed variation in SSRC number in the different microcosms which was larger than expected from the performance evaluation of the analytical method in water and sand is unexplained. Therefore the decay rate assessment was restricted to the microcosms with a statistically significant decline in number over the total period of incubation. The first order decay rate constants calculated from this microcosm study was the lowest ($0.0004\,d^{-1}$) for indigenous SSRC in the saturated sand at 15°C and the highest ($0.0023\,d^{-1}$) for spores of *C. bifermentans* in river water at 25°C. The decay rate depended on temperature and doubled in the saturated sand at a temperature rise of 10°C (Table 11.1). Another condition enhancing decay of these spores in sand was dehydration. Only few studies in literature presented decay rates for anaerobic spores. In the sediment from a lake anaerobic spore forming bacteria were counted as a function of depth and from the sediment characterization the age-depth correlation was deduced to be 1780 $year.m^{-1}$ (Rothfuss *et al.*, 1997). From the exponential profile of aerobic heat resistant bacteria (aerobic heterotrophic spores) a decay rate was calculated of $4 \times 10^{-6} - 7 \times 10^{-6}\,d^{-1}$. For the anaerobic heterotrophic spores a similar exponential profile was observed but these data did not reveal a significant fit to the decay rate equation. On basis of the similarity of number and profile of aerobic and anaerobic spores, microscopically characterized as *Clostridium* species, similarity of decay rates of aerobic and anaerobic spores was assumed. These decay rates are a factor of 100–1000 lower than the decay rates presented in the current study. In these calculations The assumption made by Rothfuss *et al.* (1997) of a similar age of detected anaerobic spores and the sampled lake sediments, however, is uncertain since spore numbers may be influenced by transport and metabolic activity at low temperatures (Molongoski and Klug, 1976). More recently, Medema *et al.* (1997) calculated higher decay rates of 0.00069 and 0.0115 d^{-1} for spores of *C. perfringens* in river water incubated 42 days at 3 and 15°C, respectively. These different decay rates demonstrate the longevity of *Clostridium* species in the environment, but also confirm the diversity in this genus with respect to the environmental persistence as hypothesized from the molecular typing results of the current study. Further research is necessary to obtain more evidence for this heterogeneity in the persistence of *Clostridium* species.

Significance of the results for SSRC as a water quality parameter. On the basis of the decay rates presented in this study and decay rates of oocysts of *Cryptosporidium* presented in literature the SSRC parameter which covers a wide range of environmental *Clostridium* species, can be regarded as a too conservative process indicator for the elimination of the oocysts in water treatment processes with the risk of accumulation (slow sand filtration, soil passage; Hijnen, 2009). Davies *et al.* (2005) presented decay rates of 0.015–0.022 d^{-1} for oocysts based on loss of viability assessed with fluorescence in situ hybridization. From cell culture-Taqman PCR assay data of King *et al.* (2005) higher decay rates of oocysts of 0.01–0.049 d^{-1} were calculated (Hijnen *et al.*, 2007) which were in the same order of magnitude as mentioned before for *C. perfringens*. In future the SSRC parameter will be used as a process indicator in the Netherlands, but for indication of faecal pollution *C. perfringens* bacteria (incl. spores) is implemented in the regulations (Anonymous, 2001). This emphasizes the need for more scientific information on the environmental behaviour of *C. perfringens* spores which includes growth and survival.

Chapter 12

Quantitative methods to assess capacity of water treatment processes to eliminate micro-organisms

Wim A.M. Hijnen

KWR Watercycle Research Institute

INTRODUCTION

Major scientific breakthroughs in drinking water safety occurred in the 19th century. The recognition of the role of water in transmission of pathogenic micro-organisms to the public and the protection of the public by means of water treatment and monitoring for faecal pollution using faecal indicator bacteria such as *E. coli* were the most prominent developments in that period. During the major part of the 20th century both water treatment and microbiological water quality monitoring to protect public health evolved gradually (Chapter 1). In the last decade of 20th century a number of observations and scientific achievements caused a more rapid development in the field of microbiological safety of drinking water. A growing amount of epidemiological studies showed the significance of persistent and highly virulent waterborne pathogens for drinking water safety and the shortcomings of both water treatment and end product monitoring for *E. coli* to guarantee microbial safety. These three aspects were exhibited in the large waterborne cryptosporidiosis outbreak in Milwaukee in 1993. At the same time, dose-response studies of relevant waterborne pathogens were published in literature which enabled a quantitative calculation of health risks of exposure to pathogens, i.e. through the consumption of drinking water.

On the basis of this information and the awareness that the current end product monitoring of faecal indicators is a curative strategy with a delay time during which consumers can be exposed to pathogens, Dutch authorities implemented a health-based target for microbial safety in the revised Drinking Water Decree (Anonymous, 2001). Water Companies using surface water as the source water should demonstrate compliance with an annual infection risk level of 10^{-4} per person for a number of index pathogens: enteroviruses, *Campylobacter* bacteria and two parasitic protozoa, *Cryptosporidium* and *Giardia* (Wetsteyn *et al.*, 2005). Compliance with this infection risk should be demonstrated by means of a Quantitative Microbial Risk Assessment (QMRA) which is part of the total system approach of Water Safety Plans, covering the whole water system from source to tap (WHO, 2004).

Edited by Wim A.M. Hijnen. ISBN: 9781843393764. Published by IWA Publishing, London, UK.

The most direct approach would be to monitor tap water for the index pathogens. However, the concentrations at which these pathogens pose an infection risk above 10^{-4} per person per year (pppy) are extremely low, much lower than the detection limit of pathogen detection methods. Hence, the concentration of pathogens in tap water has to be estimated from their concentration in source water and their removal by water treatment. The general objective of the studies presented was to develop and evaluate methods for the assessment of the elimination capacity of full-scale water treatment processes for the index pathogens.

The line of approach was to develop a method as closely related to the natural conditions in water treatment and to the daily practice of microbiological water quality monitoring. The gerneral hypothesis was that *Escherichia coli* and spores of sulphite-reducing clostridia used in routine water quality monitoring can be used as process indicators to assess the elimination capacity of water treatment processes for, respectively, pathogenic micro-organisms that are susceptible (bacteria and viruses) and pathogenic micro-organisms that are resistant (protozoan (oo)cysts) to disinfection processes under the commonly applied conditions in water treatment.

E. COLI AND SSRC AS PROCESS INDICATORS UNDER FULL-SCALE CONDITIONS

Routine water quality monitoring. Routine microbiological monitoring by water companies in all staged of water treatment yield valuable quantitative information of the first processes in water treatment on the elimination of these indicators expressed as the Decimal Elimination Capacity (DEC). These routinely collected data usually show a decreasing amount of positive samples, however, and is therefore unsuited to determine accurately DEC of a total treatment and its variability (Chapter 2).

Large volume sampling. In microbiological studies in water treatment large volume sampling has been used by others (Goyal *et al.*, 1980; Van Olphen *et al.*, 1993; Payment *et al.*, 1989, 1991, 1993 and LeChevallier and Norton, 1991). These studies used filtration and/or adsorption followed by elution and culturing. These methods are not rapid, simple, reliable and unambiguous and relatively expensive for routine monitoring. The current study showed that modification of the standard membrane filtration method to examine a larger water volumes up to 10 liters with the standard membrane filtration (mf method; Chapter 2) and an up-scaled membrane filtration method (MF sampler; Chapter 3) for larger volumes up to 1000 liters yielded a simple and reliable method with a low detection limit to monitor faecal indicator bacteria in a total water treatment chain (Chapters 4).

The MF sampler method is used in the Dutch drinking water practices by the Water Companies and in special microbiological research projects (Hoogenboezem *et al.*, 2001b; Medema and Stuyfzand, 2002 and Smeets *et al.*, 2006).

Elimination of process indicators by full-scale treatment. The study in eight Dutch treatment facilities presented in Chapter 4 clearly showed that traces of faecal contamination (Coli44 and SSRC) were occasionally found in finished waters. SSRC where detected more frequently and at higher concentrations than the susceptible Coli44. This was not only in treatments with primary chemical disinfection but also in a treatment without this process. This clearly points to differences in transport and fate of susceptible and persistent faecal micro-organisms in water treatment.

A considerable variation in the decimal elimination capacity (DEC) between the eight full-scale treatment facilities for Coli44 and SSRC was observed. For Coli44 the DEC ranged from 3.0–6.6 log and for SSRC from 1.2–4.5 log. The observed differences in the DEC between different treatment facilities for SSRC were attributed to design and operational conditions which have been tailored to the quality of the source water, conclusions also drawn from the work of others for protozoan oocysts (LeChevallier and Norton, 1991, 1992 and Smeets *et al.*, 2007).

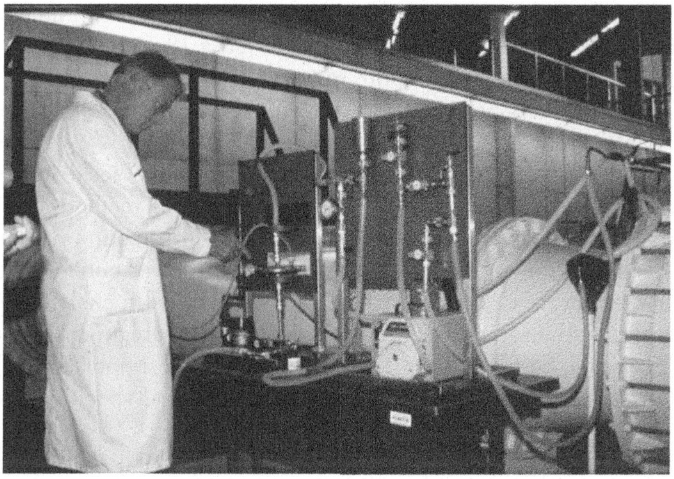

Figure 12.1 The MF-sampler used under full-scale conditions

Monitoring of Coli44 and SSRC in large volumes demonstrated also a considerable variation in DEC values of the same type of processes at the eight different facilities (Chapter 4). The largest differences were observed for chemical disinfection processes, slow sand filtration and granular activated carbon filtration. Important observation was the relatively low and variable inactivation capacity of full-scale chlorination and ozonation for Coli44, lower than expected from disinfection kinetics established in laboratory studies. This is a very significant observation for water treatment practice, as it shows that simple extrapolation from laboratory disinfection studies to full scale water treatment, as is common in the water sector, may seriously overestimate the treatment efficacy. Consequently, inactivation kinetics for indigenous micro-organisms under full-scale conditions deviates considerably from the kinetics of pure cultures assessed under well defined laboratory conditions, mostly batch tests. This was also observed in a literature study on UV (Chapter 6). Basis for this discrepancy are factors related to conditions of the studied micro-organisms such as physical protection (particle association), physiological or metabolic conditions and variation in susceptibility within one species. Different susceptibility of lab-cultured and environmental micro-organisms is demonstrated in other studies (Hijnen *et al.*, 2004b and Smeets *et al.*, 2005). Also process related factors are responsible for the different inactivation kinetics in lab-scale and full-scale disinfection studies. Low inactivation of Coli44 and SSRC observed in a local full-scale ozonation was evaluated by a Hazard Analysis Critical Control Point (HACCP) study which is an integral part of a Water Safety Plan (WHO, 2004). A team of process technicians and experts designated the high DOC content of the water, the applied ozone dosing strategy in combination with poor hydraulics of the full-scale system as major causes (Hijnen *et al.*, 2001). Evidence for the role of poor hydraulic conditions on inactivation efficacy was obtained in the studies presented by Van der Veer *et al.* (2005) and Smeets *et al.* (2006).

The study in Chapter 4 demonstrated the high efficacy of slow sand filters to remove Coli44 (2–3 log). The DEC of these filters for SSRC, however, was low and variable (0–2 log). This phenomenon was attributed to accumulation, survival and delayed breakthrough deduced from high SSRC concentrations in the sand bed (Chapter 3). Similar observations were reported for granular activated carbon GAC filters (Hijnen *et al.*, 1997) where high SSRC concentrations were detected in both the filter bed and the backwash water. More recently, low removal of aerobic and anaerobic spores by GAC filtration were observed in literature (Galofré *et al.*, 2004 and Mazoua and Chauveheid, 2005) and these authors suggested a similar explanation for their results. The validity of the proposed explanation for the low spore removal in slow sand filters and soil infiltration with no filter cleaning was supported by results presented in Chapters 7, 8 and 9. The high persistence of SSRC was clearly demonstrated in Chapter 11.

In conclusion the DEC of full-scale processes can be assessed with faecal indicator bacteria as process indicators. By increasing the sample volume of the standard analytical methods the elimination capacity of the total treatment can be assessed more accurately with additional information on the variability of the DEC. The DEC of full-scale treatment for environmental micro-organisms is potentially lower than elimination capacities assessed for the same processes under well defined conditions in challenge studies under pilot plant or laboratory conditions commonly used to assign log credits to water treatment processes (LeChevallier and Au, 2004; von Huben, 1991 and USEPA, 2006). This underlines the importance of obtaining quantitative information on elimination of micro-organisms under full-scale and site specific conditions as proposed and studied in this book.

Limitations and remaining questions. But the results also revealed some limitations and remaining questions concerning the methodology:

- The proposed MF method to assess DEC of water treatment processes with indicator bacteria Coli44 and SSRC is primarily focussed on the index pathogens *Campylobacter* and *Cryptosporidium/Giardia*. Enteroviruses, however, are also important pathogens for water safety (Regli *et al.*, 1991; de Roda Husman, 2001b and Fernandes *et al.*, 2007) and implemented in the revised Dutch Drinking Water Decree (Anonymous, 2001). These pathogens are the most critical micro-organisms in water (in)filtration (Schijven, 2001). Bacteriophages, especially F-specific RNA phages (FRNA), are described as appropriate process indicators for the removal of enteroviruses in water treatment (Havelaar, 1986; 1993 and Schijven, 2001). A standard for bacteriophages was not included in the drinking water regulations. Additional research on bacteriophages as process indicators for (entero)viruses is recommended and further explored in this book in the challenge tetsts.
- The MF method for indicator bacteria is limited to one organism at a time. Successive membrane sampling at one sampling point enables to monitor multiple bacterial species. Recently, a new method using cross-flow membrane filtration has been introduced for detection of multiple organisms, including bacteriophages. The observed recoveries for bacteriophages, bacteria and protozoa were as high as the recoveries of the MF-sampler method for bacteria and spores (Veenendaal and Brouwer, 2007). This method was first described by Leclerc *et al.* (1977) as a technique to continuously and automatically determine *E. coli* in 100 ml of water. The technique was also applied to overcome the problems with low recoveries of the dead end filtration technique used to assess concentrations of *Cryptosporidium* and *Giardia* (oo)cysts (Simmons *et al.*, 2001) and explored for multiple micro-organism detection (Morales-Morales *et al.*, 2003 and Hill *et al.*, 2005, 2007). This cross-flow filtration method was successfully applied in the experiment of Chapter 7 and further investigations are recommended to explore the use of this new LVS method in drinking water practices.
- In the course of this study and the performance of the provisional risk assessments it became apparent that for some processes site specific full-scale information on the elimination of Coli44 and SSRC was not assessable because concentrations were too low to detect (infiltration and soil passage, slow sand filters, post-disinfection) or the processes were not in operation under full-scale conditions (UV disinfection). Consequently, alternative methods are needed to assess DEC of these processes.
- The methodology with the two process indicators is valuable to establish elimination capacity and robustness of treatment for water safety, but is not suited to control the treatment processes for an optimal DEC and for momentary loss of DEC. The method has a delay time of at least 24 hours. These

temporarily 'bad day' conditions with potentially high concentrations of micro-organisms in the finished water may contribute to a large extent to the annual risk of infection (Smeets, 2008). The results of Chapter 2 and 4 have shown that Coli44 and SSRC can detect 'bad day' conditions and help to elucidate how operational process conditions such as dosing, flow rates and contact times affect the DEC. This enables the design of an on line control strategy for optimal elimination conditions and may in the long run lead to process models for predicting the elimination of micro-organisms on the basis of on line process parameters. Analyzing actual full-scale process and elimination data by statistical techniques such as multiple-regression analysis (Wiersema, 1999 and Haas *et al.*, 2001) is an example of such studies. Both studies identified pH, coagulation doses and polymers, temperature, turbidity as conditions affecting the elimination rate in coagulation processes. The outcome of such studies, however, largely depends on the quality of the collected data base on process parameters and microbial elimination data. Another example of such studies is presented by Smeets (2008) who combined process models for disinfection with full-scale elimination data and used stochastic modelling to describe elimination capacity and estimate micro-organisms concentrations in treated water.

- In the process of QMRA site specific quantitative information is required on the elimination of index pathogens which in most cases is not available as indicated before. It is unsure how well both process indicators can be translated to the elimination of these index pathogens. Some data on Coli44 as indicator for *Campylobacter* already indicated the value of this process indicator. But for the *Clostridium* spores there were observations at GAC filtration and slow sand filtration which cast doubt on the use of this parameter as process indicator for the parasitic protozoa for these processes.

Consequently, additional research was necessary to find alternative strategies for the assessment of the DEC of selected processes, to elucidate the most significant conditions that affect the variability in DEC and to compare the elimination of process indicators (PI), *E. coli* and spores of sulphite-reducing clostridia, with the elimination of the index pathogens (IP) under similar conditions.

ADDITIONAL METHODS TO ASSESS DEC, TO STUDY CONDITIONS AND TO VALIDATE PROCESS INDICATORS

To overcome the limitations of the use of faecal indicators as process indicators for DEC assessment and to address part of the remaining questions the use of challenge tests or dosing experiments was explored (Chapters 5, 7, 8, 9 and 10). Dosing micro-organisms to full-scale drinking water treatment processes is generally not allowed and feasible. Therefore dosing experiments were done at pilot or laboratory scale under controlled conditions, mimicking full-scale conditions. In this type of study elimination can be quantified precisely for selected micro-organisms, water type and process conditions. Dosing experiments with multiple organisms under the same conditions enabled comparison of the elimination of the different pathogens and their process indicators under similar conditions in order to assess their elimination ratio (IP/PI ratio) and to investigate the influence of specific process variables. The F-specific RNA phage MS2 was dosed in these studies as model for virus removal.

Challenge tests in bench-scale experiments intrinsically deviate from the full-scale conditions where multiple conditions affect the overall elimination. Consequently, the result of such tests potentially over- or under estimate the real elimination under normal practices. In the course of this study verification of the validity of these challenge tests for assessment of DEC of full-scale processes was part of the objectives. For this verification three strategies have been applied:

- when elimination data of environmental Coli44 and SSRC under full-scale conditions are available:
  - comparison of the elimination of environmental faecal indicators determined under the challenge test and under full-scale conditions;

  - comparison of elimination of dosed *E. coli* and spores of *C. perfringens* determined under the challenge test conditions with the elimination of Coli44 and SSRC determined under full-scale conditions.
- when no information of elimination of the environmental faecal indicators are available: comparison with elimination data from scientific literature of either full-scale systems or other challenge tests. Quantitative analysis of literature data is presented in Chapter 6 and a separate review (Hijnen and Medema, 2007).

DISINFECTION CHALLENGE TESTS

With respect to disinfection processes the study focussed on the two disinfection processes of major interest in the Dutch drinking water industry, ozonation and disinfection with ultraviolet radiation (UV).

Batch experiments. Disinfection studies under well defined conditions on laboratory scale are usually batch tests where pre-cultured micro-organisms are exposed to different doses of disinfectant (chlorine, ozone, UV). These inactivation kinetic studies are relatively simple and inexpensive. From the obtained dose-response curves the first order inactivation rate constant k specific for a combination of micro-organism and disinfectant is derived. These studies yield basic information on the inactivation kinetics and susceptibility of the different micro-organisms to a disinfectant (k value). Moreover, the effects of some water quality conditions such as temperature and pH can be quantified which is of interest for process modelling. With the k values and the quantified effect of temperature and water quality on these k values disinfection processes are designed to achieve a desired inactivation capacity (USEPA, 2006). Translation of these batch test results to 'real world' disinfection processes, however, will over-estimate the inactivation efficacy of these processes (Chapter 2, 3, 4 and 6). The conditions in a full-scale continuous flow system are not as optimal as the conditions in a batch laboratory system (variability in hydraulics and dosages). Furthermore, environmental micro-organisms potentially have different susceptibility to disinfectants than pre-cultured organisms.

Continuous flow systems. In continuous flow bench-scale systems process conditions can be created that are close to full-scale conditions. In these systems inactivation was less than expected from simple translation of batch study results and closer to the inactivation of micro-organisms observed under full-scale conditions for ozone and UV (Chapter 5; Hijnen *et al.*, 2004b; Chapter 6). With continuous flow systems the lower susceptibility of environmental micro-organisms for UV disinfection (Hijnen *et al.*, 2004b) and ozone (Smeets *et al.*, 2005) compared to lab-cultured species of the same group was demonstrated which accounts, in part, for the discrepancy between disinfection efficacy observed in lab and pilot tests. This emphasizes the need for further research on the difference between the susceptibility of environmental and lab-cultured micro-organisms.

FILTRATION CHALLENGE TESTS

The lack of quantitative information on removal of index pathogens by slow sand filtration and surface water infiltration and the IP/PI ratios for these processes was addressed by challenge testing with pre-cultured organisms on pilot plant and laboratory scale (Chapters 7, 8 and 9).

Laboratory column tests. Small scale column tests operated under conditions simulating full-scale filtration conditions are relatively simply and cheap to perform. The experiences in the course of this study showed that the quantitative elimination data collected which such studies, however, can not be translated directly to full-scale conditions (Chapter 7 and 9). The filtrations conditions in the columns deviate too much from the full-scale filtration conditions. This conclusion was derived from comparison of column test data with elimination data of environmental process indicators under full-scale filtration conditions. The columns were challenged with multiple micro-organisms under variable filtration conditions. This yielded comparative quantitative information on the relative elimination of viruses, bacteria and bacterial spores and *Cryptosporidium* and *Giardia* by slow sand filtration and surface infiltration. Furthermore, with these studies the relative influence of the geo-chemical and morphological properties of the sand and the infiltration rate on the overall elimination in sand beds was demonstrated. Under both conditions, slow sand filtration and surface water infiltration, the column studies showed that the average grain size of the sand is less important for microbial elimination than the geo-chemical properties of the sand. Microbial transport in filter material or soil is governed to a large extent by the size of micro-organisms. The infiltration column study (Chapter 9), however, revealed that in a soil column with fine and uniform sand *Cyptosporidium* oocysts (4 μm) were removed to a lower extent than *E. coli* and spores of *C. perfringens* (both approx. 1 μm), an observation also described by Nobel *et al.* (1999). This emphasizes the role of surface interactions in the transport of micro-organisms through sand or soil. The higher effect of filtration rate on elimination of bacteriophages compared to bacteria, bacterial spores and (oo)cysts demonstrate the additional effect of straining in the elimination of the larger micro-organisms. These results show that column studies are useful to compare removal of different micro-organisms in order to assess the PI/IP ratio and to understand the impact of process conditions on the elimination of micro-organisms. This information is necessary for further understanding and modelling of microbial transport in (in)filtration processes.

Challenge tests on pilot plant scale. Slow sand filters (SSF) of a pilot plant operated as dummy of a full-scale SSF process (Chapters 7, 8; Dullemont *et al.*, 2006 and Schijven *et al.*, 2007) were used to determine the elimination of viruses, bacteria and bacterial spores and protozoan (oo)cysts. Comparison of the results with results of elimination of environmental micro-organisms under full-scale conditions revealed that challenge tests on pilot plant scale were valid to assess the DEC of full-scale filters for bacteria (*E. coli* and *Campylobacter*) but not for SSRC (Chapter 8). The elimination of dosed spores of *C. perfingens* was much higher than the elimination of 'natural' SSRC by full-scale filters. Evidence for the suggested mechanism of delayed transport as explanation for the variable and sometimes low DEC of persistent micro-organisms by filters with no filter back wash was given in Chapter 8. The rapid decline of oocyst concentrations in the challenged filter bed suggested a lower and most likely negligible risk of delayed breakthrough of oocysts in these filters. The role of predation in this oocyst decline was hypothesised, since zooplanktonic organisms which have been described to ingest oocysts are present in these biological filters. The potential role of these organisms as transport vector to the outlet of the filter bed is an issue of further research. Preliminary studies on predation of (oo)cysts in granular activate carbon filtration recently published showed that this hypothesized process indeed may occur but QMRA modelling revealed low additional risk of infection from those ingested and transported (oo)cysts (Bichai *et al.*, 2010a,b).

In the pilot pant SSF study the large positive effect of the presence of a Schmutzdecke on a slow sand filter on the elimination of *E. coli* was demonstrated (Chapter 7). The results on MS2 bacteriophage removal showed that the Schmutzdecke has a minor effect on virus removal. Another condition that is of importance for the DEC of slow sand filtration is temperature. The DEC of the process for *E. coli* and MS2 phages assessed in challenge tests varied with 3 log in a temperature range of 7–16°C (Hijnen *et al.*, 2006).

Chapter 10 describes a challenge tests with GAC filtration on pilot plant scale which showed that a challenge test can under estimate the DEC of a process. The elimination of environmental process indicators, *E. coli* and SSRC, was higher than the elimination of the same micro-organisms dosed to the influent (Hijnen *et al.*, 2010). Additionally, this study demonstrated a high potential of the filters to eliminate protozoan (oo)cysts.

Challenge tests on pilot plant scale to assess the efficiency of conventional processes (coagulation/filtration) as a microbial barrier in water treatment have been used increasingly over the last decade. Especially studies on *Cryptosporidium* and *Giardia* removal by conventional treatment have been presented (Lodgsdon *et al.*, 1981;

West *et al.*, 1994; Patania *et al.*, 1995; Nieminski *et al.*, 1995; Dugan *et al.*, 2001; Emelko, 2001 and Huck *et al.*, 2001). In these studies effects of sub-optimal process operation on micro-organism breakthrough and water safety have been addressed since the recognition of the significant effect of such treatment inadequacies on water safety (Badenoch, 1990; Richardson *et al.*, 1991 and Craun, 1990). In challenge tests the loss of elimination capacity of such processes was clearly demonstrated and quantified at sub-optimal coagulant doses and conditions (pH, mixing) and risk moments in filtration (increasing pressure drops, backwashing moments and start up) as summarized in a review (Hijnen and Medema, 2007).

In conclusion, pilot plant studies operated as a dummy of the full-scale processes are valuable studies to assess the DEC and to elucidate the influence of process conditions on the DEC.

QUANTITATIVE ANALYSIS OF LITERATURE DATA

The international literature on microbial inactivation and elimination is a significant source of quantitative data and was extended in the last decade (LeChevallier and Au, 2004 and Hijnen and Medema, 2007). The review of the first authors described the elimination capacity of a broad range of water treatment processes without further quantitative analysis. A more quantitative method of literature reviewing was explored in this study for UV (Chapter 6) and by Hijnen and Medema (2007; third edition) for the major treatment steps currently used in treatment facilities (coagulation/floc removal, rapid granular filtration, slow sand filtration). In this method quantitative literature data were used to calculate a default value, the Microbial Elimination Capacity (MEC) of the process for viruses, bacteria and bacterial spores and prototzoan (oo)cysts. The MEC was calculated from the DEC values in those literature studies qualified for inclusion in the calculation. This qualification was based on evaluating the study design/report: experimental conditions, microbial assays and process information. The MEC-value was the weighted average value of the DEC values derived from the different studies. The weighting was based on a Full-scale index (FSI), with an increasing value of successively lab-scale, pilot plant and full-scale with either dosed or environmental organisms. With the scale of the studied processes on the y-axis and the conditions of the used micro-organisms on the x-axis, the data were ranked in an x-y matrix with values of 1–5. Drawback of this weighting system is that the same rank number between 1 and 5 represents either low conformity with the full-scale conditions or low conformity with the environmental pathogen of concern while the effect of both conditions on the DEC is probably not equal. Because data sets from full-scale studies with environmental micro-organisms were limited and in most cases evenly distributed over these data sets, the influence of this weighting on the calculated MEC was also limited (Hijnen and Medema, 2007). This diminishes the significance of the noticed drawback of this weighting on the assessed MEC values. The FSI was further used as a descriptive parameter to qualify the studies with respect to their conformity with the full-scale conditions. The range of the MEC was given by the minimum and maximum observed the DEC values. The effect of process conditions on the DEC values was presented by describing the major process conditions in the review and summarizing specific experimental data addressing the effect of selected process conditions.

The literature review yielded MEC values of conventional treatment processes (coagulation and filtration) and slow sand filtration for viruses (incl. enteroviruses and bacteriophages), bacteria (incl. indicator bacteria and *Campylobacter*) and bacterial spores (incl. *Clostridium* and *Bacillus*), *Cryptosporidium* and *Giardia*. For conventional treatment processes the highest MEC-values were observed for protozoan (oo)cysts, but in general the MEC values of the different organisms were in the same order of magnitude. The variability of the MEC values was high and only for obvious conditions low or high elimination could be attributed to process conditions (free sedimentation versus lamellae separation; pre-oxidation). The UV review (Chapter 6) yielded a required fluence table for a range of micro-organisms relevant for drinking water safety in which fluence for bacterial pathogens was corrected for the discrepancy between dose-effect observed in batch tests for pre-cultured organisms and in continuous flow systems for environmental organisms. Viruses and *Acanthamoeba* are the most persistent pathogens. Susceptibility of bacteria and *Cryptosporidium*and *Giardia* to UV were in the same order of magnitude.

These default elimination capacities derived from literature were used in QMRA calculation for specific locations with no or limited quantitative data on process indicator elimination (Medema *et al.*, 2006b). The MEC was used as an input value for a point estimate of the annual infection risk from index pathogens in drinking water of the location. In a more comprehensive stochastic calculation the uncertainty level of this annual infection risk was derived from the range in the MEC value, assuming a simple triangular distribution.

PROCESS INDICATOR VALIDATION

The IP/PI ratio. To validate the use of process indicators to quantify the elimination of the index pathogens, comparative studies on elimination are needed. The ratio between the index pathogen and the process indicator elimination (IP/PI ratio) is a parameter to evaluate this aspect of the process indicator use. The ideal IP/PI ratio is 1.0, but on the basis of the results presented in this book these ratios will be variable. The impact of these ratios on the prediction of the elimination of the index pathogen is demonstrated in Figure 12.2, assuming that this ratio is the slope of a linear relationship between the elimination of both organisms, independent from the extent of removal. This shows that the over- or under-estimation of the elimination of the index pathogen will increase with the increasing elimination capacity of the process.

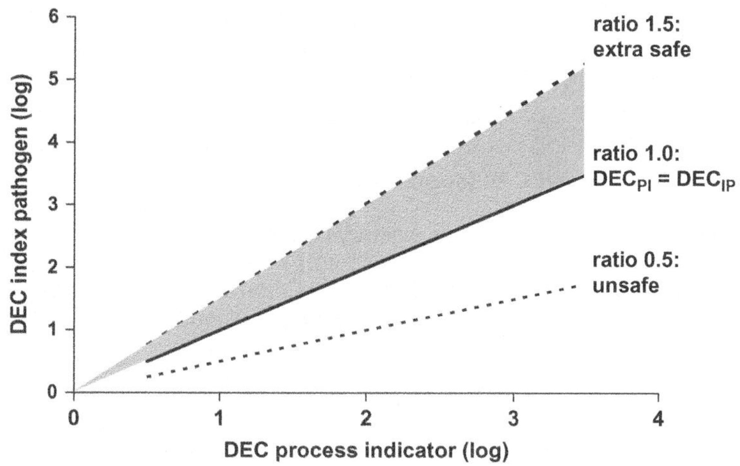

Figure 12.2 The effect of the use of the IP/PI ratios on the predicted DEC values at a ratio of respectively 0.5, 1.0 and 1.5

The IP/PI ratios assessed from the comparative studies with multiple micro-organisms presented in this book supplemented with additional studies from literature are summarized in Figure 12.3 and 12.4. In these Figures the ideal IP/PI ratio of 1.0 is depicted by a straight line.

Process indicator validation for Cryptosporidium. Figure 12.3 shows for *Clostridium* spores (SSRC and *C. perfringens*) as potential process indicator for *Cryptosporidium* removal by slow sand filtration and infiltration a wide range of ratios of <1.0 to >>1.0. Consequently, these spores are not suitable as process indicator for *Cryptosporidium* for these processes. The ratio assessed for SSRC/*Cryptosporidium* inactivation by ozone of 0.8 was close to the safe ratio of 1.0. For the use of *Clostridium* spores in conventional water treatment two IP/PI ratios of 0.9 and 1.0 were derived from a direct comparative study (Payment and Franco, 1993) and an indirect comparative study (Hijnen *et al.*, 2003; data of IP and PI elimination collected in two separate periods). The use of *Bacillus* spores as process indicator was intensively studied in the US. For a more pronounced conclusion on the IP/PI ratio of bacterial spores/ *Cryptosporidium* elimination by conventional treatment results from two external comparative studies were used (Huck *et al.*, 2001 and Prevost *et al.*, 2007, large review of data from 15 studies). These studies demonstrated that IP/PI ratio for these aerobic spores as process indicator for *Cryptosporidium* was close to the ideal value of 1.0 with the majority >1.0 (safe area).

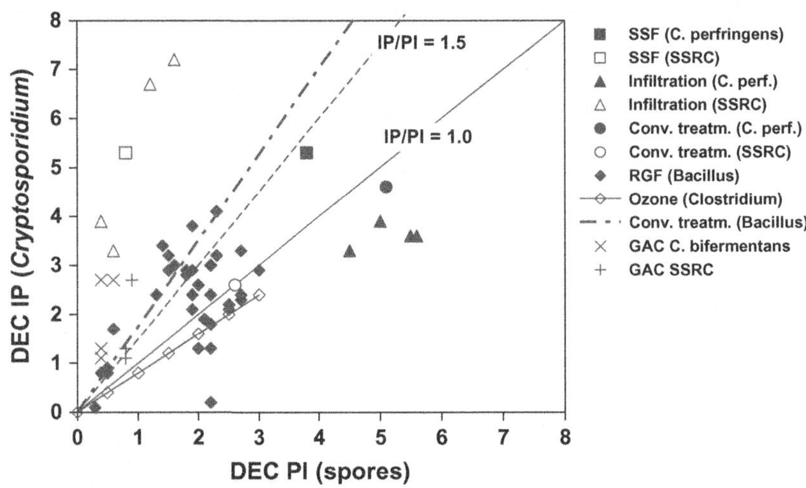

Figure 12.3 The IP/PI ratio of *Cryptosporidium* and bacterial spores *C. perfringens* (C. perf.), sulphite-reducing clostridia (SSRC) and aerobic spores (*Bacillus*) for a number of water treatment processes: Slow sand filtration (SSF; Chapters 7,8), Infiltration (Chapter 9, Nobel *et al.*, 1999), Conventional treatment (Payment and Franco, 1993; Hijnen *et al.*, 2003 and Prevost *et al.*, 2007), rapid granular filtration (Huck *et al.*, 2001) and GAC filtration (Hijnen *et al.*, 2010)

The MEC values of conventional treatment processes (coagulation and rapid granular filtration) presented in the literature review (Hijnen and Medema, 2007) can also be used to calculate IP/PI ratios. These ratios were 1.3–1.6 and in agreement with the data presented in Figure 12.3. And finally a study in 2010 (Hijnen *et al.*, 2010) compared the removal of spores of *C. bifermentans* and SSRC with the removal of *C. parvum* in GAC filtration.

Process indicator validation for Campylobacter. The IP/PI ratios for Coli44 (incl. *E. coli*) as process indicator for *Campylobacter* elimination by rapid granular filtration, slow sand filtration and ozone (Figure 12.4) showed that Coli44 complies with the requirement of a proper process indicator. Most ratios were in the safe area of >1.0, but not too high (>1.5).

Predictive value of PI for 'bad day' elimination conditions. From the results presented above it can be concluded that the IP/PI ratio exhibit a level of variation caused by the differences in characteristics of pathogen and process indicator related to the elimination process. This variation in IP/PI ratio can be included in a stochastic QMRA modelling tool used to calculate the level of uncertainty of the infection risk. Precondition is that the variation is random and not related to specific process conditions where the fate of process indicator and pathogen significantly differs. Consequently, the reaction of IP and PI in treatment process on specific conditions affecting elimination must be as similar as possible. Differences in persistence and surface properties between micro-organisms are such characteristics of interest for the fate of PI and IP in disinfection processes and filtration processes.

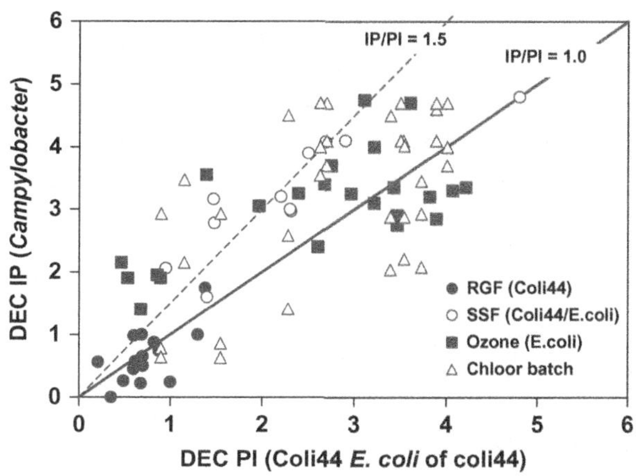

Figure 12.4 The IP/PI ratio of *Campylobacter* and Coli44 (incl. *E. coli*) for a number of water treatment processes: rapid granular filtration (RGF; Hijnen *et al.*, 1998), slow sand filtration (Hijnen *et al.*, 1995; Chapter 7, Dullemont *et al.*, 2006), ozone (Smeets *et al.*, 2005) and Chlorine (Blaser *et al.*, 1986)

Difference in persistence of micro-organisms and the distribution in disinfectant dose of full-scale disinfection processes are two major characteristics which affect the inactivation of these different micro-organisms and consequently the IP/PI ratios. Generally spoken the dose-effect relation of disinfection processes can be assessed more accurately with persistent micro-organisms than with the more susceptible micro-organisms. This was demonstrated by ozone studies with *C. parvum* oocysts and bacterial spores (Oppenheimer *et al.*, 2000; Chapter 5, Hijnen *et al.*, 2004c) and UV studies with bacterial spores and bacteriophages (Qualls and Johnson, 1983; Sommer *et al.*, 1999; Hijnen *et al.*, 2004b and USEPA, 2006). Dose assessment of UV systems is done with UV calibrated pre-cultured aerobic spores of *B. subtilis* or MS2 bacteriophages with relatively low and known susceptibility to UV. These process indicators are being used as biodosimeters for certification of full-scale UV systems. More recently *B. subtilis* is also suggested as biodosimeter for full-scale ozone systems (Broséus *et al.*, 2008). The more susceptible micro-organisms are better process indicators to study the variability in efficacy of disinfection processes related to the hydraulics of the continuous flow system. In a model study on UV biodosimetry Cabaj *et al.* (1996) demonstrated the integrated effect of the susceptibility of micro-organisms and the hydraulics of a UV reactor on the inactivation capacity. At a constant dose of 40 mJ/cm^2 their numerical modelling showed a clear decrease in inactivation capacity (or Reduction Equivalent Dose; RED) with increased susceptibility of the micro-organisms (k-value) and dose distribution. Consequently, the broadness of the dose distribution has more effect on the inactivation of susceptible micro-organisms than on the resistant ones. This was also demonstrated for ozone disinfection by Smeets *et al.* (2006). An increasing broadness of dose distribution expressed as decreasing number of continuously stirred tank reactors (CSTR), showed the largest decrease in inactivation of micro-organisms with the highest k-value. The higher impact of dose distribution of full-scale ozonation and chlorination processes on the inactivation efficacy of susceptible micro-organisms such as *E. coli* implicate that this indicator bacterium can be used as process indicator to determine if the dose distribution in the contact chambers is adequate for the required log-inactivation. The log-inactivation for pathogenic organisms that are equally or more resistant (viruses and protozoan (oo)cysts) will never exceed the log-inactivation of *E. coli.* More recently, Computational fluid dynamics (CFD) has become an engineer based tool of growing interest in design and control of full-scale water treatment disinfection and has been used to describe the dose distribution in UV systems (Ducoste *et al.*, 2005) and more recently also in full-scale ozonation contactors (Li *et al.*, 2006).

The difference in persistence of micro-organisms has also consequences for elimination and the IP/PI ratio in filtration processes with low or no filter bed cleaning as demonstrated for the elimination of SSRC and *Cryptosporidium* in slow sand filtration and surface water infiltration (Chapter 7, 8 and 9). While attached anaerobic spores will persist long under these conditions (Chapter 11), attached oocysts are susceptible to predation (Chapter 8) and potentially disappear under these conditions, although not yet everything is clear on the fate of these ingested oocysts.

On the other hand, the similar reaction of attached spores of *C. perfringens* and oocysts of *Cryptosporidium* in soil columns on a pH increase in the water presented in Chapter 9 indicates similarity in surface attachment of both micro-organisms and fate in filter beds with short contact times and frequent filter bed back washing. Similarity of reactions of spores and oocysts on coagulation and filtration conditions were also notified by others (Huck *et al.*, 2001 and Emelko, 2001). This confirms the predictive value of bacterial spore elimination for elimination of oocyst in these water treatment processes.

APPLICATION OF THE METHODS IN QMRA

An example of QMRA. During the last few years Quantitative Microbial Risk Assessment (QMRA) was applied by the Dutch Drinking Water Companies as a tool to assess drinking water safety. The provisional QMRA's used the methodologies described in this book to determine the elimination capacity of water treatment. An example of such calculations is presented in Figure 12.5.

Indicator bacteria were monitored in variable sample volumes depending on the location in treatment. For the final step in treatment (SSF), the DEC was determined by challenge tests on pilot pant scale (PI; Chapters 7, 8; Dullemont *et al.*, 2006).

On basis of these treatment data and data on the concentration of the index pathogens in the source water, the annual risk of infection for *Cryptosporidium* and *Campylobacter* in the drinking water of this locations was 1.1×10^{-6} and 6.6×10^{-5} per person per year (pppy), respectively. Using the more comprehensive method with stochastic modelling of water quality and treatment efficiency data (Medema *et al.*, 2003) the average annual risk of infection for both index pathogens was 1.9×10^{-7} and 6.0×10^{-5} with 97.5-percentile values of 1.5×10^{-6} and 1.7×10^{-4}, respectively.

This example shows that the average and range of the annual risk of infection for both index pathogens could be assessed with full-scale monitoring of both process indicators in variable sample volumes and additional pilot plant research. The average risk for *Cryptosporidium* and *Campylobacter* is below 10^{-4} pppy. The IP/PI ratios for both pathogen/process indicator combinations are usually >1.0 (Figure 12.3 and 12.4), suggesting that the risk of infection is probably lower.

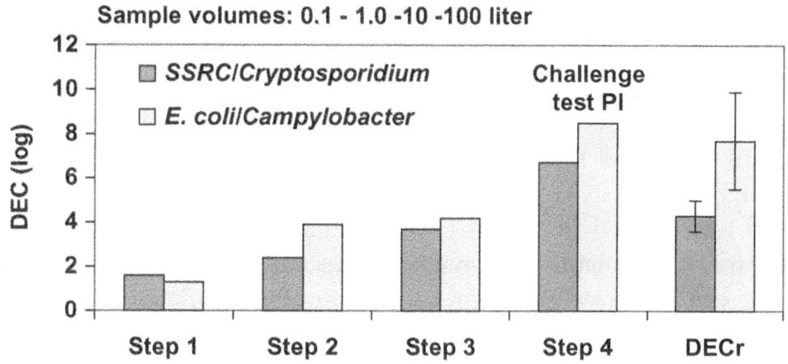

Figure 12.5 The cumulative DEC of a train of full-scale water treatment processes with four treatment steps for Coli44 and SSRC and the required DEC (DEC_r) for the index pathogens to meet the statutory infection risk of 10^{-4} pppy (error bars presents the range of DECr based on minimum and maximum IP concentrations in the source water; steps 1–4, RGF, Ozone, GAC filtration, SSF)

Significance of variability of the DEC in QMRA. The variability in the exposure assessment depends on the variability in the concentrations of index pathogens in the source water but also on the variability in treatment efficiency (Teunis *et al.*, 1997 and Medema *et al.*, 2003). Assessment of variability in the DEC by determination of *actual* Decimal Elimination (DE) with paired C_{in} and C_{out} concentrations by date as applied in the current study is common practice in microbiological risk assessment studies (LeChevallier and Norton, 1991; Nieminski *et al.*, 1995 and Teunis *et al.*, 1997). The observed variability of DEC assessed with the described microbial methods is affected by the quality of the assessed data. This quality is influenced by the accuracy of the applied methods. By decreasing the detection limit of the standard microbial methods for indicator bacteria by a factor of 1000 or more the accuracy of assessment of DEC was increased as demonstrated in Chapters 3 and 4.

Basic assumptions in this approach are that the assessed DEC is independent of the source water concentration and is predictive of the probability of breakthrough of the micro-organisms and the concentrations in the finished water. Under normal conditions treatment is continuously loaded with a low, but variable level of micro-organisms of faecal origin (faecal indicators and pathogens) in the raw water. Peak contaminations may occur infrequently, for instance after heavy rainfall. A large part of the micro-organisms that enter the treatment are initially inactivated or retained (largely in the filter beds) but the retained organisms may be remobilized over time. Therefore, transport of microbes will demonstrate a distribution of retention time or retardation depending on the mechanisms influencing their transport behaviour such as co-aggregation, and breaking of these aggregates and adsorption/desorption. These phenomena will occur in processes such as coagulation/floc-removal and the filtration processes, but also in pipes and (contact) reservoirs along the treatment. This means that the retention time of micro-organisms in water treatment may differ very significantly from the retention time of water in the system. The higher this retardation, the higher the probability of spatial and temporal clustering at the end of treatment and the less there is a direct correlation between C_{in} and C_{out} at any point in time. Moreover, in daily practices influent and effluent sampling at the processes will not be synchronized on the basis of the average water retention time.

The observed positive correlation between the DEC and C_{in} as observed in this book for SSRC and sometimes for *E. coli* (Chapter 4 and 9) and by others for oocysts (Smeets *et al.*, 2007 and Assavasilavasukul *et al.*, 2008) is a clear indication for this retardation. One possible explanation for a higher DEC at higher C_{in}

values is a positive correlation between C_{in} and the load of suspended solids causing a higher DEC due to co-aggregation or increased filter bed ripening. The observed positive correlation of C_{in} with turbidity and of turbidity removal with DEC of treatment plants for oocysts (LeChevallier and Norton, 1991 and Dugan *et al.*, 2001) and for SSRC (Chapter 4) are indications for the validity of this hypothesis. The other explanation for the correlation of DEC and C_{in} is the aspect of retardation (accumulation and delayed transport) as hypothesized for environmental SSRC in slow sand filters and GAC filters (Chapter 4), for environmental centric diatoms in slow sand filters (Chapter 8) and environmental FRNA phages, Coli44 and SSRC in soil passage (Chapter 9). Additional indications for retarded microbial transport during water treatment are i) the lower variation in elimination of overall treatment plants for environmental Coli44 and SSRC (relative standard deviation 10–13%) compared to the variation in elimination of unit processes (relative standard deviation 25–47%) in Chapter 4 and ii) the increased spatial heterogeneity and over-dispersion of dosed aerobic bacterial spores after a treatment train (Gale *et al.*, 1997, 2002) although propagation of these microbes during treatment can not be excluded.

This over-dispersion or clustering of pathogens in drinking water by treatment has two major implications on the QMRA process. It will cause an over-dispersion in the distribution of exposure to pathogens among the public. Due to this over-dispersion the average risk at a constant mean dose concentration is lower than the risk related to a Poisson distributed dose (Haas, 2002 and Havelaar *et al.*, 2004). Secondly, the actual decimal elimination assessed by paired concentrations on date as an input of stochastic modelling (Medema *et al.*, 2003) may over-estimate the variability of the DEC calculated from such data and as a result both the infection risk itself and its variability. This was confirmed by the work done by Smeets (2008) who showed that the current method using paired C_{in} and C_{out} to calculate elimination over-estimates the risk of infection by one order of magnitude. Important in this respect are the observations described in Chapter 4 and mentioned above and also described by Medema (1999), that variability of one step in treatment is dependent on the performance of the previous process. Thus, the variability in the DEC of a treatment train with multiple barriers is lower than expected from the sum of the variability of the individual processes. These observations emphasize that for a proper and realistic description of the uncertainty of the risk of infection in drinking water the variability in an overall treatment should be assessed rather than summation of the variability of the individual processes.

In conclusion the proposed methodology to quantify the DEC of specific full-scale systems by microbiological monitoring is an important input for QMRA and microbial risk management (WSP). Interpretation of these data with respect to the uncertainty of the assessed infection risk from the drinking water needs comprehensive statistical tools to account for the variability in the collected microbial data. Smeets (2008) developed and applied statistical methods with stochastic modelling by fitting the calculated distribution of concentrations with the observed distribution. This reduced the over-estimation of variability in DEC caused by the use of paired data. Besides the use of statistical methods for data interpretations, these tools can be used in the development of a microbiological monitoring strategy which accounts for the previous considerations on variability.

GENERIC METHODOLOGY TO ASSESS ELIMINATION CAPACITY OF WATER TREATMENT PROCESSES FOR PATHOGENS

On the basis of the studies presented in this book supplemented with data presented in this discussion a generic concept of site specific assessment of elimination capacity of water treatment processes is proposed and presented in Figure 12.6. Basic principle in the proposed methodology is that the elimination should be determined under local specific conditions as much as possible with simple and reliable methods which fit in the daily drinking water practices.

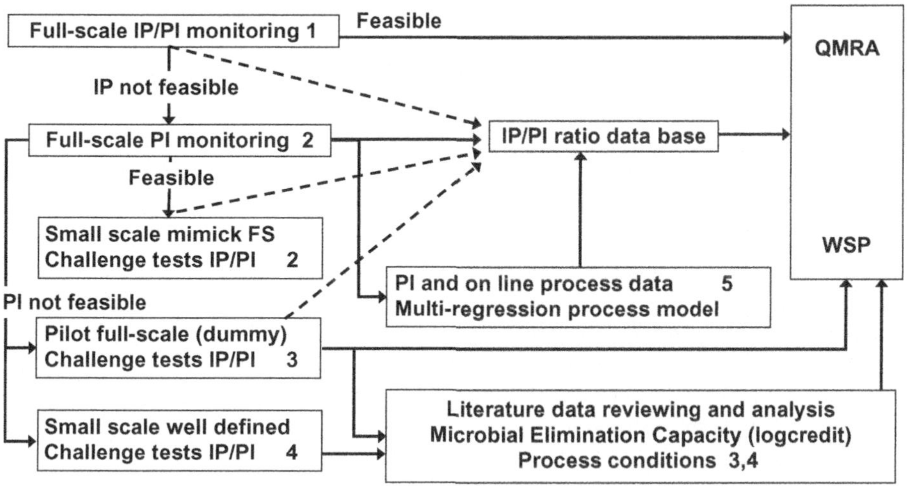

Figure 12.6 Generic methodology of assessment of elimination capacity of water treatment processes (IP and PI = index pathogen and process indicator, FS = full-scale; solid line = DEC values and dotted line = indirect information for DEC assessment)

The solid lines in this diagram represent information on elimination of micro-organisms, information which has to be collected regularly in accordance with the regulations (Anonymous, 2001). The dotted lines mean information on the IP/PI ratio which has been collected on the basis of comparative quantitative data (treatment monitoring data or challenge test results).

Description of the proposed generic methodology:

1. The methodology starts with direct monitoring for index pathogens (IP) and usage of these data in QMRA and Water Safety Plans (WSP). It was demonstrated that for some systems this was a realistic possibility (Chapter 1), but for most it is not. By measuring index pathogens (IP) and process indicators (PI) simultaneously, the elimination of process indicators IP/PI ratios can be determined. On the basis of further considerations it has to be decided to what extent pathogen monitoring in treatment has to be continued or monitoring of faecal indicator bacteria as process indicators (PI) can be used as an alternative (next best method 2; cheaper and hence more intensive).
2. From the present study it was concluded that the next best method is monitoring elimination of *E. coli* and SSRC in the full-scale water treatment facility. The historical database on faecal indicators concentrations in water treatment shows that large volume samples are needed for a reliable estimate of the DEC of downstream treatment processes and its variability (Chapters 2, 3 and 4). This method allows site specific assessment of the DEC of the PI, but requires additional information on IP/PI ratios (Chapters 5, 6, 7, 8, 9 and 10) to translate these data to the index pathogens.
3. When the elimination capacity is not assessable with process indicator monitoring, challenge tests with index pathogens and/or process indicators in a pilot of the process operated under full-scale conditions (dummy) is an alternative the method (Chapters 7, 8 and 10). The quantitative data of the index pathogen elimination can directly be used in QMRA. When for safety reasons the dummy is only challenged with the process indicator, IP/PI ratios are required to translate the observed elimination to the index pathogen removal. These ratios can be assessed with small scale experiments (Chapters 5, 7 and 9) and/or obtained from literature reviews (Chapter 6, Hijnen and Medema, 2007).

4. When both methods (2 and 3) are not applicable the use of small scale tests under well defined conditions (collimated beam, disinfection, jar-tests, column tests) is an alternative method which yields only relative quantitative data. Additional literature data assessed under more realistic process conditions (Chapter 6, Hijnen and Medema, 2007) are required to translate the results of these small tests to full-scale conditions.
5. Additional methods of interest are the use of statistical modelling of full-scale data of process conditions and elimination capacity by multi-regression analysis (Wiersema, 1999 and Haas *et al.*, 2001) or the use of process models (Schijven, 2001; Van der Wielen *et al.*, 2006 and Schijven *et al.*, 2008) in combination with stochastic modelling of elimination data as demonstrated more recently by Smeets (2008). This enables a more direct and continuous control of water safety by on line process data and can be used as supportive information for QMRA. Example of such on line process monitoring control of the DEC of processes based on microbiological elimination data is the dose (fluence) control of a full-scale UV disinfection at the drinking water production plant Berenplaat of Evides (Medema and Beerendonk, 2005). These methods are additional because application depends on the availability of elimination data assessed with the above described methods. The success of these methods depends largely on the quality of the available data bases.

FURTHER RESEARCH NEEDS

Process indicator for viruses. In the work of this book the main focus was on elimination of the index pathogens *Cryptosporidium/Giardia* and *Campylobacter*. Data on virus elimination was assessed in Chapter 6 for UV and Chapters 7, 9 and 10 on removal of the F-specific RNA phage MS2, a conservative model organism for virus transport in filtration processes (Schijven, 2001). In the microbial risk assessment of the Inspectorate (Wetsteyn *et al.*, 2005) both somatic coliphages and FRNA phages are mentioned as potential process indicators for virus removal in treatment. Microbiological water quality monitoring in drinking water practices, however, is still largely based on indicator bacteria and data on elimination of bacteriophages in full-scale treatment plants are very limited.

Two full-scale studies indicated the value of these bacteriophages as process indicators for virus removal by conventional treatment; the study of Payment and Franco (1993) and the study of Havelaar *et al.* (1995). From the elimination data of human enteric viruses and coliphages by three full-scale drinking water systems with large volume sampling an IP/PI ratio of 1.3 ± 0.5 was calculated (Payment and Franco, 1993). The study of Havelaar *et al.* (1995) yielded an IP/PI of 0.9 ± 0.3 for FRNA phages. In both studies the faceal indicator bacteria were also monitored and Payment proposed *C. perfringens* as a surrogate for virus and *Cryptosporidium/Giardia* removal in drinking water treatment. He observed an IP/PI ratio for viruses and *C. perfringens* of 1.0 ± 0.3 ($n = 3$). The IP/PI ratios for enteroviruses and faecal indicators (Coli44 and SSRC) calculated from the data presented by Havelaar was 0.7 ± 0.3. Similar ratios of 0.6–0.7 were also observed in comparative (in)filtration studies with challenge tests using MS2 phages as a model virus (Chapters 7 and 9, Dullemont *et al.*, 2006 and Nobel *et al.*, 1999). Thus, these studies indicate that faecal indicator bacteria potentially are removed to a larger extent in filtration processes than viruses and do not satisfy the requirement for a proper process indicator for virus removal by filtration processes. Because MS2 is regarded as a conservative process indicator for virus removal (Schijven, 2001) it is questionable whether this conclusion is correct, however. Using the MEC values of conventional treatment processes for viruses, bacteria and bacterial spores (Hijnen and Medema, 2007) virus/indicator ratios of >1.0 were calculated for most processes (range of 0.7–1.6). These ratios are based on indirect comparison and therefore less reliable, but they suggest that when lacking data on bacteriophage elimination, data on elimination of faecal indicators can be used as crude estimate of elimination of viruses. This hypothesis must be verified by further comparative research on the elimination of viruses, environmental bacteriophages and faecal indicator bacteria in full-scale water treatment.

Further omissions in knowledge on process indicators. In the process of QMRA it became apparent that for some processes quantitative information on elimination of micro-organisms was still missing because of lack of data on process indicator removal as well as a lack of data on IP/PI ratios. Examples of such processes are post-disinfection with chlorine dioxide and ultrafiltration.

- Post-disinfection is used at the end of a treatment train in a number of Dutch locations and is potentially an effective process in inactivation of susceptible pathogens (bacteria and viruses). The Coli44 concentrations at the end of those treatments, however, are too low to assess DEC.
- Ultrafiltration is applied both as pre-treatment step and a polishing step in water treatment. Because of the high potential of ultrafiltration as a barrier for viruses, bacteria and parasitic protozoa (DEC of 5 log or more) concentration of faecal indicators or other particulates in the source water must be high to demonstrate this DEC. Applied as a pre-treatment process such high DEC values can be demonstrated by particle counting as presented for the drinking water production plant Heemskerk PWN (Kruithof *et al.*, 2001), but used as a polishing step in treatment this is not possible.

GENERAL CONCLUSIONS AND RECOMMENDATIONS

On the basis of the studies described in this book the general conclusion is that the elimination of *E. coli* and SSRC by water treatment processes under full-scale conditions can be assessed with a reliable method with a low level of detection which fits in the daily routine of microbial water quality monitoring. The DEC of water treatment for both process indicators is determined site specific and related to the full-scale conditions as much as possible. Application of the method showed a large variability of DEC for these environmental micro-organisms which potentially differ from DEC assessed in challenge tests with pre-cultured organisms.

Challenge tests with pre-cultured organisms can be used to assess elimination of micro-organisms in water treatment processes. Challenge tests on pilot plant scale operated as a dummy of the full-scale process will generate quantitative data with a high predictive value for DEC of a full-scale system. DEC determined in small scale challenge tests (batch or columns) must be regarded as indicative for the DEC of the full-scale process. With these tests the relative DEC of different micro-organisms (pathogens and process indicators, IP/PI ratio) can be assessed and the influence of process conditions on these DEC values.

Quantitative information on elimination of micro-organisms by water treatment processes presented in literature can be used to calculate an average DEC or default value of the Microbial Elimination Capacity of these processes. These MEC values can be used when local specific conditions are missing and as reference data for the DEC assessed with challenge tests.

Spores of sulphite-reducing clostridia (SSRC; incl. *C. perfringens*) can be used as a process indicator for *Cryptosporidium* and *Giardia* (oo)cyst elimination by conventional water treatment processes (coagulation/filtration) and ozone but not for slow sand filtration, surface water infiltration and GAC filtration. The IP/PI ratios assessed for these processes were too variable and sometimes too conservative (»1.0). From the accumulated IP/PI ratios for Coli44 (incl. *E. coli*) and *Campylobacter* it was concluded that these indicator bacteria are proper process indicators for slow sand filtration, rapid granular filtration and ozonation. The IP/PI ratios for MS2 as a model virus and both faecal indicators showed that these were usually <1.0 indicating that viruses are removed to a lower extent than both indicators.

The methods described in this book has been used in practice by Water Companies to optimize their water treatment processes and to perform the mandatory Quantitative Microbial Risk Assessment to estimate the annual risk of infection with the level of uncertainty for *Campylobacter*, *Cryptosporidium* and *Giardia* in drinking water.

Finally a number of issues can be identified as subjects which need to be investigated further.

- The Hemoflow method with cross-flow filtration is a new method to determine multiple micro-organisms including bacteriophages in large water volumes. The use of this method as an alternative for the MF-sampling for both faecal indicators must be explored further in drinking water practice.
- Bacteriophages are the preferred process indicators for enterovirus elimination in water treatment. In this discussion it has been hypothesised that *E. coli* and SSRC removal are indicative for virus elimination and can be used in QMRA when data on bacteriophages are not available. This simplification of elimination assessment, however, needs further supportive evidence by collecting elimination data of indicator bacteria and bacteriophages in comparative studies under full-scale as well as under challenge test conditions.
- In the revised Dutch Drinking Water Decree (Anonymous, 2001) the statutory parameters to determine microbial safety of the drinking water are *E. coli* and spores of *C. perfringens*. The change from thermotolerant coliforms (Coli44) to *E. coli* is small since the methods are almost identical. Increasing the specificity of the analytical methods for the anaerobic spores SSRC to determine spores of *C. perfringens*, however, may have a larger impact on the use of this process indicator in QMRA and needs to be studied further.
- A process which is not addressed in this study is post-disinfection with low disinfectant doses. On the basis of the observations described in this book on chemical disinfection it seems inappropriate to use the same dose-effect data to verify inactivation efficiency as used for main disinfection processes. Especially because these processes are operated under conditions of low doses and long contact times. Challenge tests are required to evaluate the efficacy of these processes for susceptible micro-organisms (*Campylobacter* and viruses).
- Predation is an elimination mechanism which is of benefit for elimination in biological filtration processes like slow sand filtration and GAC filtration. The ingested micro-organisms which include waterborne pathogens, may survive in these zooplankton micro-organisms. Since these grazing organisms are also observed in the filtrate of these filters, these organisms can become a vector of waterborne pathogens.

References

Achalme P. (1891–1897). Compt. rend. Soc. de biol., cit. from Achalme 1897. *Ann. Inst. Past.*, **11**, 845.

Adam R. D. (2001). Biology of *Giardia lamblia*. *Clin. Microbiol. Rev.*, **14**, 447–475.

Akiba M., Kunikane S., Kim H.-S. and Kitazawa H. (2002). Algae as surrogate indices for the removal of *Cryptosporidium* oocysts by direct filtration. *Wat. Sci. Technol.: Wat. suppl.*, **2**, 73–80.

Altschul S. F., Gish W., Miller W., Myers E. W. and Lipman D. J. (1990). Basic local alignment search tool. *J. Mol. Biol.*, **215**(3), 403–410.

Anderson M. T. (1981). Improved method for separating zooplankton from detritus. Prog. Fishculturist, **43**, 42–44.

Angelotti R., Hall H. E., Foter M. J. and Lewis K. H. (1962). Quantitation of *Clostridium perfringens* in foods. *Appl. Microbiol.*, **10**, 193–199.

Anonymous (1942). Voorschriften voor de bacteriologische keuring van drinkwater: Hoofdcommissie voor de normalisatie in Nederland.

Anonymous (1956). Bacteriologisch onderzoek van drinkwater: Hoofdcommissie voor de normalisatie in Nederland. N. V. Keuringsinstituut voor Waterleidingartikelen (KIWA).

Anonymous (1962). Jaarverslag 1962. Gemeentewaterleidingen Amsterdam.

Anonymous (1975). Richtlijn 75/440/EEG van de Raad van 16 juni 1975 betreffende de vereiste kwaliteit van het oppervlaktewater dat is bestemd voor produktie van drinkwater in de Lid-Staten.

Anonymous (1980a). Kwaliteit van voor menselijke consumptie bestemd water: Publicatieblad Nr. L 229/11 van 30.08/1980.

Anonymous (1980b). Guidelines and recommendations for the protection of Groundwater intake areas, Commissie bescherming Waterwingebieden (CBW). VEWIN-RID, Rijswijk, NL.

Anonymous (1981). Bacteriologisch onderzoek van drinkwater. Onderzoek met behulp van membraanfiltratie naar de aanwezigheid van bacteriën van de coligroep, Nederlands Normalisatie Instituut NEN6553.

Anonymous (1982). Bacteriologisch onderzoek van drinkwater. Onderzoek met behulp van membraanfiltratie naar de aanwezigheid van faecale streptococcen, Nederlands Normalisatie Instituut NEN6564.

Anonymous (1984). Waterleidingbesluit Staatsblad van het Koninkrijk der Nederlanden, **220**, 1–35.

Edited by Wim A.M. Hijnen. ISBN: 9781843393764. Published by IWA Publishing, London, UK.

Anonymous (1984). Herziene Waterleidingbesluit. *Staatsblad van het Koninkrijk der Nederlanden*, 1–20, the Netherlands.

Anonymous (1985a). Bacteriologisch onderzoek van drinkwater. Onderzoek met behulp van membraanfiltratie naar de aanwezigheid van sporen van sulfiet-reduucerende clostridia, Nederlands Normalisatie Instituut NEN6567.

Anonymous (1985b). Nederlandse praktijkrichtlijn; toelichting bij Nederlands Normalisatie Instituut NEN6567, vol. NPR6568.

Anonymous (1985c). NEN6564; NEN6567.

Anonymous (1998). Council Directive 98/83/EC of 3 November 1998 on the quality of water intended for human consumption. *Official Journal of the European Communities*, **L330**, 32–54.

Anonymous (1999). Anlagen zur Desinfektion von Wasser mittels Ultraviolet-strahlen, Anforderungen und Prüfung, Önorm M 5873-1. Österreichisches Normungsinstitut, Vienna Au.

Anonymous (2001). Besluit van 9 januari 2001 tot wijziging van het waterleidingbesluit in verband met de richtlijn betreffende de kwaliteit van voor menselijke consumptie bestemd water, p. 1–53, vol. 31. Staatsblad van het Koninkrijk der Nederlanden.

Anonymous (2003). *3rd Edition of the Guidelines for Drinking Water Quality: World Health Organisation,* Geneve, Switzerland.

Atherholt T. B., Lechavallier M. W., Norton W. D. and Rosen J. S. (1998). Effect of rainfall on *Giardia* and *Cryptosporidium*. *J. Am. Water Works Assoc.*, **90**, 66–80.

Antopol S. C. and Ellner P. D. (1979). Susceptibility of *Legionella pneumophila* to Ultraviolet radiation. *Appl. Environ. Microbiol.*, **38**, 347–348.

Ashbolt N. J., Grabow W. O. K. and Snozzi M. (2001). Indicators of microbial water quality. In: Water Quality: Guidelines, Standards and Health, p. 289–316, World Health Organization WHO, London, UK.

Assavasilavasukul P., Lau B. L. T., Harrington G. W., Hoffman R. F. and Borchardt M. A. (2008). Effect of pathogen concentrations on removal of Cryptosporidium and *Giardia* by conventional drinking water treatment. *Water Res.*, **42**, 2678–2690.

Badenoch J. (1990). Cryptosporidium in water supplies, London.

Badenoch J. (1995). *Cryptosporidium* in water supplies: Second report of the group of experts. Department of the Environment, Department of Health, London UK.

Bader H. and Hoigne L. (1982). Determination of ozone in water by the Indigo Method, A submitted Standard Method. *Ozone Sci. and Eng.*, **4**, 169–176.

Bales R. C., Hinkle S. R., Kroeger T. W. and Stocking K. (1991). Bacteriophage adsorption during transport through porous media: chemical perturbations and reversibility. *Environ. Sci. Technol.*, **25**, 2088–2095.

Bales R. C. and Li S. (1993). MS-2 and Poliovirus Transport in Porous Media: Hydrophobic effect and Chemical Perturbations. *Wat. Resour. Research*, **29**(4), 957–963.

Battigelli D. A., Sobsey M. D. and Lobe D. C. (1983). The inactivation of Hepatitis A virus and other model viruses by UV irradiation. *Wat. Sci. Tech.*, **27**, 339–342.

Beakes G., Canter H. M. and Jaworski G. H. M. (1988). Zoospores ultrastructure of *Zygorhizidium affluens* Canter and *Z. planktonicum* Canter, two chytrids parasitizing the diatom *Asterionella formossa. Hassall. Can. J. Bot.*, **66**, 1054–1067.

Bellamy W. D., Silverman G. P., Hendricks D. W. and Logsdon G. S. (1985). Removing *Giardia* cysts with slow sand filtration. *J. Am. Water Works Assoc.*, **77**, 52–60.

Belosevic M., Craik S. A., Stafford J. L., Norman N. F., Kruithof J. C. and Smith D. (2001). Studies on the resistance/reactivation of *Giardia muris* cysts and *Cryptosporidium parvum* oocysts exposed to medium-pressure ultraviolet radiation. *FEMS Microbiol. Letter*, **204**, 197–203.

Bernhardt H., Hoyer O., Hengesbach B. S., Karanis D. P., Moriske H. and von Sonntag C. (1992). Desinfektion aufbereiteter Oberflächenwässers mit UV Strahlen- erste Ergebnisse des Forschungsvorhabens. GWF Wasser-Abwasser, **133**, 632–643.

Bernhardt H. (1994). Desinfektion aufbereiteter Oberflächenwässers mit UV Strahlen. GWF Wasser-Abwasser, **135**, 677–689.

Bichai F., Payment P. and Barbeau B. (2008). Protection of waterborne pathogens by higher organisms in drinking water: a review. *Can. J. Microbiol.*, **54**, 509–524.

Bichai F., Barbeau B., Dullemont Y. and Hijnen W. A. M. (2010a). Role of predation by zooplankton in transport and fate of protozoan (oo)cysts in granular activated carbon filtration. *Water Res.*, **44**, 1072–1081.

Bichai F., Hijnen W. A. M. and Barbeau B. (2010b). *Internalisation of protozoan (oo)cysts by zooplankton in granular media filtration: a quantitative microbial risk assessment in drinking water.* Submitted to J. Water and Health May, 2010.

Blanc R. and Nasser A. (1996). Effect of effluent quality and temperature on the persistence of viruses in soil. *Wat. Sci. Tech.*, **33**, 237–242.

Blaser M. J., Smith P. M., Wang W. L. and Hoff J. C. (1986). Inactivation of Campylobacter jejuni by Chlorine and Monochloramine. *Appl. Environ. Microbiol.*, **51**, 307–311.

Bolton J. R., Dussert B., Bukhari Z., Hargy T. and Clancy J. L. (1998). Presented at the Proceedings of the American Water *Works Association Water Quality Technology Conference,* San Diego, US, Nov. 1998.

Bonde G. J. (1962). *Bacterial indication of water pollution.* Teknisk Forlag, Copenhagen.

Bonde G. J. (1977). *Bacterial indication of water pollution. Adv. Aquat. Microbiol.*, **1**, 273–364.

Bouwer E. J. and Rittman B. E. (1992). Comment on use of colloid filtration theory in modeling movement of bacteria through a contaminated aquifer. *Environ. Sci. Technol.*, **26**, 400–401.

Bradford S. A., Yates S. R., Bettahar M. and Šimůnek J. (2002). Physical factors affecting the transport and fate of colloids in saturated porous media. *Water Resour. Res.*, **88**, 1327–1340.

Bradford S. A., Šimůnek J., Bettahar M. and van Genuchten M. T. (2003). Modeling colloid attachment, straining and exclusion in saturated porous media. *Environ. Sci. Technol.*, **37**, 2242–2250.

Bradford S. A., Bettahar M., Simunek J. and Van Genuchten M. T. (2004). Straining and attachment of colloids in physically heterogenous porous media. *Vadose Zone J.*, **3**, 384–394.

Bradford S. A. and Bettahar M. (2005). Straining, attachment and detachment of *Cryptosporidium* oocysts in saturated porous media. *J. Environ. Qual.*, **34**, 469–478.

Broda D. M., Lawson P. A., Bell R. G. and Musgrave D. R. (1999). *Clostridium frigidicarnis* sp. nov., a psychrotolerant bacterium associated with 'brown pack'spoilage of vacuum-packed meats. *Int. J. Syst. Bacteriol.*, **49**, 1539–1550.

Broséus R., Barbeau B. and Bouchard C. (2008). Verification of full-scale ozone contactor inactivation performance using biodosimetry. *J. Envrion. Engin.*, **4**, 314.

Brush C. F., Walter M. F., Anguish L. J. and Ghiorse W. C. (1998). Influence of pretreatment and experimental conditions on electrophoretic mobility and hydrophobicity of *Cryptosporidium parvum* oocysts. *Appl. Environ. Microbiol.*, **64**, 4439–4445.

Buchholtz ten Brink M. R., Mecray E. L. and Galvin E. L. (2000). *Clostridium perfringens* in Long Island sound sediments: an urban sedimentary record. *J. of Coast. Research*, **16**, 591–612.

Burge W. D. and Enkiri N. K. (1978). Virus adsorption in five soils. *J. Environ. Qual.*, **7**, 73–76.

Busscher H. J., Dijkstra R. J. B., Engels E., Langworthy, Collias D. I., Bjorkquist D. W., Mitchell M. D. and Van der Mei H. C. (2006). Removal of Two Waterborne Pathogenic Bacterial Strains by Activated Carbon Particles prior to and after Charge Modification. *Environ. Sci. Technol.*, **40**, 6799–6804.

Bush J. H. (1953). New bacteriological technique for testing water and sewage. Water and Sewerage Works, **100**, 151.

Butler R. C., Lund V. and Carlson D. A. (1987). Susceptibility of *Campylobacter jejuni* and *Yersinia enterocolitica* to UV radiation. *Appl. Environ. Microbiol.*, **53**, 375–378.

Cabaj A., Sommer R. and Schoenen D. (1996). Biodosimetry: model calculations for UV water disinfection devices with regard to dose distribution. *Water Res.*, **30**, 1003–1009.

Cabelli V. J. (1977). *Clostridium perfringens* as a water quality indicator. *In* A. W. Hoadley and B. J. Dutka (ed.), *Bacterial indicators: health hazards related with water*, vol. 635, p. 65–79ASTM STP.

Caplenas N. R. and Kanarek M. S. (1984). Thermotolerant Non-fecal Source *Klebsiella pneumoniae*: Validity of the Fecal Coliform Test in Recreational Waters. *Am. J. Public Health*, **74**, 1273–1275.

Camper A. K., LeChevallier M. W., Broadaway S. C. and McFeters G. A. (1985). Growth and persistence of pathogens on granular activated carbon filters. *Appl. Environ. Microbiol.*, **50**(6), 1378–1382.

Chang J. C. H., Ossoff S. F., Lobe D. C., Dorfman M. H., Dumais C. M., Qualls R. G. and Johnson J. D. (1985). UV inactivation of pathogenic and indicator micro-organisms. *Appl. Environ. Microbiol.*, **49**, 1361–1365.

Chick H. (1908). An investigation of the laws of disinfection. *J. Hyg.*, **8**, 92.

Clancy J. L., Hargy T. M., Marshall M. M. and Dyksen J. E. (1998). UV light inactivation of *Cryptosporidium* oocysts. *J. Am. Water Works Assoc.*, **90**(9), 92–102.

Clancy J. L., Bukhari Z., Hargy T. M., Bolton J. R., Dussert B. W. and Marshall M. M. (2001). Using UV to inactivate *Cryptosporidium. J. Am. Water Works Assoc.*, **92**, 97–104.

Clancy J. L., Hargy T. M., Battigelli D. A., Marshall M. M., Korich D. and Nicholson W. L. (2002). Susceptibility of multiple strains of *C. parvum* to UV light. *Am. Water Works Assoc. Res.* Found. Denver CO, US.

Clancy J. L., Fallon K., Hargy T. M., Mackey E. and Wright H. (2004). Presented at the American Water Works Association Water Quality Technology Conference, Nov. 2004, San Antonio, US.

Cleasby J. L., Hilmoe D. J. and Dimitracopoulos C. J. (1984). Slow sand and direct in-line filtration of a surface water. *J. Am. Water Works Assoc.*, **76**, 44–55.

Craik S. A., Finch G. R., Bolton J. R. and Belosevic M. (2000). Inactivation of *Giardia muris* cysts using medium-pressure ultraviolet radiation in filtered drinking water. *Water Res.*, **34**, 4325–4332.

Craik S. A., Weldon D., Finch G. R., Bolton J. R. and Belosevic M. (2001). Inactivation of *Cryptosporidium parvum* oocysts using medium- and low-pressure ultraviolet radiation. *Water Res.*, **35**, 1387–1398.

Coffey B. M., Huck P. M., Maurizo D. D., Emelko M. B., Douglas I. P. and Van den Oever J. (1999). *Presented at the Water Quality Technol. Conf.*, Tampa, Florida, US.

Colford J. M., Wade T. J., Sandhu S. K., Wright C. C., Lee S., Shaw S., Fox K., Burns S., Benker A., Brookhart M. A., van der Laan M. and Levy D. A. (2005). A randomized, controlled trial of in-home drinking water intervention to reduce gastrointestinal illness. *Am. J. Epidemiol.*, **161**, 472–482.

Craun G. F. (1988). Surface water supplies and health. *J. Am. Water Works Assoc.*, **80**(2), 40–52.

Craun G. F. (1990). Waterborne *Giardia*sis. In: E. A. Meyer (ed.), *Human Parasitic Diseases,* vol. 3, p. 267–293. Elsevier Science Publication, Amsterdam, The Netherlands.

Craun G. F., Berger P. S. and Calderon R. L. (1997). Coliform bacteria and waterborne disease outbreaks. *J. Am. Water Work Assoc.*, **89**(3), 96–104.

Craun G. F., Hubbs S. A., Frost F., Calderon C. R. L. and Via S. H. (1998). Waterborne outbreaks of cryptosporidiosis. *J. Am. Water Works Assoc.*, **90**, 81–91.

Dai X. and Hozalski R. M. (2003). Evaluation of microspheres as surrogates for *Cryptosporidium parvum* oocysts in filtration experiments. *Environ. Sci. Technol.*, **37**, 1037–1042.

Davies C. M., Long J. A. H., Donald M. and Ashbolt N. J. (1995). Survival of Faecal Microorganisms in Marine and Freshwater Sediments. *Appl. environ. microbiol.*, **61**, 1888–1896.

Davies C. M., Kaucner C., Altavilla N., Ashbolt N. J., Ferguson C. M., Krogh M., Hijnen W. A. M., Medema G. J. and Deere D. A. (2005a). *Fate and Transport of Surface Water Pathogens in Watersheds.* AWWARF, Denver CO, US.

Davies C. M., Altavilla N., Krogh M., Ferguson C. M., Deere D. A. and Ashbolt N. J. (2005b). Environmental inactivation of *Cryptosporidium* oocysts in catchment soils. *J. Appl. Microbiol.*, **98**, 308–317.

de Roda Husman A. M., Ketelaars H. A. M., Lodder W. J., Medema G. J. and Schets F. M. (2001a). De microbiologische kwaliteit van het ingenomen en afgeleverde water van Waterwinningbedrijf Brabantse Biesbosch in 2001. RIVM330250002 (in Dutch).

de Roda Husman A. M., Duizer E., Lodder W., Pribil W., Cabaj A., Gehringer P. and Sommer R. (2003). Presented at the 2nd Int. Congress on ultraviolet technologies, Vienna, Au.

de Roda Husman A. M. (2001b). *Human viruses in H2O.* H2O, **8**, 18–20.

Douglas H. W. (1955). *Trans. Faraday Soc.*, **51**, 146.

Driks A. (2007). Surface appendages of bacterial spores. *Mol. Microbiol.*, **63**, 623–625.

Drost Y. C., Groennou J. T., Hijnen W. A. M. and Van der Kooij D. (1997). Statistische methoden ter bepaling van de belasting en eliminatiecapaciteit van zuiveringsprocessesn m.b.t. micro-organismen. Kiwa Water Research SWI 97.177, Nieuwegein NL.

Drozd C. and Schwartzbrod J. (1996). Hydrophobic and electrostatic cell surface properties of *Cryptosporidium parvum. Appl. Environ. Microbiol.*, **62**, 1227–1232.

Ducoste J., Linden K., Rokjer D. and Liu D. (2005). Assessment of Reduction Equivalent Fluence Bias Using Computational Fluid Dynamics. *Environ. Eng. Sci.*, **22**, 615–628.

Dugan N. R., Fox K. R., Owens J. H. and Miltner R. J. (2001). Controlling *Cryptosporidium* oocysts using conventional treatment. *J. Am. Water Works Assoc.*, **93**(12), 64–76.

Duizer E., Bijkerk P., Rockx B., de Groot A., Twisk F. and Koopmans M. (2004). Inactivation of caliciviruses. *Appl. Environ. Microbiol.*, **70**, 4538–4543.

Dutka B. J. (1973). Coliform are an inadequate index of water quality. *J. Env. Health*, **36**, 39–46.

Dullemont Y. J., Visser A., Schijven J. F. and Hijnen W. A. M. (2004). Eliminatiecapaciteit van langzame zandfiltratie voor micro-organismen bepaald met doseerproeven. H2O, **37**(13), 25–27.

Dullemont Y. J., Schijven J. F., Hijnen W. A. M., Collin M., Magic-Knezev A. and Oorthuizen W. A. (2006). *Presented at the Recent progress in slow sand and alternative biofiltration*, Mullheim, Germany.

Duncan S. H., Louis P. and Flint H. J. (2007). Cultivable bacterial diversity from the human colon. *Lett. Appl. Microbiol.*, **44**, 343–350.

DVGW. (1997). UV-Desinfektionsanlagen für die Trinkwasserversorgung – Anforderungen und Prüfung, Technical Standard W294. In: D. V. d. G.-u. Wasserfaches (ed.), Bonn, GR.

Dykes A. C., Juranek D. D., Lorenz R. A., Sinclair S., Jakubowski W. and Davies R. B. (1980). Municipal waterborne Giardiasis: an epidemiological investigation. *Ann. Int. Med.*, **92**, 165–170.

Eijkman C. (1904). Die Gärungsprobe bei 46°C als Hilfsmittel bei der Trinkwasseruntersuchung. Centralblat für Bakteriologie Abth., **1 Orig. 37**, 742.

Ellis K. V. (1985). Slow sand filtration, Critical Reviews in Environmental Control, vol. 15, p. 315–354. CRC Press.

Emelko M. B. (2001). *Removal of Cryptosporidium parvum by granular media filtration.* University of Waterloo, Ontario, Canada.

Emelko M. B., Schmidt P. J. and Roberson J. A. (2008). Quantification of uncertainty in microbial data – reporting and regulatory implications. *J. Am. Water Work Assoc.*, **94**(3), 94–104.

Escherich T. (1885). Die Darmbakterien des Neugeborenen und Säuglings. Fortschritte des Medicin, **3**, 515–547.

Evans F. R. and Curran H. R. (1959). Influence of preheating, pH and holding temperature upon viability of bacterial spores stored for long periods in buffer substrates. *J. Bacteriol.*, **79**, 361–368.

Evers E. G. and Groennou J. T. (1999). Berekening van de verwijdering van micro-organismen bij de bereiding van drinkwater. RIVM 734301016.

Fallacara D. M., Monahan C. M., Morishita T. Y. and Wack R. F. (2001). Fecal shedding and antimicrobial susceptibility of selected pathogens and a survey of intestinal parasites in free-living waterfowl. *Avian. Dis.*, **45**, 128–135.

Farrah S. R. (1982). Chemical factors influencing adsorption of bacteriophage MS2 to Membrane Filters. *Appl. Environ. Microbiol.*, **43**(3), 659–663.

Fayer R., Trout J. M., Walsh E. and Cole R. (2000). Rotifers ingest oocysts of *Cryptosporidium parvum. J. Eukaryot. Microbiol.*, **47**, 161–163.

Fernandes T. M. A., Schout C., De Roda Husman A. M., Eilander A., Vennema H. and Van Duynhoven Y. T. H. P. (2007). Gastroenteritis associated with accidental contamination of drinking water with partially treated water. *Epidemiol. Infect.*, **135**, 818–826.

Finch G. R., Smith D. W. and Stiles M. E. (1988). Dose-response of *Escherichia coli* in ozone demand-free phosphate buffer. *Water Res.*, **22**, 1563–1570.

Finch G. R., Black E. K., Gyurék L. and Belosevic M. (1993). Ozone inactivation of *Cryptosporidium parvum* in demand-free phosphate buffer determined by in vitro excystation and animal infectivity. *Appl. Environ. Microbiol.*, **59**, 4203–4210.

Fletcher M. and Marshall K. C. (1982). Are solid surfaces of ecological significance to aquatic bacteria. *Adv. Microb. Ecol.*, **6**, 199–236.

Fogel D., Isaac-Renton J., Guasparini R., Moorehead W. and Ongerth J. (1993). Removing *Giardia* and *Cryptosporidium* by Slow Sand Filtration. *J. Am. Water Works Assoc.*, **85**, 77–84.

Foster S. J. and Johnstone K. (1990). Pulling the triggers: The mechanism of bacterial spore germination. *Mol. Microbiol.*, **4**, 137–141.

Frankland P. and Frankland P. (1894). *Micro-organisms in water. Their significance, identification and removal.* Longmans, Green and co., London/New York.

Fujioka R. S. and Shizumura L. K. (1985). *Clostridium perfringens*, a reliable indicator of stream water quality. *Jour. Wat. Pol. Control Fed.*, **57**, 986–992.

Galbraith N. S., Barrett N. J. and Stanwell-Smith R. (1992). *Water and disease after Croydon: a review of water-borne and water-associated disease in the UK 1937–86.* J NEWWA, 169–185.

Gale P., van Dijk P. A. H. and Stanfield G. (1997). Drinking water treatment increases micro-organism clustering; the implications for microbiological risk assessment. *J Water SRT-Aqua*, **46**, 117–126.

Gale P., Pitchers R. and Gray P. (2002). The effect of drinking water treatment on the spatial heterogeneity of micro-organisms: implications for assessment of treatment efficiency and health risk. *Water Res.*, **36**, 1640–1648.

Galofre B., Israel S., Dellunde J. and Ribas F. (2004). Aerobic bacterial spores as process indicators for protozoa cysts in water treatment plants. *Wat. Sci Techn.*, **50**(1), 165–72.

Gehr R. and Nicell J. (1996). Pilot studies and assessment of down stream effects on UV and ozone disinfection of a physicochemical wastewater. *Water Qual. Res. J. Can.*, **31**, 263–281.

Gehr R., Wagner M., Veerasubramanian P. and Payment P. (2003). Disinfection efficiency of peracetic acid, UV and ozone after enhanced primary treatment of municipal wastewater. *Water Res.*, **37**, 4573–4586.

Gerba C. P. and Rose J. B. (1990). Viruses in source and drinking water. In: G. A. McFeters (ed.), *'Drinking water Microbiology: progress and recent developments'*, Springer-Verlag New-York Inc.

Gerba C. P. and Rose J. B. (1993). Estimating viral disease risk from drinking water. In: C. R. Cothern (ed.), *Comparative Environmental Risk Assessment*, p. 117–135, Lewis Publishers.

Gerba C. P., Gramos D. M. and Nwachuku N. (2002). Comparative inactivation of enteroviruses and adenovirus 2 by UV light. *Appl. Environ. Microbiol.*, **68**, 5167–5169.

Gibbs P. A. (1964). Factors affecting the germination of spores of *Clostridium bifermentans. J. Gen. Microbiol.*, **37**, 41–8.

Godfree A. F., Kay D. and Wyer M. D. (1997). Faecal streptococci as indicators of faecal contamination in water. *J. of Appl. Microbiol. Symp. Suppl.*, **83**, 110S–119S.

Goldschmid J., Zohar D., Argaman J. and Kott Y. (1972). Effect of dissolved salts on the filtration of coliform bacteria in sand dunes. In: P. Jenkins (ed.), *Advances in water pollution research.* Pergamon Press, Oxford, UK.

Gould G. W. and Hurst A. (1969). The Bacterial Spore. Academic Press, London and New York.

Gould G. W. (2006). History of science – spores. Lewis B Perry Memorial Lecture 2005. *J. Appl. Microbiol.*, **101**, 507–513.

Goyal S. M., Hanssen H. and Gerba C. P. (1980). Simple method for the concentration of influenza virus from allantoic fluid on microporous filters. *Appl. Environ. Microbiol.*, **39**, 500–4.

Graaf de W. C. (1922). Een en ander over het bacteriologisch wateronderzoek. Tijdschrift voor Microbiologie en Gezondheidsleer, Vergelijkende en Tropische Geneeskunde, Parasitaire en Infectieziekten, **7**, 1–22.

Grabow W. O. K., Gauss-Muller V., Prozesky O. W. and Deinhardt F. (1983). Inactivation of Hepatitis A virus and indicator organisms in water by free chlorine residuals. *Appl. Environ. Microbiol.*, **46**, 619–624.

Gray P., Pitchers R., Stanfield G., Camm R., Connor K. and Hall T. (1997). *Quantifying the removal of microbes during water treatment,* WRC, Report PT 2073, Medmenham UK.

Guy M. D., McIver J. D. and Lewis M. J. (1977). Removal of virus by a pilot treatment plant. *Water Res.*, **11**, 421–428.

Haag W. R. and Hoigné J. (1983). Ozonation of bromide-containing waters: kinetics of formation of hypobromous acid and bromate. *Environ. Sci. Technol.*, **17**, 261–267.

Haas C. N. (1983). Estimation of risk due to low doses of micro-organisms: A comparison of alternative methodologies. *Am. J. Epidemiol.*, **118**, 573–82.

Haas C. N., Joffe J., Anmangandla U., Jacangelo J. G. and Heath M. (1996). Water quality and disinfection kinetics. *J. Am. Water Works Assoc.*, **88**, 95–103.

Haas C. N., Rose J. B., Gerba C. P. and Regli S. (1993). Risk assessment of virus in drinking water. *Risk analysis*, **13**, 545–522.

Haas C. N., Rose J. B. and Gerba C. P. (1999). *Quantitative microbial risk assessment.* John Wiley & Sons, New York, USA.

Haas C. N. and Eisenberger J. N. S. (2001). *Risk Assessment.* In: L. Fewtrell and J. Bartram (ed.), Water Quality: Guidelines, Standards and Health, p. 161–183. IWA Publishing, London, UK.

Haas C. N., French K., Finch G. R. and Guest R. K. (2001). Data review on the physical/chemical removal of Cryptosporidium. Am. Water Works Assoc. Research Foundation, Denver CO, US.

Haas C. N. (2002). Conditional dose–response relationships for microorganisms: development and application. *Risk analysis*, **22**, 455–463.

Harden V. P. and Harris J. O. (1953). The isoelectric point of bacterial cells. *J. Bacteriol.*, **65**, 198–202.

Harris G. D., Dean Adams V. D., Sorensen D. L. and Curtis M. S. (1987). Ultraviolet inactivation of selected bacteria and viruses with photoreactivation of the bacteria. *Water Res.*, **21**, 687–692.

Harris J. R. and Petry F. (1999). *Cryptosporidium parvum*: Structural components of the Oocyst wall. *J. Parasitol.*, **85**, 839–849.

Harter, T., Wagner S. and Atwill E. R. (2001). Colloid transport and filtration of *Cryptosporidium* in sandy soils and aquifer sediments. *Environ. Sci. Technol.*, **34**, 62–70.

Harvey R. W., Kenner N. E., Bunn A., MacDonald D. and Metge D. (1995). Transport behavior of groundwater protozoa and protozoan-sized microspheres in sandy aquifer systems. *Appl. Environ. Microbiol.*, **61**, 209–217.

Harwood V. J., Levine A. D., Scott T. M., Chivukula V., Lukasik J., Farrah S. R. and Rose J. B. (2005). Validity of the indicator organisms paradigm for pathogen reduction in reclaimed water and public health protection. *Appl. Environ. Microbiol.*, **71**, 3163–3170.

Hashimoto A., Hirata T. and Kunikane S. (2001). Occurrence of *Cryptosporidium* oocysts and *Giardia* cysts in a conventional water purification plant. *Wat. Sci. Techn.*, **43**(12), 89–92.

Havelaar A. H. (1983). Microbiologisch onderzoek van drinkwater. *H2O*, **16**, 105–112.

Havelaar A. H., Pot R., van Kranen H. J., van Welie E., Hoekstra A. C., Visser A., van Breemen L. W. C. A. and van der List C. (1983). *Een methode voor het tellen van sporen van sulfietreducerende clostridia in drinkwater en andere soorten water*. Report No. 148112001, RIVM, Bilthoven, the Netherlands.

Havelaar A. H. (1986). *F-specific RNA bacteriophages as model viruses in water treatment processes.* University of Utrecht, RIVM Bilthoven, the Netherlands.

Havelaar A. H., Pot-Hogeboom W. M., Koot W. and Pot R. (1987). F-specific bacteriophages as indicators of the disinfection efficiency of secondary effluent with ultraviolet radiation. *Ozone Sci. and Eng.*, **9**, 353–368.

Havelaar A. H., Meulemans C. C. E., Pot-Hogeboom W. M. and Koster J. (1990). Inactivation of bacteriophage MS2 in wastewater effluent with monochromatic and polychromatic ultraviolet light. *Water Res.*, **24**, 1387–1393.

Havelaar A. H., van Olphen M. and Drost Y. C. (1993). F-specific RNA bacteriophages are adequate model organisms for enteric viruses in fresh water. *Appl. environ. Microbiol.*, **59**, 2956–2962.

Havelaar A. H. (1993). Bacteriophages as models of enteric viruses in the environment. *ASM News*, **59**, 614–619.

Havelaar A. H. (1994). Application of HACCP to drinking water supply. *Food Control*, **5**, 145–152.

Havelaar A. H., van Olphen M. and Schijven J. F. (1995). Removal and inactivation of viruses by drinking water treatment processes under full scale conditions. *Wat. Sci. Tech.*, **31**(5–6), 55–62.

Havelaar A. H., De Hollander A. E. M., Teunis P. F. M., Evers E. G., Van Kranen H. J., Johanna, Versteegh F. M., Van Koten J. E. M. and Slob W. (2000). Balancing the risks and benefits of drinking water disinfection: Disability adjusted life-years on the scale. *Environ. Health Perpect.*, **108**, 315–321.

Havelaar A. H., Nauta M. J. and Jansen J. T. (2004). Fine-tuning food safety objectives and risk assessment. *Int. J. Food Microbiol.*, **93**, 11–29.

Heijnen L. and Medema G. J. (2006). Quantitative detection of *E. coli, E. coli* O157 and other shiga toxin producing *E. coli* in water samples using a culture method combined with real-time PCR. *J. Water and Health*, **04**(4), 487–498.

Hellard M. E., Sinclair M. I., Forbes A. B. and Fairley C. K. (2001). A randomized, blinded, controlled trial investigating the gastrointestinal health effects of drinking water quality. *Environ. Health Perpect.*, **109**, 773–778.

Henry V., Helbronner A. and Recklinghausen M. (1910). Nouvelles recherches sur la sterilization de grandes quantited d'eau par les rayons ultraviolets. *Comp. Rend. Acad. Sci.*, **151**, 677–680.

Hernández-Eugenio G., Fardeau M.-L., Cayol J.-L., Patel B. K. C., Thomas P., Macarie H., Garcia J.-L. and Ollivier B. (2002). *Clostridium thiosulfatireducens* sp. nov., a proteolytic, thiosulfate- and sulfur-reducing bacteriom isolate from an upflow anaerobic sludge blanket (UASB) reactor. *Int. Syst. Evol. Microbiol.*, **52**, 1461–1468.

Herwaldt B. L., Craun G. F., Stokes S. L. and Juranek D. D. (1992). Outbreaks of waterborne disease in the United States: 1989–90. *J. Am. Water Works Assoc.*, **84**(4), 129–135.

Herzig J. P., Leclerc D. M. and LeGoff P. (1970). Flow of suspensions through porous media: application to deep filtration. *Ind. Eng. Chem.*, **62**, 8–35.

Hijnen W. A. M., Visser A. and Van der Kooij D. (1995). Verwijdering van *Campylobacter* bacteriën bij de drinkwaterbereiding op productielocatie Scheveningen, DZH. SWO 95.272, Kiwa Water Research, Nieuwegein NL.

Hijnen W. A. M., van der Speld W. M. H., Houtepen F. A. P. and van der Kooij D. (1997). Presented at the International Symposium on Waterborne *Cryptosporidium*, Newport Beach, California, US.

Hijnen W. A. M., Visser A. and Medema G. J. (1998). Verwijdering van indicatorbacterien bij de drinkwaterbereiding op productielocatie Scheveningen, NV Duinwaterbedrijf Zuid-Holland. SWI 98.216 Kiwa Water Research Nieuwegein NL.

Hijnen W. A. M., Veenendaal. D., Van der Speld W. M. H., Visser A., Hoogenboezem W. and Van der Kooij D. (2000a). Enumeration of faecal indicator bacteria in large water volumes using on site membrane filtration to assess water treatment efficiency. *Water Res.*, **34**, 1659–1665.

Hijnen W. A. M., Willemsen-Zwaagstra J., Hiemstra P., Medema G. J. and van der Kooij D. (2000b). Removal of sulphite-reducing clostridia spores by full scale water treatment processes as a surrogate for protozoan (oo)cysts removal. *Wat. Sci. Tech.*, **41**, 165–171.

Hijnen W. A. M., Bosklopper T. G. J., Hofmann J. A. M. H., Bosch A. D. and Medema G. J. (2001). Presented at the 10th Int. Ozone Assoc. Congres, IOA, London.

Hijnen W. A. M., Van der Veer A. J., Van Beveren J. and Medema G. J. (2002). Spores of sulphite-reducing clostridia (SSRC) as surrogate for verification of the inactivation capacity of full-scale ozonation for *Cryptosporidium. Wat. Sci. Technol.: Wat. suppl.*, **2**, 163–170.

Hijnen W. A. M., Vrouwenvelder J. S. and Manolarakis S. A. (2003). Mcrobiologische aspecten van de levering van huishoudwater: beleidsonderbouwende monitoring. KWR 02.095B. Kiwa Water Research, Nieuwegen, the Netherlands.

Hijnen W. A. M., Schijven J. F., Bonné P., Visser A. and Medema G. J. (2004a). Elimination of viruses, bacteria and protozoan oocysts by slow sand filtration. *Wat. Sci. Technol.*, **50**, 147–154.

Hijnen W. A. M., van der Veer A. J., Beerendonk E. F. and Medema G. J. (2004b). Increased resistance of environmental anaerobic spores to inactivation by UV. *Wat. Sci. Technol.: Water Supply*, **4**, 54–61.

Hijnen W. A. M., Medema G. J., Baars E., Bosklopper T. G. J., van der Veer A. J. and Meijers R. T. (2004c). Influence of DOC on the Inactivation Efficiency of Ozonation Assessed with *Clostridiumperfringens* and a Lab-Scale Continuous Flow System. *Ozone Sci. Eng.*, **26**, 465–473.

Hijnen W. A. M., Medema G. J. and van der Kooij D. (2004d). Quantitative assessment of the removal of indicator bacteria in full-scale treatment plants. *Wat. Sci. Technol.: Water Supply*, **4**, 47–54.

Hijnen W. A. M., Brouwer-Hanzens A. J., Charles K. and Medema G. J. (2005). Transport of MS2 phage, *Escherichia coli, Clostridium perfringens, Cryptosporidium parvum* and *Giardia intestinalis* in a gravel and a sandy soil. *Environ. Sci. Technol.*, **39**, 7860–7868.

Hijnen W. A. M. (2006). Verslag van de BTO workshop 'Langzame zandfiltratie als veiligheidsbarrière' 25 oktober 2006. BTO2007.002, Kiwa Water Research, Nieuwegein NL.

Hijnen W. A. M. and Medema G. J. (2007). Elimination of micro-organisms by drinking water treatment processes: A review BTO 2003.13-3 (Third edition). Kiwa Water Research, Nieuwegein, the Netherlands.

Hijnen W. A. M., Dullemont Y. J., Schijven J. F., Brouwer-Hanzens A. J., Rosielle M. and Medema G. J. (2007). Removal and fate of *Cryptosporidium parvum*, *Clostridium perfringens* and small-sized centric diatoms (*Stephanodiscus hantzschii*) in slow sand filters. *Water Res.*, **41**, 2151–2162.

Hijnen W. A. M., Suylen G. M. H., Bahlman J. A., Brouwer-Hanzens A. J. and Medema G. J. (2010). GAC adsorption filters as barriers for viruses, bacteria and protozoan (oo)cysts in water treatment. *Water Res.*, **44**, 1224–1234.

Hill R. T., Knight I. T., Anikis M. S. and Colwell R. R. (1993). Benthic distribution of sludge indicated by *Clostridium perfringens* at a deep-ocean dump site. *Appl. Environ. Microbiol.*, **59**, 47–51.

Hill V. R., Kahler A. M., Jothikumar N., Johnson T. B., Hahn D. and Cromeans T. L. (2007). Multistate evaluation of an ultrafiltration-based procedure for simultaneous recovery of enteric microbes in 100-liter tap water samples. *Appl. Environ. Microbiol.*, **73**, 4218–4225.

Hill V. R., Polaczyk A. L., Hahn D., Narayanan J., Cromeans T. L., Roberts J. M. and Amburgey J. E. (2005). Development of a rapid method for simultaneous recovery of diverse microbes in drinking water by ultrafiltration with sodium polyphosphate and surfactants. *Appl. Environ. Microbiol.*, **71**, 6878–6884.

Hoogenboezen W., Medema G. J., Ketelaars H. A. M., Rijs G. B. J. and Schijven J. F. (2001a). *Cryptosporidium* and *Giardia*: occurrence in sewage, manure and surface water. RIWA report.

Hoogenboezem W., Donker F., Gijsbers E. and Strating S. (2001b). Beëindiging veiligheidsdesinfectie op pompstation Mensink? H2O, **34**, 21–24.

Houston A. C. (1899). Bacterioscopic Examination of Drinking-water with particular reference to the relations of Streptocooci and Staphylococci with Water of this Class. Supplement to the Twenty-eighth Annual Report of the Local Government Board Containing the Report of the Medical Officer for 1898–99, 467.

Hofer J. W. and Davis J. (1972). Survival and dormancy. *Texas Medicine*, **68**, 80.

Hoyer O. (1998). Testing performance and monitoring of UV systems for drinking water disinfection. *Wat. Supply*, **16**, 424–429.

Hoyer O. (2004). Presented at the European Conference UV Karlsruhe, UV radiation, Effects and Technologies, Karlruhe, GR, 2003.

Hrudey S. E., Payment P., Huck P. M., Gillham R. W. and Hrudey E. J. (2003). A fatal waterborne disease epidemic in Walkerton, Ontario: comparison with other waterborne outbreaks in the developed world. *Wat. Sci. Tech.*, **47**(3), 7–14.

Huck P. M., Emelko M. B., Coffey B. M., Maurizo D. D. and O'Melia C. R. (2001). *Filter operation effects on pathogen passage*. American Water Works Assoc. Reserach Foundation, Denver CO US.

Huisman J. and Nobel P. J. (1981). Enkele epidemiologische gegevens over de gevolgen van de faecale drinkwaterverontreiniging in het Scheepsvaartkwartier te Rotterdam in maart 1981. H2O, **6**(8), 199–200.

Hsu B.-M. and Huang C. (2002). Influence of ionic strength and pH on hydrophobicity and zeta potential of *Giardia* and *Cryptosporidium*. *Colloids Surf. A*, **201**, 201–206.

Hutchison M. L., Walters L. D., Moore T., Thomas D. J. I. and Avery S. M. (2005). Fate of pathogens present in livestock wastes spread onto fescue plots. *Appl. Environ. Microbiol.*, **71**, 691–696.

Jacangelo J. G., Loughran P., Petrik B., Simpson D. and McIloy C. (2004). Removal of enteric viruses and selected microbial indicators by UV irradiation of secondary effluent. *Wat. Sci. Tech.*, **47**, 193–198.

Jin Y., Yates M. V., Thompson S. S. and Jury W. A. (1997). Sorption of viruses during flow through saturated sand columns. *Environ. Sci. Technol.*, **31**, 548–555.

Jofre J., Ollé E., Ribas F., Vidal A. and Lucena F. (1995). Potential usefulness of bacteriophages that infect *Bacteroides fragilis* as model organisms for monitoring virus removal in drinking water treatment plants. *Appl. Environ. Microbiol.*, **61**, 3227–3231.

Johnson A. M., Linden K., Ciociola K. M., De Leon R., Widmer G. and Rochelle P. A. (2005). UV inactivation of *Cryptosporidium hominis* as measured in cell culture. *Appl. Environ. Microbiol.*, **71**, 2800–2802.

Johnstone K. (1994). The trigger mechanism of spore germination: Current concepts. *Soc. Appl. Bacteriol. Symp. Ser.*, **23**, 17S–24S.

Karmali M. A., Simor A. E., Roscoe M., Flemming P. C., Smith C. S. and Lane J. (1986). Evaluation of a blood-free, Charcoal-based, selective medium for the isolation of *Campylobacter* organisms from feces. *J. Clin. Microbiol.*, **23**, 456–459.

Ketelaars H. A. M., Medema G. J., Van Breemen L. C. W. A., van der Kooij D., Nobel P. J. and Nuhn P. (1995). Occurrence of *Cryptosporidium* oocysts and *Giardia* cysts in the river Meuse and removal in the Biesbosch reservoirs. *Aqua*, **44**(1), 108–111.

Keynan A. and Evenchik Z. (1969). *Activation*. In: G. W. Gould and A. Hurst (ed.), The Bacterial Spore. Academic Press, London and New York.

King B. J., Keegan A. R., Monis P. T. and Saint C. P. (2005). Environmental temperature controls *Cryptosporidium* oocyst metabolic rate and associated retention of infectivity. *Appl. Environ. Microbiol.*, **71**, 3848–3857.

Klein E. (1895). Centralbl. f. Bakt., **18**, 737.

Klein E. (1897–1898). Report on the Morphology and Biology of *Bacillus enteritidis* sporogens; on the association of this microbe with infantile diarrhoea and with Choler Nostras; and its relation with milk, with sewage, and with manure. Supplement to the Twenty-seventh Annual Report of the Local Government Board Containing the Report of the Medical Officer for 1897–98, 210.

Klein E. and Houston A. C. (1898–1899). Further reports on bacteriological evidence of recent and therefore dangerous sewage pollution of elsewise potable waters. Supplement to the Twenty-eigth Annual Report of the Local Government Board Containing the Report of the Medical Officer for 1898–99, 498.

Knappett P. S. K., Emelko M. B., Zhuang J. and McKay L. D. (2008). Transport and retention of a bacteriophage and microspheres in saturated, angular porous media: Effects of ionic strength and grain size. *Water Res.*, **42**, 4368–4378.

Knudson G. B. (1985). Photoreactivation of UV irradiated *Legionella pneumophila* and other *Legionella* species. *Appl. Environ. Microbiol.*, **49**, 975–980.

Knorr N. (1937). Die Schutzzonenfrage in der Trinkwater-hygiene. Das Gas- und Wasserfach, **80**, 330–355.

Koch R. (1884). Die Aetiologie der Tuberkulose. Mitt Kaiser Gesundh., **2**, 1–88.

Koch R. (1893). Wasserfiltration und Cholera. Gesammelte Werke, vol. Bd 2 Teil 1, 1912. Georg Thieme, Leipzig.

Konert M. and Vandenberghe J. (1997). Comparison of laser grain size analysis with pipette and sieve analysis: A solution for the underestimation of the clay fraction. *Sedimentology*, **44**, 523–535.

Kramer M. H., Herwaldt B. L., Craun G. F., Calderon R. L. and Juranek D. D. (1996). Waterborne Disease: 1993 and 1994. *J. Am. Water Works Assoc.*, **88**, 66–80.

Kruithof J. C., Van der Leer R. C. and Hijnen W. A. M. (1992). Practical experiences with UV disinfection in The Netherlands. *J. Water SRT-Aqua*, **41**, 88–94.

Kruithof J. C., Kamp P. C., Folmer H. C., Nederlhof M. M. and van Hoof S. C. J. M. (2001). Development of a membrane integrity monitoring strategy for the UF/RO Heemskerk drinking water treatment plant. *Wat Supply*, **1**(5–6), 261–271.

Kruithof J. C., Kamp P. C., Martijn B. J., Belosevic M. and Williams G. (2005). Presented at the Third Int. Congress on Ultraviolet Technologies, Whistler, BC Canada, May 24–27.

Kuznar Z. A. and Elimelech M. (2005). Role of Surface Proteins in the Deposition Kinetics of *Cryptosporidium parvum* Oocysts. *Langmuir*, **21**, 710–716.

Lechavallier M. W., Hassenauer T., Camper A. K. and McFeters G. A. (1984). Disinfection of bacteria attached to granular activated carbon. *Appl. Environ. Microbiol.*, **48**(5), 918–923.

LeChevallier M. W., Norton W. D. and Lee R. G. (1991). Occurrence of *Giardia* and *Cryptosporidium* spp. in surface water supplies. *Appl. Environ. Microbiol.*, **57**, 2610–2616.

LeChevallier M. W. and Norton W. D. (1992). Examining relationships between particle counts, and *Giardia* and *Cryptosporidium* and turbidity. *J. Am. Water Works Assoc.*, **84**(12), 54–60.

LeChevallier M. W. and Au K. (2004). *Water treatment and pathogen control: process efficiency in achieving safe drinking water: World Health Organisation*, IWA Publishing, London, UK.

Leclerc F., Mossel D. A., Trinel P. A. and Gavini F. (1977). Microbiological monitoring: a new test of fecal contamination. In A. W. Hoadley and B. J. Dutka (ed.), Bacterial indicators: health hazards associated with water. American society for testing and materials, Philadelphia, US.

Leuschner R. G. K., Weaver A. C. and Lillford P. J. (1999). Rapid particle size distribution analysis of *Bacillus* spore suspensions. *Colloids and Surfaces B: Biointerfaces*, **13**, 47–57.

Lewis J. C. (1969). *Dormancy.* In: G. W. Gould and A. Hurst (ed.), The Bacterial Spore. Academic Press, London and New York.

Li J., Zhang J., Miao J., Ma J. and Dong W. (2006). Application of computational fluid dynamics (CFD) to ozone contactor optimization. *Wat Sci Technol: Wat Supply*, **6**(4), 9–16.

Linden K. G., Shin G., Faubert G., Cairns W. and Sobsey M. D. (2002). UV Disinfection of *Giardia lamblia* cysts in water. *Environ. Sci. Technol.*, **36**, 2519–2522.

Lindenauer K. G. and Darby J. L. (1994). Ultraviolet disinfection of wastewater: effect of dose on subsequent photoreactivation. *Water Res.*, **28**, 805–817.

Lloyd B. J. (1996). *The significance of protozoal predation and adsorption for the removal of bacteria by slow sand filtration.* In: N. Graham and R. Collins (ed.), Advances in slow sand filtration and alternative biological filtration. John Wiley & Sons, West Sussex, UK.

Logan B. E., Jewett D. G., Arnold R. G., Bouwer E. J. and O'Melia C. R. (1995). Clarification of clean-bed filtration models. *J. Environ. Engineer.*, **121**, 869–873.

Logan J. L., Stevik T. K., Siegrist R. L. and Ronn R. M. (2001). Transport and fate of *Cryptosporidium* parvum oocysts in intermittent sand filters. *Water Res.*, **35**, 4359–4369.

Logsdon G. S., Symons J. M., Hoye R. L. and Arozarena M. M. (1981). Alternative filtration methods for removal of *Giardia* cysts and cyst models. *J. Am. Water Works Assoc.*, **2**, 111.

Logsdon G. S., Thurman V. C., Frindt E. S. and Stoecker J. G. (1995). Evaluating sedimentation and various filter media for removal of Giardia cysts. *J. Am. Water Works Assoc.*, **75**(2), 61–66.

Lytle C. D. (1971). Host-cell reactivation in mammalian cells I. survival of ultra-violet-irradiated Herpes virus in different cell-lines. *Int. Jour. of Rad. Biol.*, **19**, 329–337.

MacConkey A. (1905). Lactose-fermenting bacteria in faeces. *Journal of Hygiene*, **5**, 333–379.

MacKenzie W. R. H., Hoxie N. J., Proctor M. E., Gradus S., Blair K. A., Peterson D. E., Kazmierczak J. J., Addiss D. G., Fox K. R., Rose J. B. and Davis J. P. (1994). A massive outbreak in Milwaukee of *Cryptosporidium* infection transmitted through the public water supply. *New. Engl. J. Med.*, **331**, 161–167.

Magic-Knezev A. and van der Kooij D. (2004). Optimisation and significance of ATP analysis for measuring active biomass in granular activated carbon filters used in water treatment. *Water Res.*, **38**, 3971–3979.

Maier A., Tougianidou D., Wiedenmann A. and Botzenhart K. (1995). Detection of poliovirus by cell culture and by PCR after UV disinfection. *Wat. Sci. Tech.*, **1**, 141–145.

Malley J. P., Ballester N. A., Margolin A. B., Linden K. G., Mofidi A., Bolton J. R., Crozes G., Laine J. M. and Janex M. L. (2004). *Inactivation of pathogens with innovative UV technologies.* AWWARF, Denver CO, US.

Mamane-Gravetz H. and Linden K. G. (2004). Presented at the American Water Works Association Water Quality Technology Conference, San Antonio, US.

Mamane-Gravetz H. and Linden K. G. (2005a). Relationship between physiochemical properties, aggregation and UV inactivation of isolated environmental spores in water. *J. Appl. Microbiol.*, **98**, 351–363.

Mamane-Gravetz H., Linden K. G., Cabaj A. and Sommer R. (2005b). Spectral sensitivity of *Bacillus subtilis* spores and MS2 coliphage for validation testing of ultraviolet reactors for water disinfection. *Environ. Sci. Technol.*, **39**, 7845–7852.

Marin$$as B. J., Rennecker J. L., Teefy S. and Rice E. W. (1999). Assessing ozone disinfection with nonbiological surrogates. *J. Am. Water Works Assoc.*, **91**, 79–89.

Martiny H., Schubert T. and Rüden H. (1990). Use of UV radiation for the disinfection of water V. Communication: Microbiological investigations with bacteria of the exponential and the stationary phase in cold and in warm drinking water. *Zbl. Hyg.*, **190**, 380–394.

Matz L. L., Beaman T. C. and Gerhardt P. (1970). Chemical Composition of Exosporium from Spores of *Bacillus cereus. J. Bacteriol.*, **101**, 196–201.

Maya C., Beltrán N., Jiménez B. and Bonilla P. (2003). Evaluation of the UV disinfection process in bacteria and amphizoic amoeba inactivation. *Wat. Sci. Technol.: Wat. suppl.*, **3**, 285–291.

Mazoua S. and Chauveheid E. (2005). Aerobic spore-forming bacteria for assessing quality of drinking water produced from surface water. *Water Res.*, **39**, 5186–5198.

McCaulou D. R., Bales R. C. and McCarthy J. F. (1994). Use of short-pulse experiments to study bacteria transport through porous media. *J. Contam. Hydrol.*, **15**, 1–14.

McTigue N. E., LeChevallier M. and Clancy J. (1996). *Findings of the national particle project.* In: Annual Conference AWWA. Toronto, Canada.

McTigue N. E., et al. (1998). *National assessment of particle removal by filtration: AWWARF,* Denver CO US.

Medema G. J. and Havelaar A. H. (1994a). *Micro-organismen in water: een gezondheidsrisico.* RIVM 289202002.

Medema G. J. and Schets F. M. (1994b). *Campylobacter* en *Salmonella* in open reservoirs voor de drinkwaterbereiding. RIVM 149103002.

Medema G. J., Teunis P. F., Havelaar A. H. and Haas C. N. (1996). Assessment of the dose-response relationship of *Campylobacter jejuni. Int. J. Food Microbiol.*, **30**, 101–111.

Medema G. J., Bahar M. and Schets F. M. (1997). Survival of *Cryptosporidium parvum*, *Escherichia coli*, faecal streptococci and *Clostridium perfringens* in river water. *Wat. Sci. Tech.*, **35**(11–12), 249–252.

Medema G. J., Schets F. M., Teunis P. F. M. and Havelaar A. H. (1998). Sedimentation of free and attached *Cryptosporidium* oocysts and Giardia cysts in water. *Appl. Environ. Microbiol.*, **64**, 4460–4466.

Medema G. J. (1999). *Cryptosporidium and Giardia: new challenges to the water industry.* University of Utrecht, Utrecht, NL.

Medema G. J. (2002). Effecten nieuwe wetgeving BTO 2002.123 (c). Kiwa Water Research.

Medema G. J. and Stuyfzand P. J. (2002). Presented at the 4th international symposium on artificial recharge, Adelaide, Australia, September 22–26.

Medema G. J., Hoogenboezem W., van der Veer A. J., Ketelaars H. A. M., Hijnen W. A. M. and Nobel P. J. (2003). Quantitative risk assessment of *Cryptosporidium* in surface water treatment. *Wat. Sci. Tech.*, **47**(3), 241–247.

Medema G., Teunis P. F. M., Blokker M., Deere D., Davison A., Charles P. and Loret J. F. (2006a). *WHO Guidelines for Drinking Water Quality*: Cryptosporidium. World Health Organization, London UK.

Medema G. J., Loret J. F., Stenström T. and Ashbolt N. J. (2006). Quantitative Microbial Risk Assessemnt in Water Safety Plans. Report for the European Community, EC. Kiwa Water Research, Nieuwegein NL; Suez Environmental-CIRSEE, LePecq, FR; Swedish Institute for Infectious Disease Control, Stockholm, SW; University of New South Wales, Sydney, AU.

Medema G. J. and Beerendonk E. F. (2005). *Testing UV-fluence monitoring of full-scale UV at Berenplaat: Kiwa Water Reserach,* KWR 05.028, Nieuwegein, NL.

Meng Q. S. and Gerba C. P. (1996). Comparative inactivation of enteric adenoviruses, poliovirus and coliphages by ultraviolet irradiation. *Water Res.*, **30**, 2665–2668.

Mofidi A. A., Rochell P. A., Chou C. I., Mehta H. M., Linden K. G. and Malley J. P. (2002). Presented at the American Water Works Association Water Quality Technology Conference, Seattle US, November 10–14.

Molongoski J. J. and Klug M. J. (1976). Characterization of anaerobic heterotrophic bacteria isolated from freshwater lake sediments. *Appl. Environ. Microbiol.*, **31**, 83–90.

Moore A. C., Herwaldt B. L., Craun G. F., Calderon R. L., Highsmith A. K. and Juranek D. D. (1994). Waterborne disease in the United States, 1991 and 1992. *J. Am. Water Works Assoc.*, **86**(2), 87–99.

Morales-Morales H. A., Vidal G. O., J., Rock C. M., Dasgupta D., Oshima K. H. and Smith G. B. (2003). Optimization of a Reusable Hollow-Fiber Ultrafilter for Simultaneous Concentration of Enteric Bacteria, Protozoa and Viruses from Water. *Appl. Environ. Microbiol.*, **69**, 4098–4102.

Morales C. F. L., Strathmann M. and Flemming H.-C. (2007). Influence of biofilms on the movement of colloids in porous media. Implications for colloid facilitated transport in subsurface environments. *Water Res.*, **41**, 2059–2068.

Morita S., Namikoshi A., Hirata T., Oguma K., Katayama H., Ohgaki S., Motoyama N. and Fujiwara M. (2002). Efficacy of UV irradiation in inactivating *Cryptosporidium parvum* oocysts. *Appl. Environ. Microbiol.*, **68**, 5387–5393.

Mossel D. A. A. (1982). Marker (index and indicator) organisms in food and drinking water. Semantics, ecology, taxonomy and enumeration. *Ant. Leeuwenhoek*, **48**, 609–611.

Murrell W. G. (1969). *Chemical composition of spores and spore structures*. In: G. W. Gould and A. Hurst (ed.), The bacterial spore. Academic Press, London and New York.

Nieminski E. C. and Ongerth J. E. (1995). Removing *Giardia* and *Cryptosporidium* by conventional treatment and direct filtration. *J. Am. Water Works Assoc.*, **87**(9), 96–106.

Nobel P. J., et al. (1999). *Effecten van begrazing op waterkwaliteit.* KOA 98.204, Kiwa Water Research, Nieuwegein NL.

Novak J. S., Juneja V. K. and McClane B. A. (2003). An ultrastructural comparison of spores from various strains of *Clostridium perfringens* and correlations with heat resistance parameters. *Int. J. Food Microbiol.*, **86**, 239– 247.

Nwachuku N., Gerba C. P., Oswald A. and Mashadi F. D. (2005). Comparative inactivation of adenovirus serotypes by UV light disinfection. *Appl. Environ. Microbiol.*, **71**, 5633–5636.

Oesterholt F., Martijnse G., Medema G. J. and D. Van der Kooij D. (2007). Health risk assessment of non-potable domestic water supplies in the Netherlands. *J Wat Sup: Aqua*, **56**, 171–179.

Oguma K., Katayama H. and Ohgaki S. (2002). Photoreactivation of *Escherichia coli* after low- or medium-pressure UV disinfection determined by an endonuclease sensitivity site assay. *Appl. Environ. Microbiol.*, **68**, 6029–6035.

Oguma, K., Katayama H. and Ohgaki S. (2004). Photoreactivation of *Legionella* pneumophila after inactivation by low- or medium pressure ultraviolet lamp. *Water Res.*, **38**(11), 2757–2763.

O'Melia C. R. and Ali W. (1978). The role of retained particles in deep bed filtration. *Prog. Water Technol.*, **10**, 167–182.

Oppenheimer J. A., Hoagland J. E., Laine J.-M., Jacangelo J. G. and Bhamrah A. (1993). Presented at the Water Environmental Federation Specialty Conference Series: Planning, design & operation of effluent disinfection systems, Alexandria, VA.

Oppenheimer J. A., Jacangelo J. G., Laîné J. M. and Hoagland J. E. (1995). Presented at the American Water Works Association Water Quality Technology Conference, New Orleans US, November 12–16.

Oppenheimer J. A., Aieta E. M., Trussell R. R., Jacangelo J. G. and Najm I. N. (2000). Evaluation of *Cryptosporidium* inactivation in natural waters. American Water Works Association Research Foundation, Denver CO US.

Örmeci B. and Linden K. G. (2002). Comparison of UV and chlorine inactivation of particle and non-particle associated coliforms. *Wat. Sci. Technol.: Wat. suppl.*, **2**, 403–410.

Patania N. L., Jacangelo J. G., Cummings L., Wilczak A., Riley K. and Oppenheimer J. (1995). Optimization of Filtration for Cyst Removal. American Water Works Association Research Foundation, Denver CO US.

Payment P., Trudel M. and Plante R. (1985). Elimination of viruses and indicator bacteria at each step of treatment during preparation of drinking water at seven water treatment plants. *Appl. Environ. Microbiol.*, **49**, 1418–1428.

Payment P., Bérubé A., Perrefault D., Armon R. and Trudel M. (1989). Concentration of *Giardia lamblia* cysts, *Legionella pneumophila*, *Clostridium perfringens*, human enteric viruses, and coliphages from large volumes of drinking water, using a single filtration. *Can. J. Microbiol.*, **35**, 932–935.

Payment P. (1991). Fate of human enteric viruses. coliphages, and *Clostridium perfringens* during drinking-water treatment. *Can. J. Microbiol.*, **37**, 154–157.

Payment P., Richardson L., Semiatycki J., Dewar R., Edwardes M. and Franco E. (1991). A randomized trial to evaluate the risk of gastrointestinal disease due to consumption of drinking water meeting current microbiological standards. *Am. J. Public Health*, **81**, 703–708.

Payment P. and Franco E. (1993). *Clostridium perfringens* and somatic coliphages as indicators of the efficiency of drinking water treatment for viruses and protozoan cysts. *Appl. Environ. Microbiol.*, **59**, 2418–2244.

Payment P., Siemiatycki J., Richardson L., Renaud G., Franco E. and Prevost M. (1997). A prospective epidemiological study of gastrointestinal health effects due to the consumption of drinking water. *Int. J. Environ. Health Res.*, **7**, 5–31.

Penrod S. L., Olson T. M. and Grant S. B. (1996). Deposition kinetics of two viruses in packed beds of quartz granular media. Langmuir, **12**, 5576–5587.

Persson F., Långmark J., Heinicke G., Hedberg T., Tobiason J., Stenström T. and Hermansson M. (2005). Characterisation of the behaviour of particles in biofilters for pre-treatment of drinking water. *Water Res.*, **39**, 3791–3800.

Prescott S. C., Winslow C. A. and McCrady M. H. (1945). *Water Bacteriology with special reference to sanitary water analysis,* 6th ed. John Wiley & Sons, New York.

Prévost M., Jaidi K., Payment P. and Barbeau B. (2007). Presented at the Water Quality and Technology Conference, Charlotte, US.

Powell T., Brion G. M., Jagtoyen M. and Derbyshire F. (2000). Investigating the Effect of Carbon Shape on Virus Adsorption. *Environ. Sci. Technol.*, **34**, 2779–2783.

Poynter S. F. B. and Slade J. S. (1977). The removal of viruses by slow sand filtration. *Prog. Water Tech.*, **9**, 75–88.

Qian S. S., D. M., Schmelling D. C., M. M., Linden K. G. and Cotton C. (2004). Ultraviolet light inactivation of protozoa in drinking water: a Bayesian meta-analysis. *Water Res.*, **38**, 317–326.

Qualls R. G. and Johnson J. D. (1983a). Bioassay and dose measurement in UV disinfection. *Appl. Environ. Microbiol.*, **45**, 872–877.

Qualls R. G., Flynn M. P. and Johnson J. D. (1983b). The role of suspended particles in ultraviolet disinfection. *J. Water Pollut. Control Fed.*, **55**, 1280–1285.

Rajagopalan R. and Tien C. (1976). Trajectory analysis of deep-bed filtration with the sphere-in-a-cell porous media model. *AIChE J.*, **22**, 869–873.

Regli S., Rose J. B., Haas C. N. and Gerba C. P. (1991). Modeling the risk from *Giardia* and viruses in drinking water. *J. Am. Water Works Assoc.*, **83**, 76–84.

Ribeiro C. D. and Price T. H. (1984). The use of Preston Enrichment broth for the isolation of thermophilic *Campylobacters* from water. *J. Hyg. Camb.*, **92**, 45–51.

Rice E. W., Fox K. R., Miltner R. J., Lytle D. A. and Johnson C. H. (1996). Evaluating plant performance with endospores. *J. Am. Water Works Assoc.*, **88**(9), 122–130.

Richardson A. J., Frankenberg R. A., Buck A. C., Selkon J. B., Colbourne J. S., Parsons J. W. and Mayon-White R. T. (1991). An outbreak of waterborne cryptosporidiosis in Swindon and Oxfordshire. *Epidemiol. Infect.*, **107**, 485–95.

Rivera-Utrilla J., Bautistat-Toledo I., Ferro-Garcia M. A. and Moreno-Castilla C. (2001). Activated carbon surface modifications by adsorption of bacteria and their effect on aqueous lead adsorption. *J. Chem. Technol. Biotechnol.*, **76**, 1209–1215.

Rijnaarts H. H. M., Norde W., Bouwer E. J., Lyklema J. l. and Zehnder A. J. B. (1995). Reversibility and mechanism of bacterial adhesion. *Colloids Surf. B.*, **5**, 5–22.

Roberts T. A. and Hitchins A. D. (1969). *Resistance of spores.* In: G. W. Gould and A. Hurst (ed.), The Bacterial Spore. Academic Press, London and New York.

Robles S., Rodríguez J. M., Granados I. and Guerrero M. C. (2000). Sulfite-reducing clostridia in the sediment of a high mountain lake (Laguna Grande, Gredos, Spain) as indicators of fecal pollution. *Internat. Microbiol.*, **3**, 187–191.

Rochelle P. A., Mofidi A. A., Marshall M. M., Upton S. J., Montelone B., Woods K. and Di Giovanni G. (2004). An investigation of UV disinfection and repair in *Cryptosporidium* parvum. AWWARF, Denver CO, US.

Rontó G., Gáspár S. and Bérces A. (1997). Phage T7 in biological UV dose measurement. *J. Photochem. Photobiol. B Biol.*, **12**, 285–294.

Rook J. J. (1974). Formation of haloforms during chlorination of natural water. *Water Treatment Exam.*, **23**, 234–245.

Rose J. B., Haas C. N. and Gerba C. P. (1993). Waterborne pathogens: assessing health risks. *Health and Environment Digest.*, **7**, 1–8.

Rosef O. and Mork A. V. (1985). Kontaminert brφnnvann som årsak til *Campylobacter*-diaré (Contaminated well water as cause of *Campylobacter*-diarrhoea). *Nor. Vet. Todsskr.*, **97**, 891–832.

Rosef O., Rettedal G. and Lageide L. (2001). Thermophilic *Campylobacters* in surface water: a potential risk of campylobacteriosis. Int. *J. Environ. Health Res.*, **11**, 321–327.

Rothfuss F., Bender M. and Conrad R. (1997). Survival and activity of bacteria in a deep, aged lake sediment (Lake Constant). *Microb. Ecol.*, **33**, 69–77.

Ryan J. N. and Elimelech M. (1996). Colloid mobilization and transport in groundwater. *Colloid. surf. A*, **107**, 1–56.

Ryan J. N., Elimelech M., Ard R. A., Harvey R. W. and Johnson P. R. (1999). Bacteriophage PRD1 and silica colloid transport and recovery in an iron oxide-coated sand aquifer. *Environ. Sci. Technol.*, **33**, 63–73.

Saiers J. E. and Ryan J. N. (2005). Colloid deposition on non-ideal porous media: The influences of collector shape and roughness on the single-collector efficiency. Geophysical Research Letters 32, L21406.

Sakthivadivel R. (1966). *Theory and Mechanism of Filtration of Non-colloidal Fines through a Porous Medium*, p. 110, University of California, Berkeley, US.

Sakthivadivel R. (1969). *Clogging of a Granular Porous Medium by Sediment*, p. 110, University of California, Berkeley, US.

Sartory D. P. (1988). Faecal clostridia and indicator bacteria levels in an eutrophic impoundment. *Water SA*, **14**, 115–117.

Schardinger F. (1892). Ueber das Vorkommen Gährung erregender Spaltpilze im Trinkwasser und ihre Bedeutung fiir die hygienische Beurteilung desselben. *Wien. Klin. Wochenschr.*, **5**, 421–423.

Schijven J. F., Hoogenboezem W., Hassanizadeh S. M. and Peters J. H. (1999). Modelling removal of bacteriophages MS2 and PRD1 by dune recharge at Castricum, Netherlands. *Water Resour. Res.*, **35**, 1101–1111.

Schijven J. F., Medema G. J., Vogelaar A. J. and Hassanizadeh S. M. (2000). Removal of microorganisms by deep well injection. *J. Contam. Hydrol.*, **44**, 301–327.

Schijven J. F. and Hassanizadeh S. M. (2000). Removal of viruses by soil passage: overview of modeling, processes and parameters. *Crit. Rev. Envrion. Sci. Tech.*, **31**, 49–125.

Schijven J. F. (2001). *Virus removal from groundwater by soil passage.* Technische Universiteit Delft, Delft, the Netherlands.

Schijven J. F., de Bruin H. A. M., Hassanizadeh S. M. and de Roda Husman A. M. (2003). Bacteriophages and clostridium spores as indicator organisms for removal of pathogens by passage through saturated dune sand. *Water Res.*, **37**, 2186–2194.

Schijven J. F., Colin M., Dullemont Y. J., Hijnen W. A. M., Magic-Knezev A. and Oorthuizen W. (2007). *Removal of micro-organisms by slow sand filtration.* RIVM 703719019/2007 (in Dutch).

Schijven J. F., Colin M., Dullemont Y. J., Hijnen W. A. M., Magic-Knezev A., Oorthuizen W., Wubbels G., Rutjes B. J. S. A., and de Roda Husman A. M. (2008). Effect filtratiesnelheid, temperatuur en korrelgrootte op de verwijdering van micro-organismen door langzame zandfiltratie. RIVM 330204002/2008, Bilthoven (in Dutch).

Schijven J. F. and De Roda Husman A. M. (2009). Analyse microbiologische veiligheid drinkwater, RIVM 703719038/2009, Bilthoven, NL.

Schoenen D., Zemke V. and Kolch A. (1991). The influence of reflection of the UV-rays on the disinfection of drinking water. *Zbl. Hyg.*, **191**, 396–405.

Schoenen D. K., Gebel A., J. and Hoyer O. (1995). UV-Desinfektion von Trinkwasser, Grundlagen und Anvorderungen für die praktische Anwendung. DVGW Schriftenreihe Wasser 86.

Scholl M. A., Mills A. L., Herman J. S. and Hornberger G. M. (1990). The influence of mineralogy and solution chemistry on the attachment of bacteria to representative aquifer materials. *J. Contam. Hydrol.*, **6**, 321–336.

Schuler P. F., Ghosh M. M. and Gopalan P. (1991). Slow Sand and Diatomaceous Earth Filtration of Cysts and Other Particulates. *Water Res.*, **25**, 995–1005.

Scott T. M., Sabo R. C., Lukasik J., Boice C., Shaw K., Barroso-Giachetti L., El-Shall H., Farah S. R., Park C., Moudgil B. and Koopman B. (2002). Performance and cost effectiveness of ferric and aluminium hydrous metal oxide coating on filter media to enhance virus removal. *KONA*, **20**, 159–164.

Setlow P. (1988). Small, acid-soluble spore proteins of *Bacillus* species: structure, synthesis, genetics, function, and their degradation. *Annu. Rev. Microbiol.*, **42**, 319–338.

Severin B. F., Suidan M. T. and Engelbrecht R. S. (1983). Effects of temperature on ultraviolet light disinfection. *Environ. Sci. Technol.*, **17**, 717–721.

Sheppard D. A. E. (1995). *John Snow: Anesthetist to a Queen and Epidemiologist to a Nation: A Bibliography York Point:* Cornwell Edward Island, Canada.

Sherman J. M. (1937). The Streptococci. *Bacteriological Reviews*, **1**, 3.

Shin G., Bohrerova Z., Linden K. G. and Faubert G. (2005). Presented at the Third Int. Congress on Ultraviolet Technologies, Whistler, BC Canada, May 24–27.

Shin G., Linden K. G., Arrowood M. J. and Sobsey M. D. (2001). Low-pressure UV inactivation and DNA repair potential of *Cryptosporidium parvum* oocysts. *Appl. Environ. Microbiol.*, **67**, 3029–3032.

Signor R. S., Rosen D. J., Ashbolt N. J. and Ball J. E. (2005). Quantifying the impact of runoff events on microbiological contaminant concentrations entering surface drinking source waters. *J. Wat. Health*, **03**, 453–468.

Simmons O. D., Sobsey M. D., Heaney C. D., Schaefer F. W. and Francy D. S. (2001). Concentration and detection of *Cryptosporidium* oocysts in surface water samples by method 1622 using Ultrafiltration and capsule filtration. *Appl. Environ. Microbiol.*, **67**, 1123–1127.

Simpson D., Jacangelo J. G., Loughran P. and McIlroy C. (2003). Investigation of potential surrogate organisms and public health risk in UV irradiated secondary effluent. *Wat. Sci. Tech.*, **47**, 37–43.

Šim nek J., Šejna M. and van Genuchten M. T. (1998). The HYDRUS-1D software package for simulating one-dimensional movement of water, heat, and multiple solutes in variably-saturated media. Version 2.0, IGWMC-TPS-70. Groundwater Modeling Center, Colorado School of Mines, Golden, Colorado, 251 pp.

Slade J. S. (1978). Enteroviruses in slow sand filtered water. *J. Inst. Water Eng. Sci.*, **32**, 530–536.

Slade J. S., Harris N. R. and Chisholm R. G. (1986). Disinfection of chlorine resistant enteroviruses in groundwater by ultraviolet irradiation. *Wat. Sci. Tech.*, **18**, 115–123.

Smeets P. W. M. H., Dullemont Y. J. and Medema G. J. (2005). Presented at the 17th IOA conference, Strasbourg, France.

Smeets P. W. M. H., Van der Helm A. W. C., Dullemont Y. J., Rietveld L. C., Van Dijk J. C. and Medema G. J. (2006). Inactivation of *Escherichia coli* by ozone under bench-scale plug flow and full-scale hydraulic conditions. *Water Res.*, **40**, 3239–3248.

Smeets P. W. M. H., Van Dijk J. C., Stanfield G., Rietveld L. C. and Medema G. J. (2007). How can the UK statutory Cryptosporidium monitoring be used for Quantitative Risk Assessment of *Cryptosporidium* in drinking water? *J. Wat. & Health*, **05**, 107–118.

Smeets P. M. (2008). *Stochastic modelling of drinking water treatment in quantitative microbial risk assessment.* Technical University Delft, Delft NL.

Smith T. (1895). Ueber den Nachweis des *Bacillus coli communis* im Wasser. Zentralblatt für Bakteriologie, **18**.

Smoluchowski M. (1917). Versuch einer mathematischen Theorie der Koagulations kinetic kolloider Losunger. Zeitschrift Physicalische Chemie, **92**, 129.

Sobsey M. D., Dean C. H., Knuckles M. E. and Wagner R. A. (1980). Interaction and survival of enteric viruses in soil materials. *Appl. Environ. Microbiol.*, **40**, 92–101.

Sobsey M. D. (1989). Inactivation of health-related microorganisms in water by disinfection processes. *Wat. Sci. Tech.*, **21**, 179–195.

Sommer R., Weber G., Cabaj A., Wekerle J., Keck G. and Schauberger G. (1989). UV inactivation of micro-organisms in water. *Zbl. Hyg.*, **189**, 214–224.

Sommer R., Cabaj A. and Haider T. (1996). Microbiocidal effect of reflected UV radiation in devices for water disinfection. *Wat. Sci. Tech.*, **34**, 173–177.

Sommer R., Cabaj A., Pribil W. and Haider T. (1997). Influence of lamp intensity and water transmittance on the UV disinfection of water. *Wat. Sci. Tech.*, **35**, 113–118.

Sommer R., Haider T., Cabaj A., Pribil W. and Lhotsky M. (1998). Time fluence reciprocity in UV disinfection of water. *Wat. Sci. Tech.*, **38**, 145–150.

Sommer R., Cabaj A., Sandu T. and Lhotsky M. (1999). Measurement of UV radiation using suspensions of micro-organisms. *J. Photochem. Photobiol. B:Biol.*, **53**, 1–6.

Sommer R., Lhotsky M., Haider T. and Cabaj A. (2000a). UV inactivation, liquid-holding recovery, and photoreaction of *Escherichia coli* OH157 and other pathogenic *Escherichia coli* strains in water. *Jour. Food Prot.*, **63**, 1015–1020.

Sommer R., Cabaj A., Pribil W., Haider T. and Hirschmann G. (2000b). Presented at the Wasser Berlin, IOA, European-African-Asian-Australian Group, Berlin GR, October 23–26.

Sommer R., Pribil W., Appelt S., Gehringer P., Eschweiler H., Leth H., Cabaj A. and Haider T. (2001). Inactivation of bacteriophages in water by means of non-ionizing (UV-253.7 nm) and ionizing (gamma) radiation: A comparative approach. *Water Res.*, **35**, 3109–3116.

Song Y., Liu C., Molitoris D., Tomzynski T. J., Mc Teague M., Read E. and Finegold S. M. (2002). Use of 16S-23S rRNA spacer-region (SR)-PCR for identification of intestinal Clostridia. *System Appl. Microbiol.*, **25**, 528–535.

Sorensen D. L., Eberl S. G. and Dicksa R. A. (1989). *Clostridium perfringens* as a point source indicator in non-point polluted streams. *Water Res.*, **23**, 191–197.

Strange R. E. and Hunter J. R. (1969). *Outgrowth and synthesis of macromolecules.* In: G. W. Gould and A. Hurst (ed.), The Bacterial Spore. Academic Press, London and New York.

States S. (1997). Protozoa in River Water: Sources, Occurrence and Treatment. *J. Am. Water Works Assoc.*, **89**(9), 74–83.

Stetler R. E., Waltrip S. C. and Hurst C. J. (1992). Virus removal and recovery in the drinking water treatment train. *Water Res.*, **26**, 727–731.

Stott R., Matsushita E. and Warren A. (2001). Protozoan predation as a mechanism for the removal of *Cryptosporidium* oocysts from wastewaters in constructed wetlands. *Wat. Sci. Tech.*, **44**, 194–198.

Stott R., May E., Ramirez E. and Warren A. (2003). Predation of *Cryptosporidium* oocysts by protozoa and rotifers: implications for water quality and public health. *Wat. Sci. Tech.*, **47**, 77–83.

Stumm W. and Morgan J. J. (1981). *Aquatic chemistry, An Introduction Emphasizing Chemical Equilibria in Natural Waters.* (Second edition). Wiley, New York US.

Stuyfzand P. J. and van der Jagt H. (1997). Toelichting op de anorganische-chemische analysemethoden voor grond in het LAM van Kiwa. Kiwa, KOA96.116, Nieuwegein, NL.

Swertfeger J., Metz D. H., DeMarco J., Braghetta A. and Jacangelo J. G. (1997). Effect of filter media on cyst and oocyst removal. *J. Am. Water Works Assoc.*, **91**(9), 90–100.

Taylor E. W. (1949). Thresh, Beale & Suckling's The examination of waters and water supplies (6th ed). Churchill, London, UK.

Teunis P. F. M., Evers E. G., Havelaar A. H. and Medema G. J. (1996). Presented at the American Water Quality Technology Conference, Boston, November 1996.

Teunis P. F. M., Medema G. J., Kruidenier L. and Havelaar A. H. (1997). Assessment of the risk of infection by *Cryptosporidium* and *Giarda* in drinking water from a surface water source. *Water Res.*, **31**, 1333–1346.

Theunissen J. J. H., Nobel P. J., van de Heide R., de Bruin H. A. M., van Veenendaal D., Lodder W. J., Schijven J. F., Medema G. J. and van der Kooij D. (1998). Enterovius concentraties bij de innamepunten van oppervlaktewater voor de bereiding van drinkwater. RIVM 289202013.

Thiercelin M. E. (1899). Sur un diploque saprophyte de líntestin susceptible de devenir pathogene. *Comptes rendus Société de Biologie*, **5**, 269–271.

Thompson S. S., Jackson J. L., Suva-Castillo M., Yanko W. A., El jack Z., Kuo J., Chen C., Williams F. P. and Schnurr D. P. (2003). Detection of infectious human Adenoviruses in tertiary-treated and ultraviolet-disinfected wastewater. *Wat. Environ. Res.*, **75**, 163–170.

Thurston-Enriquez J. A., Haas C. N., Jacangelo J. G., Riley K. and Gerba C. P. (2003). Inactivation of Feline Calicivirus and Adenovirus Type 40 by UV radiation. *Appl. Environ. Microbiol.*, **69**, 577–582.

Timms S., Slade J. S., Fricker C. R., Morris R., Grabow W. O. K., Botzenhart K. and Wyn Jones A. P. (1995). Removal of *Cryptosporidium* by Slow Sand Filtration. *Wat. Sci. Tech.*, **31**, 81–84.

Tisa L. S., Koshikawa T. and Gerhardt P. (1982) Wet and dry bacterial spore densities determined by buoyant sedimentation. *Appl. Environ. Microbiol.*, **43**, 1307–1310.

Toride N., Leij F. J. and van Genuchten M. T. (1997). The CXTFIT code for estimating transport parameters from laboratory or field tracer experiments. Version 2.0 US Salinity Laboratory, Agricultural Research Service US Department of Agriculture, Report 137, Riverside, California US.

Tong M. and Johnson W. P. (2007). Colloid Population Heterogeneity Drives Hyperexponential Deviation from Classic Filtration Theory. *Environ. Sci. Technol.*, **41**, 493–499.

Tufenkji N. and Elimelech M. (2004a). Correlation equation for predicting single-collector efficieny in physicochemical filtration in saturated porous media. *Environ. Sci. Technol.*, **38**, 529–536.

Tufenkji N., Miller G. F., Ryan J. F., Harvey R. W. and Elimenech M. (2004b). Transport of *Cryptosporidium* oocysts in porous media: role of straining and physicochemical filtration. *Environ. Sci. Technol.*, **38**, 3620–3629.

Tufenkji N. and Elimelech M. (2005). Spatial distribution of *Cryptosporidium* oocysts in porous media: evidence for dual mode deposition. *Environ. Sci. Technol.*, **39**, 3620–3629.

Tufenkji N., Dixon D. R., Considine R. and Drummond C. J. (2006). Multi-scale *Cryptosporidium*/sand interactions in water treatment. *Water Res.*, **40**, 3315–3331.

Tufenkji N. (2007). Modeling microbial transport in porous media: Traditional approaches and recent developments. Ad. *Water Resources*, **30**, 1455–1469.

USEPA. (2001). Method 1623: *Cryptosporidium* and *Giardia* in Water by Filtration/IMS/FA. USEPA, Washington DC, US.

USEPA. (2003). Ultraviolet disinfection guidance manual. EPA 815-D-03-007. EPA, Washington DC, US.

USEPA. (2006). LT2ESWTR Long Term Second Enhanced Surface Water Treatment Rule; Final Rule. USEPA.

van Breemen L. C. W. A. and Waals J. M. J. (1998). Storage of surface water in the Netherlands: challenges of the future. *Water Supply*, **16**, 375–381.

Van der Kooij D., Drost Y. C., Hijnen W. A. M., Willemsen-Zwaagstra J., Nobel P. J. and Schellart J. A. (1995). Multiple barriers against micro-organisms in water treatment and distribution in the Netherlands. *Wat. Supply*, **13**, 13–23.

Van der Kooij D. (2002). Presented at the 54e vakantiecursus in drinkwatervoorziening & 21e Vakantiecursus in Riolering en Afvalwaterbehandeling, Delft NL.

Van der Veer A. J., Schotsman R. M., Martijn B., Schurer R. and van Dijk J. C. (2005). Presented at the 17th IOA Conference, Strasbourg Fr.

Van der Wielen P. W. J. J., Blokker M. and Medema G. J. (2006). Modelling the length of microbiological protection zones around phreatic sandy aquifers in The Netherlands. *Wat. Sci. Technol.*, **54**, 63–69.

Van Olphen M., van der Baan E. and Havelaar A. H. (1993). Virus removal at river bank filtration. H_2O (in Dutch), **26**, 63–66.

Veenendaal H. R. and Brouwer-Hanzens A. J. (2007). *A method for the concentration of microbes in large volumes of water. Kiwa Water Research*, Nieuwegein NL.

Vigneswaran S. and Chang J. S. (1989). Experimental testing of mathematical models describing the entire cycle of filtration. *Water Res.*, **23**, 1413–1421

Visser A., Hijnen W. A. M., Dullemont Y. and Medema G. J. (2004). Langzame zandfilters als effectieve barriére voor micro-organismen. H2O, **37**(13), 28–30.

von Huben H. (1991). *Surface Water: the new rules.* American Water Works Association, Denver CO, US.

von Sonntag C., Kolch A., Gebel J., Oguma K. and Sommer R. (2004). Presented at the European Conference UV Karlsruhe, UV radiation. Effects and Technologies, Karlsruhe GR, September 22–24.

Walker J. R, Gnanam A. J., Blinkova A. L., Hermandson M. J., Karymov M. A., Lyubchenko Y. L., Graves P. R., Haystead T. A. and Linse K. D. (2007). *Clostridium taeniosporum* spore ribbon-like appendage structure, composition and genes. *Mol. Microbiol*, **63**, 629–643.

Wang D. S., Gerba C. P. and Lance J. C. (1981). Effect of soil permeability on virus removal through soil. *Appl. Environ. Microbiol.*, **42**, 83–88.

Watanabe K. and Takesue S. (1976). Colloid titration for determining the surface charge of bacterial spores. *J. Gen. Microbiol.*, **96**, 221–223.

Watercare. (2002). *Pilot plant investigations, surrogate study, results and recommendations.* Disinfection review group report Water Care Services Ltd., New Sealand.

Watson H. E. (1908). A note on the variation of the rate of disinfection with change in the concentration of the disinfectant. *J. Hyg.*, **8**, 536–592.

Weber-Shirk M. L. and Dick R. I. (1997). Biological mechanisms in slow sand filters. *J. Am. Water Works Assoc.*, **89**, 72–83.

Weelden V. and Mingelen. (1868). Commissie tot onderzoek van drinkwater, 1868. Rapport aan de Koning.

Welch W. H. and Nuttall G. H. F. (1892). A gas-producing *Bacillus* (*Bacillus aerogenes capsulatus*, N. S.) capable of rapid development in the blood vessels after death. Bulletin of the Johns Hopkins Hospital, July, 1892.

West T. P., Meyerhofer D. P., DeGraca A., Leonard S. and Gerba C. P. (1994). Presented at the Annual Conf. American Water Works Assoc., Denver CO, US.

Wetsteyn F. J. (2005). Analyse microbiologische veiligheid drinkwater. Artikelcode: 5318. VROM-Inspectorate, Haarlem NL.

Whitby G. E. and Palmateer G. (1993). The effect of UV transmission, suspended solids, wastewater mixtures and photoreactivation on micro-organisms in wastewater treated with UV light. *Wat. Sci. Tech.*, **27**, 379–386.

White G. C. (1999). *Handbook of chlorination and alternative disinfectants.* John Wiley & Sons, inc., New York.

WHO. (2004). *Guidelines for Drinking Water Quality, (third edition).* World Health Organization, Geneva, SW.

Wiedenmann A., Fischer B., Straub U., Wang C. H., Flehmig B. and Schoenen D. (1993). Disinfection of Hepatitis A virus and MS2 coliphage in water by ultraviolet irradiation: comparison of UV susceptibility. *Wat. Sci. Tech.*, **27**, 335–338.

Wiencek K. M., Klapes N. A. and Foegeding P. M. (1990). Hydrophobicity of *Bacillus* and *Clostridium* Spores. *Appl. Environ. Microbiol.*, **56**, 2600–2605.

Wiersema I. (1999). Influence of process factors on the removal of indicator bacteria by coagulation processes during drinking water production. Doctoral report 97–35. Wageningen university, WUR, NL.

Wijmer S. (1992). Water om te drinken. Vereniging van Exploitanten van Waterbedrijven in Nederland VEWIN, Rijswijk NL.

Willis A. T. (1956). Anaerobes as an index of faecal pollution in water. *J. appl. Bact.*, **19**, 105–107.

Willis A. T. (1957). Anaerobic bacilli in a treated water supply. *J. appl. Bact.*, **20**, 61–64.

Wilson B. R., Roessler P. F., van Dellen E., Abbaszadegan M. and Gerba C. P. (1992). Presented at the American Water Works Association Water Quality Technology Conference, November 15–19, Toronto CN.

Wilson W. J. and Blair E. M. M. (1925). Correlation of the sulphite reduction test with other tests in the bacteriological examination of water. *Jour. of Hyg.*, **XXIV**, 111–119.

Wilson W. J. and Blair E. M. M. (1931). Official Circular 96. British Waterworks Association.

Windle Taylor E., Burman N. P. and Oliver C. W. (1955). Membrane filtration technique applied to the routine bacteriological examination of water. *Journal of the institution of Water Engineers*, **9**, 248–263.

Windle Taylor E. (1969). *The removal of viruses by slow sand filtration. Rep. Results bac. chem. Biol. Exam.* London Waters 1969–1970.

Wolfe R. L. (1990). Ultraviolet disinfection of potable water. *Environ. Sci. Technol.*, **24**, 768–773.

Wubbels G. H. (1996). Inventarisatie van Thermofiele *Campylobacter* bacterien No. Code 945. Haren: Gemeentelijk Waterbedrijf Groningen.

Xu S., Gao B. and Saiers J. E. (2006). Straining of colloidal particles in saturated porous media. *Water Resour. Res.* **42**, W1216.

Yao K. M., Habibia M. and O'Melia C. R. (1971). Water and wastewater filtration: Concepts and applications. *Environ. Sci. Technol.*, **5**, 1105–1112.

Yolton D. P., Huettel R. N., Simpson D. K. and Rode L. J. (1972). Isolation and partial chemical characterization of the spore appendages of *Clostridium taeniosporum*. *J. Bacteriol.*, **109**, 881–885.

Yolton D. P., Pope L., Williams M. G. and Rode L. J. (1968). Further electron microscope characterization of spore appendages of *clostridium bifermentans*. *J. Bacteriol.*, **95**, 231–238.

Yoo H. S., Lee S. U., Park K. Y. and Park Y. H. (1997). Molecular typing and epidemiological survey of prevalence of *C. perfringens* types by multiplex PCR. *J. Clin. Microbiol.*, **35**, 228–232.

Zimmer J. L. and Slawson R. M. (2002). Potential repair of *Escherichia coli* DNA following exposure to UV radiation from both medium- and low-pressure UV sources used in drinking water treatment. *Appl. Environ. Microbiol.*, **68**, 3293–3299.

Zimmer J. L., Slawson R. M. and Huck P. M. (2003). Inactivation and potential repair of *Cryptosporidium parvum* following low- and medium-pressure ultraviolet irradiation. *Water Res.*, **37**, 3517–3523.

IWA Publishing's authorised EU representative for General Product Safety Regulations is Diane D'Arras, 15 rue Duret, 75116 Paris, France, e-mail: safety@iwap.co.uk.

Printed and bound by CPI Group (UK) Ltd, Croydon, CR0 4YY
16/06/2026
02140385-0001